Traveler Behavior and Values 2018

Transportation Research Record: Journal of the Transportation Research Board Volume 2672, Issue 47

ISBN: 978-1-5108-8253-9

Printed from e-media with permission by:

Curran Associates, Inc.
57 Morehouse Lane
Red Hook, NY 12571

proceedings.com

Some format issues inherent in the e-media version may also appear in this print version.

Published by Sage Publications USA

ISBN (Print) 978-1-5108-8253-9
ISBN 2018 Printed Set (All Issues) 978-1-5108-7735-1

Printed by Curran Associates, Inc. (2019)

Additional copies of this publication are available from:

Curran Associates, Inc.
57 Morehouse Lane
Red Hook, NY 12571 USA
Phone: 845-758-0400
Fax: 845-758-2633
Email: curran@proceedings.com
Web: www.proceedings.com

Volume 2672 Number 47 December 2018

TRR
JOURNAL OF THE TRANSPORTATION RESEARCH BOARD

Contents

Traveler Behavior and Values

2018 Thomas B. Deen Distinguished Lecture

Articles

2018 PYKE JOHNSON AWARD: Outstanding Paper in Planning and Environment

Articles

* Membership as of February 2019.

TRB

TRANSPORTATION RESEARCH BOARD

CORE SPONSORS

Transportation Departments of the 50 States and the District of Columbia

Federal Government

U.S. Department of Transportation
- Federal Aviation Administration (FAA)
- Federal Highway Administration
- Federal Motor Carrier Safety Administration (FMCSA)
- Federal Railroad Administration (FRA)
- Federal Transit Administration
- National Highway Traffic Safety Administration (NHTSA)
- Office of the Assistant Secretary of Transportation for Research and Technology (OST-R)

U.S. Air Force Civil Engineer Center (AFCEC)
U.S. Coast Guard (USCG)
U.S. Department of Energy
U.S. Department of the Interior (DOI)
U.S. Environmental Protection Agency (EPA)

Nongovernmental Organizations

American Public Transportation Association (APTA)
Association of American Railroads (AAR)
California Air Resources Board

The National Academies of
SCIENCES · ENGINEERING · MEDICINE

The **National Academy of Sciences** was established in 1863 by an Act of Congress, signed by President Lincoln, as a private, nongovernmental institution to advise the nation on issues related to science and technology. Members are elected by their peers for outstanding contributions to research. Dr. Marcia McNutt is president.

The **National Academy of Engineering** was established in 1964 under the charter of the National Academy of Sciences to bring the practices of engineering to advising the nation. Members are elected by their peers for extraordinary contributions to engineering. Dr. C. D. Mote, Jr., is president.

The **National Academy of Medicine** (formerly the Institute of Medicine) was established in 1970 under the charter of the National Academy of Sciences to advise the nation on medical and health issues. Members are elected by their peers for distinguished contributions to medicine and health. Dr. Victor J. Dzau is president.

The three Academies work together as the National Academies of Sciences, Engineering, and Medicine to provide independent, objective analysis and advice to the nation and conduct other activities to solve complex problems and inform public policy decisions. The Academies also encourage education and research, recognize outstanding contributions to knowledge, and increase public understanding in matters of science, engineering, and medicine.

Learn more about the National Academies of Sciences, Engineering, and Medicine at **www.national-academies.org.**

The **Transportation Research Board** is one of seven major programs of the National Academies of Sciences, Engineering, and Medicine. The mission of the Transportation Research Board is to increase the benefits that transportation contributes to society by providing leadership in transportation innovation and progress through research and information exchange, conducted within a setting that is objective, interdisciplinary, and multimodal. The Board's varied activities annually engage about 7,000 engineers, scientists, and other transportation researchers and practitioners from the public and private sectors and academia, all of whom contribute their expertise in the public interest. The program is supported by state transportation departments, federal agencies including the component administrations of the U.S. Department of Transportation, and other organizations and individuals interested in the development of transportation.

Learn more about the Transportation Research Board at **www.TRB.org**.

Transportation Research Record
2018, Vol. 2672(47) 1–11

DOI: 10.1177/0361198118798602
journals.sagepub.com/home/trr
SAGE

The Times They Are A-Changin': What Do the Expanding Uses of Travel Time Portend for Policy, Planning, and Life?

Patricia L. Mokhtarian[1]

Abstract
Personal travel is becoming increasingly "passengerized," with new travel products and services being enabled by the same technology that is expanding the ways in which travel time can be used. This paper explores the implications of these changes from the traveler's perspective. A typology of possible changes in travel time is presented, followed by an extended discussion of reasons why, and mechanisms by which, travel time could, in particular, increase. Some key research needs are suggested and concluding reflections offered.

Time is one of life's most precious resources, a currency we expend on every act we perform. It is a key element in the organization of modern society, not only in the work domain (where punctuality is generally expected, and its lack is frequently punished) but often also in the social and sport/recreation domains (consider the role of the clock in races and most ball games—baseball and cricket being notable exceptions) and, certainly, in the travel domain. Unlike money, we all accrue this currency at precisely the same *rate* of 24 hours per day although the subjective *quality, effectiveness*, and *perceived rate of passage* of the time we spend will vary. Also, unlike money, "time savings and time losses cannot easily be transferred and exchanged" (*1*, p. 119). The passage of time and our time-use choices are classic themes of poetry (Ecclesiastes 3, *The Bible*) and song (Jim Croce, "Time in a Bottle," 1973) as well as the visual arts (Salvador Dali, *The Persistence of Memory*, 1931) and science fiction (H.G. Wells, *The Time Machine*, 1895).

Today's Information Age is characterized by the increasing fragmentation of time. Couclelis (*2*, p. 346) describes the "disintegration of activities" across space and time. Whereas, in the past, activities such as work, shopping, and play tended to occur at a single location in a continuous block of time, it is now increasingly common for such activities to occur across multiple locations and across times that are alternated, interleaved, or overlaid (*3*) with other activities. Some studies have suggested that 95% of the population multitasks every day, and some authors have found that activities are multitasked for an average of about 7 hours per day (*4*). Evidently, a multitasking mentality has now become the rule rather than the exception, such that most of us may be in a state of "continuous partial attention" (https://lindastone.net/2009/11/30/beyond-simple-multi-tasking-continuous-partial-attention/).

Not surprisingly, time spent traveling has long been of importance to our profession: it would be difficult to exaggerate the centrality of travel time to transportation policy, planning, and models. Although alternative views have gained traction in recent years (*5–7*), it is still the case that travel is almost universally treated as a cost of participating in spatially separated activities and, thus, that travel time is considered a disutility that travelers wish to minimize.

There is little doubt that the ways people use their travel time are evolving. In the past, travel largely served as the conveyance between "continents" containing complete activities. Increasingly, however, the complete activities are disintegrating, and travel is becoming not only a bridge joining "islands" of spatially and temporally fragmented activity, but also a spatio-temporal locus for the activity fragments themselves. Although far from an unmitigated blessing (e.g., *8*), advances in information and communication technology (ICT) are broadening the range of activities that can be conducted while traveling, thereby increasing the effectiveness with which

[1]School of Civil and Environmental Engineering, Georgia Institute of Technology

Corresponding Author:
Address correspondence to Patricia L. Mokhtarian: patmokh@gatech.edu

traditional activities can be performed and stimulating the widespread adoption of devices that support new and better activities. Simultaneously, ICT is supporting more efficient management of the transportation system and enabling the creation of new transportation products and services. As a result, we are in a period of nearly unprecedented, ICT-enabled ferment with respect to transportation vehicles, networks, services, operations – and traveler behavior.

It is the perspective of the traveler that undergirds the present paper: with all of the changes in transportation options swirling around, how will the traveler react? Specifically, how will her uses of travel time change? How will her *valuation* of travel time change? How will those changes influence her travel-related decisions of all kinds? For the sake of concreteness, I will focus much of the discussion on the prospect of a future in which fully automated vehicles (AVs) are ubiquitous. I want to stress, however, that many of the issues to be discussed apply even to the present day and, certainly, to an evolving future that will assuredly involve the increasing "passengerization" of travel, by which I mean drivers more often becoming passengers (e.g., through the use of smartphone-enabled ride-hailing services) and/or being relieved of more of the cognitive load of driving (through greater amounts of automation). Air travel and more conventional (but also evolving) transit and carpooling modes also qualify for the term. While technically not passengerization, personal concierge and errand services, such as grocery and meal delivery, fall into a related realm in which the customer's travel time is freed for other uses and, as such, will also be included in the discussion.

In the remainder of this paper, I explore these issues at a largely conceptual level, calling on some of the limited but growing empirical evidence as appropriate. The next section 2 presents a typology of *potential* time-related impacts of increasingly passengerized travel and some of their likely implications. In the following section, I question how likely the most troublesome of these impacts are, discuss how they might be predicted, and raise some associated issues. The final section contains some concluding reflections.

A Typology of Potential Time-Use Impacts of Increasingly Hands-Free Travel

Six Types of Potential Impacts

At the individual level, it is possible to identify at least six potential time-use impacts of the increased passengerization of travel (Figure 1). In three of them, travel time remains constant; in one of them travel time shrinks; and in two of them, it expands. Across people and time, all of these effects will likely be true to some extent, and thus, one of our challenges will be to assess the *net* impact of partly counteracting influences. Again, for the purposes of discussion, I will assume a fully AV context, which can serve as an extreme-case scenario. Note carefully, however, that in the discussion, I am focusing on the time use of the *traveler* and not of the *vehicle*. If the future permits zero-occupant AVs to be sent on errands, the erstwhile *traveler* saves time, but the *vehicle* may not. Conversely, if people become more inclined to share rides with others, the *traveler* may not save time, but in the aggregate, *vehicle* travel could decrease or, at least, increase much less (e.g., *9*).

Travel Time Remains Constant

Type 1. As a degenerate case, travel time remaining constant could mean that no changes in time use occur. Alternatively, a minimal change could be that essentially the *same activities* occur during travel as before, except that the expanding capabilities of ICT (magnified by increasingly hands-free travel) may allow those activities to be performed with *greater quality* and/or in *higher quantity*. It could also happen that an increasing fragmentation of attention may result in a *degradation of the quality* and/or *quantity* of those activities.

Type 2. Again, because of the expanding capabilities of ICT, increasingly hands-free travel, or the combination of both, *new within-trip (WT) activities may be generated*, overlaid onto the travel time, and possibly displacing some previous WT activities.

Type 3. For the same reasons, it may become possible to *bring formerly outside-the-trip (OT) activities into the trip, thereby freeing time for new or expanded OT activities.*

Travel Time Shrinks

Type 4. In a fully AV world, personal (albeit not vehicular) travel time could shrink to the extent that a *driverless car could be sent on chauffeuring and errand-running trips* (e.g., to pick up/drop off children or elderly dependents or to pick up groceries, dinner, or the dry cleaning). Shrinkage could also occur through more efficient trip chaining being enabled (*10*) or through the price of travel increasing substantially (whether by market forces or regulation [*11*]). Activities would be generated or expanded to fill the freed-up time.[1] Under the assumption of a shrinking of travel time, these activities would take place at least partly OT, but some of them may involve travel, thereby partly counteracting the personal travel time savings.[2]

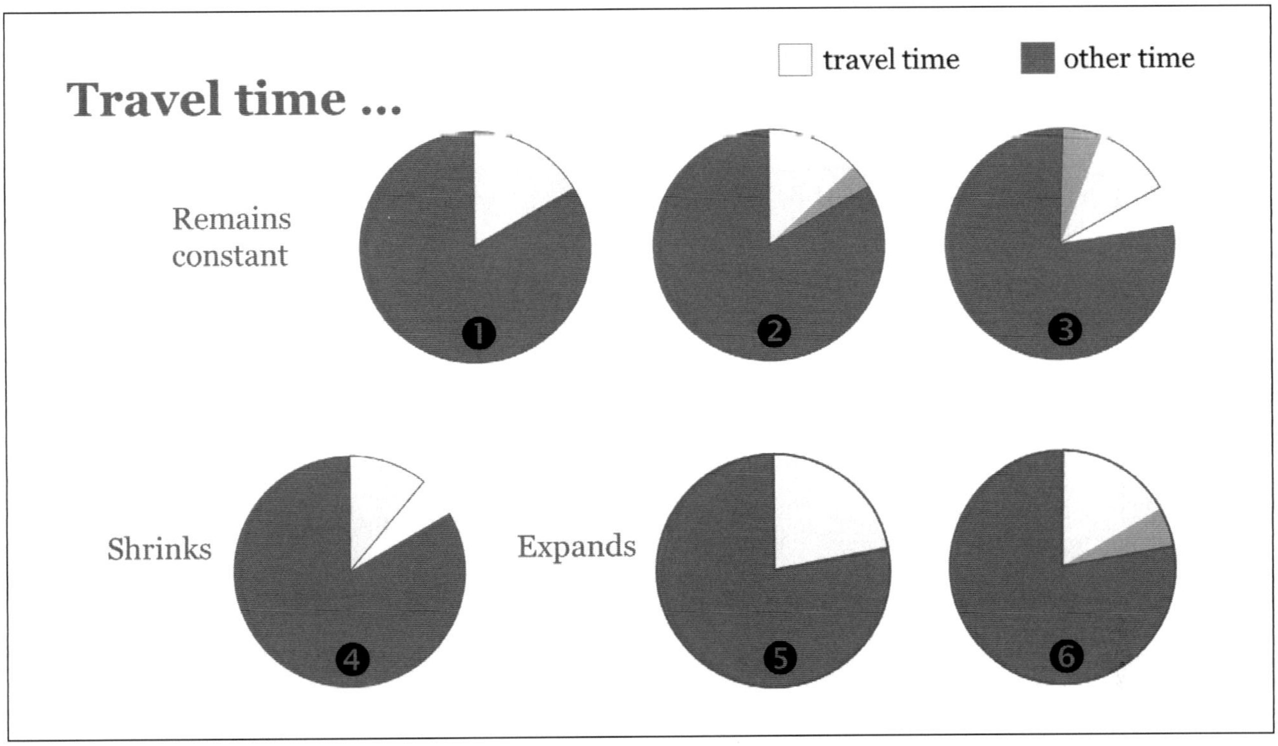

Figure 1. A visual typology of potential time use impacts of the increasing passengerization of travel.
Note: The medium gray wedges represent the newly-occurring overlay of other activities onto travel time, which occurs either with the generation of entirely new within-trip activities (2), or by bringing formerly outside-the-trip activities inside the trip (3 and 6). The blank wedges represent newly-freed time, which occurs either by travel time shrinking (4), or by activities formerly conducted outside the trip being brought inside (3).

Travel Time Expands. The prospect of travel time expanding calls for more discussion because this is the most sinister type of outcome with respect to resource consumption and congestion. Here, I briefly mention two types of outcomes, and in the next subsection, I explore reasons why travel time might increase and the mechanisms by which it may do so.

> Type 5. *Total travel time increases* (for reasons explained in the following subsection), *thereby crowding out some former OT activities.*
> Type 6. Some *formerly OT activities become WT activities, thereby freeing some* (formerly OT) *time that is then (at least partly) used to travel.* Note carefully the distinctions between Types 3, 4, and 6. In Type 4, *WT* time is freed, and not all of that new time is again spent WT (since travel time shrinks). In Type 6, *OT* time is freed, and not all of that new time is again spent OT (since travel time increases). In Type 3, *OT* time is freed, and all of the new time is again spent OT (since travel time remains constant).

The distinction between Types 5 and 6 can be subtle. However, Type 5 can occur even with little or no change in WT activities (aside from lengthening the time engaged in them), whereas Type 6 depends on some OT activities becoming WT activities. In Type 5, some OT activities (or some of the time spent on them) are (is) forgone altogether, because either they cannot be done WT or the traveler does not wish to do so. In Type 6, by contrast, through changing formerly OT activities to WT activities, the traveler creates a hole in the activity schedule *without* sacrificing activities (entirely), and that hole is then filled (at least in part) by travel.

Travel Time Expansion: Reasons and Manifestations

As discussed elsewhere (*12*; Singleton and Mokhtarian, 2018), actions occur when the following two conditions are met: (1) constraints are weak enough and/or facilitators are strong enough, and (2) the motivation to act is strong enough. Each condition alone is necessary but not sufficient: even in the absence of any binding constraints, the action will not occur unless a strong enough motivation is present, and even if motivation is strong, constraints may be insurmountable. Both conditions together are sufficient.[3]

In the current context, this suggests the following major avenues by which travel time might expand in an increasingly passengerized world: (1) constraints may be weakened, and/or facilitators may be strengthened; and (2) motivations may be heightened. With respect to the first (constraints) avenue, it is not hard to imagine that there is an *existing* unrealized latent demand (non-negligible motivation) for travel, which is currently being suppressed because of various constraints or barriers. Current and emerging technologies are likely to lower some of these barriers, e.g., to

- reduce the generalized costs of travel (potentially including monetary cost but also including inconvenience, physical energy, stress, and other factors)
- reduce or eliminate the physical and mental requirements of driving, in particular, but of other forms of travel as well;
- release various schedule constraints through the means identified in the preceding subsection;
- and increase the facilitation of travel by making it more pleasant/fun or productive.

Note, again, that none of these outcomes are sufficient in themselves to generate more travel if the motivation to travel is not strong enough.

With respect to the second (motivations) avenue, however, it is pertinent to ask whether the new ways to use travel time could actually *increase* the motivation to travel. At least on occasion, this outcome seems credible.[4] For one thing, as in-vehicle comfort and entertainment technologies continue to advance, there could be a joyriding effect, in which the trip itself more often becomes an entertaining activity (or an integral part of a bundle of entertainment activities) that is desirable in its own right. In other circumstances, the activities that are planned to be conducted while traveling may influence the length of the trip (e.g., the chosen destination or the route taken); e.g., it may "need" to be "long enough to get some work done" or "to watch my favorite TV show on the way."[5] In yet other cases, it is easy to imagine a moving vehicle sometimes becoming the venue of choice for normally stationary activities such as a business meeting or social event (much as has occasionally been the case in the past with ships, trains, and planes). Some ideas along these lines (e.g., the yoga vehicle or the social networking vehicle) are offered in a discussion of "the wow factor" by Nordhoff et al. (*14*).

However, as long as there is net unrealized latent demand (which may be debatable for some—see, e.g., Section 4.2 of Gruel and Stanford (*15*)—but which is likely to be the case *at least* among currently mobility-limited sectors of society), reducing constraints and increasing facilitation will have much the same effect as strengthening motivation, i.e., the generation of more travel. Regardless of which of those two reasons apply, how might this increased generation be manifested? Some likely pathways are listed in Table 1 (also see *16–19*) together with illustrative vignettes. The first manifestation, more time in congestion, is a byproduct of the direct choices that comprise the rest of the table.

Putting the Possibilities into Perspective

How Likely are These Impacts, and How Can We Tell?

If the kinds of changes described in the previous section materialize on a large scale, the impacts will obviously be profound. But how likely are such changes? After all, the waste bin is full of dramatic predictions of technology impacts for which reality has fallen far short, such as the "electronic cottage" (in which working at home becomes predominant [*21*]) and the "death of distance" (in which ICT makes distance irrelevant [*22*]). Are the prophesied impacts of increasing passengerization just another instance of technology hype?

We do not really know.

Well, shouldn't we be able to predict what will happen? After all, time (or its close correlates distance, accessibility, impedance, and generalized cost) has a prominent place in many (if not most) of our travel-demand models (see Table 2 and *23*); therefore, doesn't that signify that we know a lot about the role of travel time in travel choices? In truth, these models have always been flawed, with some rather deep cracks that have been papered over. For example, practitioners have acknowledged that, surprisingly often, initial estimations of mode-choice models have yielded *positive* coefficients of travel time, counter to that most firmly held tenet that travel is a disutility to be minimized, as mentioned in the Introduction.[7] Clearly, in such cases travel time is standing in for some unmeasured variables that are correlated with it, such as "unobserved travel-experience attributes or conjoint activities" (*24*, p. 221), which most models are currently unable to capture. The disruptive changes that the transportation system is now undergoing, in which the influence precisely of the *travel experience* and *accompanying activities* are gaining more prominence in the traveler's decision-making process, are serving to shine a spotlight on those cracks in our existing models and throw them into sharp relief.

In recent analyses of the prospective impacts of AVs on travel demand (e.g., *11*, *23*), it is common to make an assumption about the diminishing disutility of travel time. But this raises the question of what assumption should be made, and on what basis? My colleagues and I have been investigating ways to remove some of the guesswork from that assumption. To my knowledge, Malokin et al. (*33*) is the first study to incorporate the influence of a propensity for productive travel-based

Table 1. Manifestations of Increased Time in Travel

Manifestation	Before	After
More time in congestion	"I'm willing to spend *x* minutes driving in congestion"	"Since I can now use the time more productively, I'm certainly not ***less*** willing than before to spend *x* minutes driving in congestion – indeed, I am now willing to spend ***more than*** *x* minutes in congestion"
New trip generation	"I'm unable to drive (e.g., at night, or at all), but taxis are too uncertain in price, availability, and driver skill/compatibility" "I don't have time to go there because I have to do *x*"	"I can have my own AV in the garage, and not have to worry about those things – I have my independence back"[6] "I can do *x while* I go there", OR "The time I saved by doing *other* things while traveling allows me to do *x* and *still have time* to go there"
More distant destinations	"It's a week night – let's eat nearby"	"Let's go to this special place 50 miles away – we can do our last hour of work in the car on the way there, and watch our favorite TV show on the way back (as we would have done at home)"
Longer routes	"The travel time is wasted, so I want to get there as quickly as possible"	"If we take the scenic route, the trip becomes long enough to get something useful done, and we can enjoy the nicer scenery too"
Mode changes	"A 45-minute transit ride is better than a 30-minute car drive because I can use the time productively"	"Even if the car drive becomes 45 minutes, I'm no worse off than I was on transit, because now I can use *car* time productively – and enjoy all the advantages of the automobile (privacy, ubiquity...)"
Residential/job relocation	"I'm willing to waste *30 minutes* (each way) on my commute"	"Since I can • work productively, and • eat, sleep, exercise in the car, I'm willing to spend *1.5 hours* (each way) on my commute – and now I can live on the lake/in the mountains/ by the shore!"
More/longer long-distance trips	"I just can't afford to take the time away from work to see my elderly parents (in another state) as often as I'd like" "It's impossible to travel with the kids for more than 2–3 hours at a time"	"Since I can work and sleep in the car, I'm no longer really losing any work time during the travel itself" "Since the kids can sleep and play in the car much like they were at home, we can drive to see Grandma and Grandpa much more frequently now"

Table 2. Influence of Travel Time in Behavior/Demand Models

Behavior	Example roles of travel time (TT)	Example references
Trip generation	Trip frequency = f(accessibility, etc.) Preference to telecommute = f(commute time, etc.)	*25, 26*
Destination choice	TT (or distance) traded off against destination attractiveness, as in aggregate gravity models or disaggregate destination choice models	*25, 27, 28*
Mode choice	TT traded off against travel cost and other variables, e.g., the multinomial logit model	*29*
Route choice	"Minimize generalized travel cost" (a weighted combination of monetary cost and time) considered the most common objective function for travelers	*30*
Residential location	Commute cost or employment accessibility traded off against housing cost in an aggregate model; commute time and/or accessibility traded off against neighborhood attractiveness and other variables in a disaggregate model	*31, 32*

multitasking into a *revealed preference* mode choice model. Findings suggest that in a fully-AV environment, this propensity alone (i.e., not accounting for other benefits and disadvantages of AV travel) could increase the commute mode share of the "drive" alone mode by 1.5 percentage points – non-negligible, albeit hardly staggering. A follow-on study (*34*) empirically confirms, again using revealed preference models, that compared to their older counterparts, millennial commuters have lower values of travel time savings for both in-vehicle and out-of-vehicle travel time, even after controlling for personal attitudes and the propensity to multitask.

Recent work (*35*) proposes to model the travel time coefficient in a mode-choice model as a function of the propensity to travel multitask (among other variables). The analysis of the resulting individual- and mode-specific discounts on travel time disutility (i.e., the percentages by which the magnitude of the travel time coefficient is reduced) offers a way to rigorously inform the assumptions that need to be made in the context of large-scale regional travel-demand forecasting models. Preliminary evidence indicates that the discount for the commuter-rail mode in 2011–2012 averaged 16–18%, which is roughly in line with the lower end of the hypothetical discounts that were assumed in future AV scenarios analyzed by Auld et al. (*23*) and others. Of course, any specific results obtained now are subject to change as technology continues to reshape nearly every aspect of the transportation system. However, the methodology demonstrated in the Etezady et al. study (*35*) can be repeatedly applied to investigate how the estimate of the discount evolves over time.

At the same time, work is in progress on ways to directly incorporate variables such as a propensity to travel multitask (as well as numerous other attitudes) into regional demand forecasting models, using machine learning methods to transfer attitudinal information from smaller-scale specialized datasets into the large-scale travel/activity survey datasets that constitute the source for estimating such regional models. Although this new direction is still in a very early stage, Malokin et al. (*36*) showed a first proof-of-concept.

Another fruitful approach to assessing the transportation impacts of increasing passengerization is to update classic time-allocation models to account for the fragmentation of activities in general and the transfer of activity (fragments) to travel in particular. Pawlak et al. (*37*, *38*) offered recent constructive developments in this direction with a utility-maximization model in which joint activity/travel combinations are characterized in part by the degree to which travel time can be used productively. Along similar lines, Pudane et al. (*10*) proposed a model allowing for the joint choice of activities and their locations, including combinations of stationary locations and travel modes. Such models could also enable the assessment of how out-of-trip time that is freed up (either by travel time shrinking or by bringing out-of-trip activities within the trip) will be used, including the extent to which it will result in additional trips.

Some Directions for Future Research

The volatility of the current phase of transportation technology development produces considerable uncertainty about future volumes of travel such as their distribution across time, space, and mode and their implications for land use, safety, consumption of non-renewable resources, and other realms of life (e.g., physical activity and subjective well-being). This state of uncertainty is likely to persist for some time, which means that thoughtful, proactive, ongoing research and planning have never been more important.

To that end, and maintaining the present paper's focus on the individual traveler, a number of research needs present themselves. The potential reasons for increased travel suggested above serve as a backdrop for discussing some of these needs.

Most fundamentally, we need to rethink most, or all, of our key travel-demand models. The present models, understandably enough, focus on the objectively measurable constraints of travel time and monetary cost to the virtual exclusion of other relevant variables such as the subjective constraints of stress, schedule inflexibility, motion sickness, physical and mental abilities, and negative expectations of the travel experience; facilitating factors such as the expectation of a pleasant, entertaining, and/or productive travel experience; and motivations to travel in the first place. Although the "usual suspects" of time and cost may have been adequate as first-order approximations in the past, they are becoming increasingly *inadequate* for characterizing the utility of various transportation choices (e.g., *11*). Even the choice set itself has become more complex; e.g., an Uber ride differs from a taxi ride in meaningful ways, and sharing a ride with only an Uber driver is different from also sharing it with friends, which is different from sharing it with strangers. In addition, we need to explore the use of functional forms that account for the dual necessity of high-enough motivations and low-enough constraints/strong-enough facilitators, whereby the presence of one set of these conditions does not necessarily compensate for the absence of the other. Initially, model improvements along these lines will inevitably fall into the realm of basic research, but eventually, our applied regional travel-demand forecasting models must better reflect the more complex reality that is emerging as well.

As a corollary, we also need to rethink how we compare and evaluate transportation improvement alternatives. The role of the individual's value of travel time savings in such assessments has long had its critics (e.g., *39–41*). However, the monetization of (often very small) travel time savings as the vastly dominant benefit to be achieved by infrastructure improvements becomes even less defensible to the extent that travel time becomes more like "ordinary" time (*42*, *43*) or more transparent time onto which ordinary activities can be overlaid and, accordingly, becomes *less* in need of being saved.

Beyond these fundamental needs, some other (illustrative, but necessarily selective) research questions seem especially engaging (also see *16*, *44*):

- No matter how effectively or pleasantly the time can be spent, for most people, there is presumably an upper bound on the amount of time that they are willing to spend in a vehicle: "travel time budget effects or in-vehicle-activity satiation effects … could make … drastic travel time increases infeasible" (*23*, p. 7). What is that upper bound, and how does it differ by person and circumstance?
- To what extent (for whom, and under what circumstances) is a productive or hedonic use of travel time a meaningful factor in a travel-related decision, i.e., a facilitator or motivator, as opposed to an incidental byproduct of a decision made for entirely other reasons? It is easy for a work-oriented, time-starved academic to project her own desire to use travel time productively onto the entire human race, but the subset of travelers (1) with information-based jobs that can be performed on-the-go; (2) who want or need to work while traveling; (3) who possess and (4) travel with the requisite productivity tools; and (5) know how to use them effectively may not be very large. Cyganski et al. (*45*, pp. 10–11), e.g., found that only 10–13% of their sample saw the ability to work while traveling as predominantly or absolutely "the particular advantage" of an AV. Even granting the inability to fully envision one's time use in an AV, the presence of additional groups who wish to be productive in personal rather than work-related ways, and the existence of those for whom hedonic uses of travel time are actively valued rather than merely passively experienced, are the expanding uses of travel time really compelling reasons for traveling more for very many people?
- Granting that the opportunities and benefits of productive travel multitasking are not evenly distributed, are there social equity implications of the disparities? For example, to the extent that transit operators expend scarce resources on improving mobile ICT services to attract affluent choice riders (such as the senior managers whom Pawlak et al. [*38*] found to be more likely than others to engage in productive work activities on the train and tended to do so for longer durations), do they fail to improve, or even diminish, services and amenities for less affluent captive riders?
- For whom, and under what circumstances, are the expanding uses of travel time a curse rather than a blessing (or both; see, e.g., *46*)? How will institutional and social expectations for individuals' availability while traveling evolve as technology continues to stretch current boundaries of availability, thereby permitting an even greater fragmentation of time and blurring of boundaries between realms of life than we already experience? To what extent will individuals be able to maintain the boundaries they desire to have?
- Given that even boredom, idleness, daydreaming, and other "non-productive leisure activities" can be beneficial (*47*, p. 273; also see *48–51*) and that the motion and changing scenery of travel can offer comforts of their own (e.g., *52*), then even if only passively experienced rather than actively valued, to what extent is time spent traveling restorative, especially once the stress and attentional demands of operating a vehicle are removed?
- In short, it seems that "making time" is not an unalloyed good, nor is "killing time" an unalloyed bad (*53*). In terms of value to the individual and to the society at large, how are we to evaluate this complex intermixture of blessing and curse?

These and other questions will require our full arsenal of research methods, both qualitative approaches (including focus groups and interviews) as well as quantitative ones (*16*). Because of the disruptive nature of the technological developments underway, these methods will necessarily continue to include hypothetical scenarios, whether explored in qualitative discussions, created for input to aggregate models, or designed for disaggregate stated response studies. At the disaggregate level, e.g., the "teleportation test" (*5*, *54*) was devised to identify the circumstances under which respondents would *not* prefer to be instantly transported to a desired destination if such an action were possible, thereby indicating circumstances under which travel possesses some utility beyond that of providing access to spatially separated activities. It would be illuminating to investigate how the answers to the teleportation question might vary with the degree to which travel time were hands-free. Along the same lines, we could ask respondents to consider how much they would travel by car (in terms of hours per day for local travel, and frequency and length for long-distance travel), if it were absolutely free and in a vehicle that was equipped with various amenities, to begin to ascertain an upper bound on the amount of time people wish to spend in a moving vehicle under various circumstances. As a complement to simply asking people to imagine life in a moving vehicle, we can deploy "passengering simulators," including virtual and augmented reality techniques, which replicate the internal and external environments of a moving AV that has various amenities.

But we must also continue to discover and refine ways to leverage what current reality can tell us about the future, whether through novel applications of revealed preference methods, such as the ones I describe above (applications to the currently little-studied context of time use during airline travel would be of particular interest); through creative naturalistic experiments, such as providing a chauffeur to ordinary families to simulate the availability of hands-free and zero-occupant trips in the future (*55*); or through some other clever means. Standard cross-sectional surveys can still be useful, but longitudinal data collection and analysis have never been more vital. The current reality is a rapidly-moving target, and it is essential to keep updating our knowledge of it as a baseline for the future. Passive and continuous data-collection methods that are enabled by ICT (e.g., Global Positioning System traces of travel patterns and social media postings) are important, but I maintain that those cannot yet replace the measurement of the attitudes, motivations, and facilitators/constraints that are essential for understanding observed behaviors and, thence, predicting evolving ones.

In a similar vein, no single discipline has a monopoly on the framework and methods needed to understand the complexities associated with travel-related choices in the present and coming eras. Quite the contrary, numerous disciplines bring diverse and meaningful perspectives to bear on these issues, and we need to increase the opportunities for cross-fertilization of these perspectives (*44*).

Concluding Reflections

We live in an age of unprecedented advancements in transportation-related and other technologies. Consequently, the behavioral changes that will be brought about by these advancements are certainly unknown and, to a great extent, unpredictable. Yet predict them we must, if we are to maximize the societal gain and mitigate the negative unintended consequences of new technology. Assuredly, the prospective gains are enormous, in terms of, e.g., improved safety, greater mobility for individuals who are currently immobile or dependent on other people, and more efficient system utilization. In contrast, the potential downsides are also massive: particularly with respect to the possibility of stimulating even more travel, there are major implications for fuel and vehicle materials consumption, time spent in congestion delays, and land consumption, to name just a few.

The potential magnitudes of these impacts compel our attention. On the technical side, they require us to attempt to estimate what those magnitudes would be under a number of reasonable scenarios involving various degrees and types of policy intervention. However, equally importantly, on the policy side, they make a public discussion of societal values and priorities more imperative than ever. What do we want our land-use patterns to look like (i.e., with what distribution of high, medium, and low densities and use mixtures)? How do we balance the negative externalities imposed by travel against its demonstrable benefits to economic prosperity and subjective well-being (e.g., *56*)? Getting the prices right (i.e., ensuring that the cost of travel includes proper compensation for the negative externalities it causes) is certainly an important element of the answer. But we should pay equal attention to getting the benefits right, because only then can we put the costs in proper context.

Acknowledgments

This paper was substantially influenced by several activities funded by the Center for Teaching Old Models New Tricks (TOMNET), which is a University Transportation Center that is sponsored by the U.S. Department of Transportation through Grant No. 69A3551747116. It is also a pleasure to acknowledge the contributions to my thinking made by the numerous outstanding students and colleagues with whom I have been privileged to interact across the years. More specifically, comments and assistance by Atiyya Shaw, Sung Hoo Kim, Yongsung Lee, Ram Pendyala, and Frank Koppelman have improved earlier drafts of this paper. Any remaining flaws are my own responsibility.

Author's Note

In the article title, "The Times They Are A-Changin'" is an homage to the Bob Dylan (1964) song of the same name.

Notes

1. It is also possible that the expansion of OT activities, for reasons that are unrelated to passengerization, crowds out some travel (i.e., that such expansion is the cause rather than the effect of the shrinkage of travel time).
2. To the extent that the original shrinkage occurs because of dispatching driverless cars to run errands, a partial rebound effect would signify a multiplication of the number of trips that could be generated by a single person *at the same point in time*.
3. In the Fogg (*13*, p. 3) Behavior Model, "motivation and ability are trade-offs of a sort[; p]eople with low motivation may perform a behavior if [it] is simple enough," although "motivation alone – no matter how high – may not get people to perform a behavior if they don't have the ability." These somewhat contradictory statements reflect two possibilities for the relationship between motivation and ability: compensatory and non-compensatory. I have chosen to emphasize the possible non-compensatory nature of motivations and constraints because I believe that this aspect is often underappreciated, especially with our penchant for estimating linear-in-parameters compensatory models.
4. The examples that follow loosely call on the "tripartite" sources of the positive utility of travel described by

Mokhtarian and Salomon (*5*); i.e., the intrinsic utility of the trip itself, the utility of activities conducted while traveling, and the utility of reaching a destination, respectively.

5. Just as most people, even now, would prefer a somewhat longer transit trip to a shorter one requiring a transfer that "breaks up the trip," as shown by the "transfer penalty" included in most mode-choice models (*20*). Of course, the transfer penalty accounts not only for the generalized cost of fragmenting a block of time, but also for the additional physical and mental effort of changing vehicles (typically also involving a walk and a wait).
6. Of course, AV travel will still cost money, but in suggesting that the traveler will not worry (much) about cost, I refer to the well-known phenomenon that most people, if they consider price at all, consider only the relatively inexpensive out-of-pocket cost for a trip using a vehicle that they personally own, whereas the total cost of a trip in a vehicle-for-hire both (1) includes some amortization of the capital costs of owning that vehicle, and (2) makes that additional component of the total cost quite explicit to the traveler.
7. In 1968, the eminent urban planner/economist William Alonso noted "Commonly, in the early runs the model will produce some outrageous results, and the modelers will use their necromantic powers to have the black boxes give reasonable results" (*57*, p. 253).

References

1. Leclerc, F., B. H. Schmitt, and L. Dube. Waiting Time and Decision Making: Is Time Like Money? *The Journal of Consumer Research*, Vol. 22, No. 1, 1995, pp. 110–119.
2. Couclelis, H. From Sustainable Transportation to Sustainable Accessibility: Can We Avoid A New "Tragedy Of the Commons"? In *Information, Place, and Cyberspace: Issues In Accessibility* (Janelle, D. G., and D. C. Hodge, eds.), Springer, Berlin, 2000, pp. 342–356.
3. Circella, G., P. L. Mokhtarian, and L. K. Poff. A Conceptual Typology of Multitasking Behavior and Polychronicity Preferences. *electronic International Journal of Time Use Research (eIJTUR)*, Vol. 9, No.1, 2012, pp. 59–107.
4. Kenyon, S. What Do We Mean By Multitasking? Exploring the Need for Methodological Clarification in Time Use Research. *electronic International Journal of Time Use Research* (*eIJTUR*), Vol. 7, No. 1, 2010, pp. 42–60.
5. Mokhtarian, P. L., and I. Salomon. How Derived is the Demand for Travel? Some Conceptual and Measurement Considerations. *Transportation Research Part A: Policy and Practice*, Vol. 35, No. 8, 2001, pp. 695–719.
6. Mokhtarian, P. L., I. Salomon, and M. E. Singer. What Moves Us? An Interdisciplinary Exploration of Reasons for Traveling. *Transport Reviews*, Vol. 35, No. 3, 2015, pp. 250–274.
7. Singleton, P., and P. L. Mokhtarian. Is There a Positive Utility of Travel? A Review of Concepts, Measures, and Evidence. 2018 (in progress).
8. Reinecke, L., S. Aufenanger, M. E. Beutel, M. Dreier, O. Quiring, B. Stark, K. Wölfling, and K. W. Müller. Digital Stress Over the Life Span: The Effects of Communication Load and Internet Multitasking on Perceived Stress and Psychological Health Impairments in a German Probability Sample. *Media Psychology*, Vol. 20, No. 1, 2017, pp. 90–115.
9. Fagnant, D. J., and K. M. Kockelman. Dynamic Ride-sharing and Fleet Sizing for a System of Shared Autonomous Vehicles in Austin, Texas. *Transportation*, Vol. 45, No. 1, 2018, pp. 143158.
10. Pudane, B., E. J. E. Molin, T. A. Arentze, Y. Maknoon, and C. G. Chorus. *A Time-Use Model for the Automated Vehicle Era*. Transportation Research Part C: Emerging Technologies, Vol. 93, 2018, pp. 102–114.
11. Childress, S., B. Nichols, B. Charlton, and S. Coe. Using An Activity Based Model to Explore Possible Impacts of Automated Vehicles. *Transportation Research Record: Journal of the Transportation Research Board*, 2015. 2493: 99–106.
12. Mokhtarian, P. L., and I. Salomon. Modeling the Choice of Telecommuting: Setting the Context. *Environment and Planning A*, Vol. 26, 1994, pp. 749–766.
13. Fogg, B. J. A Behavior Model for Persuasive Design. *Proc., 4th International Conference on Persuasive Technology*, Claremont, California, USA, April 26 – 29, 2009. https://Dl.Acm.Org/Citation.Cfm?Id=1541999. Accessed March 6, 2018. 10.1145/1541948.1541999.
14. Nordhoff, S., B. van Arem, and R. Happee. Conceptual Model to Explain, Predict, and Improve User Acceptance of Driverless Podlike Vehicles. *Transportation Research Record: Journal of the Transportation Research Board*, 2016. 2602: 60–67.
15. Gruel, W., and J. M. Stanford. System Effects of Widespread Use of Fully Automated Vehicles – Three Scenarios. In *Disrupting Mobility: Impacts of Sharing Economy and Innovative Transportation On Cities* (Meyer, G., and S. Shaheen, eds.), Springer, Cham, Switzerland, 2017, ch. 9.
16. Milakis, D., B. van Arem, and B. van Wee. Policy and Society Related Implications of Automated Driving: A Review of Literature and Directions for Future Research. *Journal of Intelligent Transportation Systems*, Vol. 21, No. 4, 2017, pp. 324–348.
17. LaMondia, J., D. J. Fagnant, H. Qu, J. Barrett, and K. Kockelman. Shifts in Long-Distance Travel Mode Due to Automated Vehicles: Statewide Mode-Shift Simulation Experiment and Travel Survey Analysis. *Transportation Research Record: Journal of the Transportation Research Board*, 2016. 2566: 1–10.
18. Harper, C. D., C. T. Hendrickson, S. Mangones, and C. Samaras. Estimating Potential Increases in Travel with Autonomous Vehicles for the Non-Driving, Elderly and People with Travel-Restrictive Medical Conditions. *Transportation Research C*, Vol. 72, 2016, pp. 1–9.
19. Truong, L. T., C. De Gruyter, G. Currie, and A. Delbosc (2017) Estimating the Trip Generation Impacts of Autonomous Vehicles On Car Travel In Victoria, Australia. *Transportation*, Vol. 44, No. 6, 2017, pp. 1279–1292.
20. Guo, Z., and N. H. M. Wilson. Assessment of the Transfer Penalty For Transit Trips: Geographic Information System-Based Disaggregate Modeling Approach. *Transportation Research Record: Journal of the Transportation Research Board*, 2004. 1872: 10–18.

21. Toffler, A. *The Third Wave*. Bantam Books, New York, 1980.
22. Cairncross, F. *The Death of Distance: How the Communications Revolution Will Change Our Lives*. Harvard Business School Press, 1997.
23. Auld, J., V. Sokolov, and T. S. Stephens. Analysis of the Effects of Connected-Automated Vehicle Technologies on Travel Demand. *Transportation Research Record: Journal of the Transportation Research Board*, 2017. 2625: 1–8.
24. Hess, S., M. Bierlaire, and J. W. Polak. Estimation of Value of Travel-Time Savings Using Mixed Logit Models. *Transportation Research Part A: Policy and Practice*, Vol. 39, No. 1/2, 2005, pp. 221–236.
25. Ortuzar, J. de D., and L. G. Willumsen. *Modelling Transport*, 4th ed. John Wiley and Sons, Chichester, 2011.
26. Mokhtarian, P. L., and I. Salomon. Modeling the Desire to Telecommute: The Importance of Attitudinal Factors In Behavioral Models. *Transportation Research Part A: Policy and Practice*, Vol. 31, No. 1, 1997, pp. 35–50.
27. Bhat, C., A. Govindarajan, and V. Pulugurta. Disaggregate Attraction-End Choice Modeling: Formulation and Empirical Analysis. *Transportation Research Record: Journal of the Transportation Research Board*, 1998. 1645: 60–68.
28. Bernardin, V. Jr, F. Koppelman, and D. Boyce. Enhanced Destination Choice Models Incorporating Agglomeration Related to Trip Chaining While Controlling For Spatial Competition. *Transportation Research Record: Journal of the Transportation Research Board*, 2009. 2132: 143–151.
29. Ben-Akiva, M. and S. R. Lerman. *Discrete Choice Analysis: Theory and Application to Travel Demand*. MIT Press, Cambridge, MA, 1985.
30. Xu, H., H. Yang, J. Zhou, and Y. Yin. A Route Choice Model with Context-Dependent Value of Time. *Transportation Science*, Vol. 51, No. 2, 2017, pp. 536–548.
31. Ahlfeldt, G. If Alonso Was Right: Modeling Accessibility and Explaining the Residential Land Gradient. *Journal of Regional Science*, Vol. 51, No. 2, 2011, pp. 318–338.
32. Guo, J. Y., and C. R. Bhat. Operationalizing the Concept of Neighborhood: Application to Residential Location Choice Analysis. *Journal of Transport Geography*, Vol. 15, No. 1, 2007, pp. 31–45.
33. Malokin, A., G. Circella, and P. L. Mokhtarian. How Do Activities Conducted While Commuting Influence Mode Choice? Testing Public Transportation Advantage and Autonomous Vehicle Scenarios. Revision Under Peer Review, 2018.
34. Malokin, A., G. Circella, and P. L. Mokhtarian. Do Multitasking Millennials Value Travel Time Differently? A Revealed Preference Study of Northern California Commuters. Paper #17-00891. Presented at the 96th Annual Meeting of the Transportation Research Board, Washington, D.C., January 2017.
35. Etezady, A., A. Malokin, P. L. Mokhtarian, and G. Circella. Not All Minutes are Created Equal: How Does the Impact of Travel Time on Mode Choice Differ by Demographics and Propensity to Multitask? 2018 (in progress).
36. Malokin, A., P. L. Mokhtarian, and G. Circella. An Investigation of Methods to Enrich National Household Travel Survey Data with Attitudinal Variables. 2018 (in progress).
37. Pawlak, J., J. W. Polak, and A. Sivakumar. Towards a Microeconomic Framework for Modelling the Joint Choice of Activity – Travel Behaviour and ICT Use. *Transportation Research A*, Vol. 76, 2015, pp. 92–112.
38. Pawlak, J., J. W. Polak, and A. Sivakumar. A Framework for Joint Modelling of Activity Choice, Duration, and Productivity while Traveling. *Transportation Research B*, Vol. 106, 2017, pp. 153–172.
39. Mackie, P. J., S. Jara-Díaz, and A. S. Fowkes. The Value of Travel Time Savings in Evaluation. *Transportation Research E*, Vol. 37, 2001, pp. 91–106.
40. Daly, A., F. Tsang, and C. Rohr. The Value of Small Time Savings for Non-Business Travel. *Journal of Transport Economics and Policy*, Vol. 48, No. 2, 2014, pp. 205–218.
41. Næss, P. Inaccurate and Biased: Cost-Benefit Analyses of Transport Infrastructure Projects. In *Crisis System: A Critical Realist and Environmental Critique of Economics and the Economy* (Naess, P., and L. Price, eds.), Routledge, New York, 2016.
42. Gripsrud, M., and R. Hjorthol. Working on the Train: From 'Dead Time' to Productive and Vital Time. *Transportation*, Vol. 39, 2012, pp. 941–956.
43. Lyons, G., and J. Urry. Travel Time Use in the Information Age. *Transportation Research A*, Vol. 39, No. 1/2, 2005, pp. 257–276.
44. Ben-Elia, E., G. Lyons., and P. L. Mokhtarian. Epilogue: The New Frontiers of Behavioural Research on the Interrelationships between ICT, Activities, Time Use and Mobility. *Transportation*, Vol. 45, No. 2, 2018, pp. 479–497.
45. Cyganski, R., E. Fraedrich, and B. Lenz. Travel Time Valuation for Automated Driving: A Use-Case-Driven Study. Paper #15-4259. Presented at the 94th Annual Meeting of the Transportation Research Board. Washington, D.C., January 2015.
46. Shaw, F. A., A. Malokin, P. L. Mokhtarian, and G. Circella. It's Not All Fun and Games: Modeling the Reported Benefits and Disadvantages of Conducting Activities While Commuting. 2018 (in progress).
47. Wilhoit, E. D. 'My Drive Is My Sacred Time': Commuting As Routine Liminality. *Culture and Organization*, Vol. 23, No. 4, 2017, pp. 263–276.
48. Baird, B., J. Smallwood, M. D. Mrazek, J. W. Y. Kam, M. S. Franklin, and J. W. Schooler. Inspired by Distraction: Mind Wandering Facilitates Creative Incubation. *Psychological Science*, Vol. 23, No. 10, 2012, pp. 1117–1122.
49. Bench, S. W., and H. C. Lench. On the Function of Boredom. *Behavioral Sciences*, Vol. 3, 2013, pp. 459–472.
50. Mann, S., and R. Cadman. Does Being Bored Make Us More Creative? *Creativity Research Journal*, Vol. 26, No. 2, 2014, pp. 165–173.
51. Holley, D., J. Jain, and G. Lyons. Understanding Business Travel Time and Its Place in the Working Day. *Time and Society*, Vol. 17, No. 1, 2008, pp. 27–46.
52. Larsen, J. Tourism Mobilities and the Travel Glance: Experiences of Being on the Move. *Scandinavian Journal of Hospitality and Tourism*, Vol. 1, No. 2, 2001, pp. 80–98.
53. Jain, J., and G. Lyons. The Gift of Travel Time. *Journal of Transport Geography*, Vol. 16, 2008, pp. 81–89.

54. Russell, M., and P. L. Mokhtarian. How Real is a Reported Desire to Travel for Its Own Sake? Exploring the 'Teleportation' Concept in Travel Behaviour Research. *Transportation*, Vol. 42, 2015, pp. 333–345.
55. Harb, M., Y. Xiao, G. Circella, P. L. Mokhtarian, and J. Walker. Projecting Travelers into a World of Self-driving Vehicles: Estimating Travel Behavior Implications via a Naturalistic Experiment. *Transportation*, in press. https://doi.org/10.1007/s11116-018-9937-9
56. Friman, M., T. Gärling, D. Ettema, and L. E. Olsson. How Does Travel Affect Emotional Well-Being and Life Satisfaction? *Transportation Research A*, Vol. 106, 2017, pp. 170–180.
57. Alonso, W. Predicting Best with Imperfect Data. *Journal of the American Institute of Planners*, Vol. 34, 1968, pp. 248–255.

Author Biography

Patricia L. Mokhtarian, the Susan G. and Christopher D. Pappas Professor of Civil and Environmental Engineering (CEE) at the Georgia Institute of Technology, is the 2018 recipient of the Thomas B. Deen Distinguished Lectureship.

Dr. Mokhtarian joined the Georgia Institute of Technology in August 2013, after 23 years on the CEE faculty at the University of California, Davis (UCD). At UCD she was the Associate Director for Education of the Institute of Transportation Studies and founding Chair of the interdisciplinary M.S./Ph.D. program in Transportation Technology and Policy, as well as a professor of CEE. While at UCD, she also served as Principal Investigator on a $2.7 million National Science Foundation Integrative Graduate Education and Research Traineeship award. Prior to joining UCD in 1990, she spent nine years in regional planning and consulting in southern California.

The Thomas B. Deen Distinguished Lectureship recognizes the career contributions and achievements of an individual in one of the areas covered by TRB's Technical Activities Division. Originally known as the Transportation Research Board Distinguished Lectureship, the award was renamed in 2002 in honor of the eighth TRB Executive Director who served with distinction from 1980 to 1994. Honorees are invited to present overviews of their technical areas, including the evolution, present status, and prospects for the future. TRB also publishes each lecture in a volume of the *Transportation Research Record: Journal of the Transportation Research Board.* Dr. Mokhtarian's lecture is titled "The Times They Are A-Changin': What Do the Expanding Uses of Travel Time Portend for Policy, Planning, and Life?"

Regarding her academic career, Dr. Mokhtarian regularly teaches graduate classes on transportation survey methods and discrete choice modeling, as well as undergraduate probability and statistics, the latter taken by more than 2,000 students during the course of her career. She has been credited with making important and substantial contributions in the areas of transportation planning and policy making, information and communications technologies' impacts on travel, and human attitudes and perceptions toward travel. She has authored or co-authored more than 200 refereed journal articles, technical reports, and other publications. Dr. Mokhtarian is one of the top five most highly cited scholars in the field of transportation, and accordingly has shaped the nature of discourse in the field for generations to come.

Since 1988, Dr. Mokhtarian has provided continued service to TRB as a chair or a member of six different TRB standing committees or subcommittees. In recognition of her service, she is an Emeritus Member of the Standing Committee on Effects of Information and Communication Technologies on Travel Choices and the Standing Committee on Traveler Behavior and Values.

Dr. Mokhtarian was also a member of the committee that produced TRB Special Report 298: Driving and the Built Environment: The Effects of Compact Development on Motorized Travel, Energy Use, and CO2 Emissions. The report estimates the contributions that changes in residential and mixed-use development patterns and transit investments could make to reducing vehicle miles traveled by 2030, and then also by 2050, and the impact this could have on meeting future transportation-related greenhouse gas reduction goals.

She earned her master's degree and Ph.D. in Industrial Engineering/Management Science from Northwestern University and her bachelor's degree in Mathematics (summa cum laude) from Florida State University.

Article

TRR
JOURNAL OF THE TRANSPORTATION RESEARCH BOARD

Transportation Research Record
2018, Vol. 2672(47) 12–22

DOI: 10.1177/0361198118790124
journals.sagepub.com/home/trr
SAGE

A Stated Preference Survey Approach to Understanding Mobility Choices in Light of Shared Mobility Services and Automated Vehicle Technologies in the U.S.

Hamidreza Asgari[1], Xia Jin[1], and Terrence Corkery[2]

Abstract

This paper presents the preliminary results of a recent nationwide survey that focuses on mode choice behavior in view of emerging mobility options such as ride-sourcing and automated vehicles (AV). The survey provides a comprehensive scan of current mode choice patterns and the influencing factors. Then it presents stated preference (SP) choices to understand how travelers measure the trade-offs among different mode alternatives. The choice experiments focused on four potential user markets: drivers who usually drive for daily activities, passengers who depend on other household members or friends, transit users or users who do not have access to a private vehicle on a regular basis, and visitors or who do not have access to a private vehicle occasionally. The results suggest that on-demand services incorporating AV technologies (with lower operating costs) may become a viable option for many travelers. Most drivers and passengers preferred single ride than shared ride regardless of whether it is on a daily or occasional basis. However, for transit users, shared rides showed higher potential than exclusive services, which may indicate that cost is a primary consideration in the mode choice decisions for these users.

Today's world is deeply influenced by the way new technology evolves. Advances in information and communication technologies have changed how we live and travel, and will continue to do so. As an example, rapidly emerging mobile apps and web payment methods (such as wechat pay) have contributed to the quick expansion of bike-sharing, car-sharing, ride-sourcing, and various other on-demand services around the world. Similarly, connected and autonomous vehicle technologies are expected to bring a paradigm shift in how we define mobility. Several research efforts have predicted the market penetration of automated vehicles (AV) (*1–4*). Although the details in the assumptions and projections vary, there is a general consensus that AVs would represent a significant portion of vehicles driving on the road around year 2040.

In light of this timeline, it is essential to incorporate AV considerations into current long-range transportation planning efforts, which usually extend to the next 20 to 30 years. However, there are a lot of uncertainties with respect to technology development, regulations and policies, and user acceptance that make it challenging to draw a clear picture of how shared mobility and AVs may affect our daily travel and the potential implications on the society as a whole. In this regard, various surveys have been conducted to measure public acceptance of AVs. However, most of the surveys focused on perceptions, attitudes, or simple choices. There is a need to engage in carefully designed stated preference (SP) choice experiments to measure the likelihood and extent of behavioral changes.

Given the above motivations, this paper presents an effort in providing new insights into mode choice in the upcoming age of shared mobility and AVs. An SP survey was designed with the aim of exploring how travelers evaluate the trade-offs between emerging modes (ride-sourcing, AVs) and conventional modes (private vehicles and public transit). Stated choice experiments are able to

[1]Department of Civil and Environmental Engineering, Florida International University, Miami, FL
[2]Forecasting and Trends Office, Florida Department of Transportation, Tallahassee, FL

Corresponding Author:
Address correspondence to Xia Jin: xjin1@fiu.edu

assess the independent influence of various factors on mobility decisions made by individuals facing an array of options. These preference experiments will provide better insight into potential behavioral changes in response to emerging mobility options. The findings will serve as the foundation to construct scenarios for further impact analysis.

The next section provides a brief overview of existing studies and surveys that focused on AV adoption and potential mobility impacts, followed by the description of the survey design elements. Survey results are presented next. The last section summarizes the major findings from the survey and concludes the paper with further analysis needs.

Literature Review

Regarding potential influences on vehicle ownership and mobility options, the literature believes that self-driving technologies would enable more efficient use of cars and expand opportunities for vehicle sharing (*5–8*), which could redefine vehicle ownership entirely.

Among the alternative modes, mass transit has been the subject of many discussions. Some argued that transit ridership would decrease, as AVs would offer the same advantages with respect to eliminating the stress of driving, allowing multitasking, and providing service for the disabled and other transit-captive population, especially for inter-city travel when AVs become affordable (*5*). On the other hand, others thought that AV technologies might bring new opportunities for transit, such as autonomous feeder bus services, demand-responsive services with flexible routes, or even Uber-like shared-ride services (*9*, *10*). It should be noted that AV adoption and penetration would depend highly on the cost of AVs, which has not been explicitly addressed in existing impact studies.

A modified four-step model was employed to address how AVs might affect transit demand during highly congested peak hours (*11*). A nested logit model was used to study the choice between driving and transit, parking and repositioning of AVs. Travelers were divided into various value of time classes ranging from $1.15 to $22.00 per hour, and a multi-class assignment was used to distinguish AVs and other vehicles. The model results based on an Austin downtown sub-network showed that when AVs become financially available to low and middle-income classes transit demand decreased by 61.4%, and AV demand reached 83% at full market penetration. However, the total number of trips increased by 271.4%, and speed decreased modestly.

Similar findings were indicated in a report that focused on business marketing guidelines for automobile companies (*12*). It predicted that there would be an increasing desire for mobility options as well as a large increase in person-miles traveled. Other factors including safety needs, weather considerations, enhanced mobility options, and travel for leisure would also lead to high desirability of driverless mobility-on-demand alternatives.

In this regard, transportation network companies (TNC) (e.g., Uber and Lyft) along with automobile manufacturers have begun testing autonomous taxi (aTaxi) services. A simulation framework for an aTaxi service in New Jersey was developed to explore its demand potential and the impacts on congestion and mobility (*13*). Aggregating the results for the whole state, it showed that around 92.7% of the demand could be addressed by an aTaxi system with an average trip length of 33.6 miles. The study pointed out that a major limitation was the lack of information on attitudinal aspects and personal preferences that might impede drivers from shifting to aTaxis.

Along this line, Malokins et al. conducted a revealed preference (RP) study that investigated the impact of multitasking attitudes and behavior on the utility of various modes (driving alone, shared driving, local transit, commuter rail, and biking) (*10*). The results indicated that engaging in productive activities (i.e., electronic reading/writing and using a laptop/tablet) significantly influenced mode utility and could account for a small but statistically significant impact on observed mode shares. The model estimated that commuter rail and car/vanpool shares decreased by 0.38 and 3.22 percentage points, respectively, and the drive-alone share increased by 3.00 percentage points, when the option of multitasking was not available. In a hypothetical scenario when people could engage in highly productive activities, drive-alone and car/vanpool shares increased by 0.95 and 1.08 percentage points, respectively.

In the absence of RP data on driverless cars, many other studies relied on stated preferences to identify user behavior and attitudes toward AVs and the influential factors on AV adoption. Overall, there was general knowledge (i.e., at least heard of) on AV technologies throughout the nation (*14–20*); however, this knowledge might be too general and therefore lead to high expectations, which need to be moderated through education and advertisement. Higher familiarity with the technology led to higher willingness of AV adoption.

When it comes to perceived benefits and concerns of AV technologies, interestingly, most people did not view travel time savings or congestion relief as major benefits; instead, reduced stress, high productivity, and fewer crashes were expected (*14*, *18*, *19*, *21*). Safety, cost, data privacy, vehicle performance in unexpected situations, surrendering full control to the vehicle, and trust were the major concerns (*14*, *18*, *20*, *22–24*).

Regarding AV adoption, positive feedback was received from the respondents (*14, 17–19, 22, 25*); however, very few were willing to pay extra for the technology: 54.5% respondents in the U.S., 59.8% in the UK, and 55.2% in Australia were willing to pay $ 0 for completely self-driving technology (*14*). The surveys also showed that socio-economic and demographic information had an important role in AV adoption (*14–18, 21, 24, 26*).

Although these surveys provided useful information on public perceptions and potential responses toward new technologies, in most cases the respondents were presented with simple multiple-choice questions without any context. They were likely to consider ideal or unrealistic situations or hidden hypotheses/attitudes that were unaccounted for. In this regard, a carefully designed choice experiment environment is essential to allow the analyst to quantify the behavioral changes and assess the impacts of influential factors on mobility decisions. The next section presents detailed considerations in the SP survey design for this study.

Survey Design

This section provides an overview of the key elements and considerations in the survey design. For more detailed discussion of the SP survey approach and the complete survey questionnaire, please refer to Jin and Asgari (*27*).

General Context

Some researchers have argued that AV adoption may begin as a shared service, because of affordability or operational issues at the initial stage (*28–31*). Given this consideration, this survey does not specifically focus on the choice behavior toward AVs, but rather the trade-offs between different mobility options with consideration of AV technologies. In this context, AV was incorporated into shared mobility options by reducing the cost associated with human drivers. This approach could avoid potential biases or attitudes toward AVs that may influence respondents' choices. Also, as participants are already familiar with on-demand services (such as Uber and Lyft), more realistic choices could be expected from the respondents. This approach also avoids the need to educate the participants (and potentially introduce biases) on the features and concerns of AV technologies. A separate section was included in the survey that measures travelers' attitudes and acceptance of AVs specifically.

Choice Experiments

Travelers' choice behavior and preferences may be different depending on the circumstances. To better reflect the trade-offs among the travel modes and capture the preferences under different conditions, this survey considers four distinct scenario types as illustrated in Figure 1:

- S1: focuses on the choice between driving and ride-sourcing. This reflects the situation where the traveler has access to a car for daily activities. The main trade-offs depend on the service attributes (travel time, cost) and level of multitasking.
- S2: captures the choice behavior between passenger ride and ride-sourcing. The difference is that a passenger ride depends on the availability and schedule of other household members or friends. A most common example would be escorting trips.

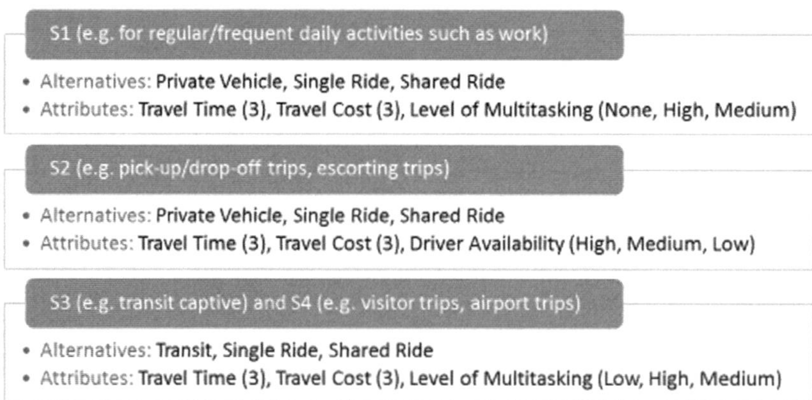

Figure 1. Choice alternatives and attributes for different scenarios.

The main trade-offs would be dependency on other members, in addition to service attributes.

- S3: considers the choice between transit and ride-sourcing when private vehicles are not available on a regular basis, for example, transit-captive users. The main trade-offs would depend on the service attributes of different travel modes.
- S4: considers the choice between transit and ride-sourcing when private vehicles are not available, but for occasional situations, for example, when visiting a place, or traveling to/from an airport. The main trade-offs are the same as S3.

Base Values for Travel Cost and Travel Time

In this study each scenario has three alternatives defined by attributes in terms of travel time, travel cost, level of multitasking or driver availability (for passengers), as presented in Figure 1. An example of a choice scenario presented in the SP section is shown in Figure 2.

The essence of a SP survey is to expose the respondents to a series of scenarios consisting of alternatives defined by attributes with varying values. In this study, the attributes of interest, travel time and travel cost, both have three levels, ranging from −30% to + 30% from the base value. The base values for travel time and cost were determined through extensive research of the literature regarding existing operating time and cost of various modes (*32–39*).

Specifically, the American Automobile Association reported average driving cost per mile (*32*), which considers the costs for fuel, tire, maintenance, insurance, license and taxes, depreciation, and finance. The costs depend on the driver's annual driving mileage; that is, the lower the annual mileage, the higher the cost per mile. Considering 15,000 miles per year as a benchmark for an average American driver, the average cost per mile for a private vehicle trip was about $ 0.57.

For ride-sourcing fares, multiple sources were taken into account, including Uber fare estimator and various other studies (*33–38*). Considering that AV technologies eventually may bring the cost down by 30–40%, a $ 0.5 per mile fare was used for a single-person ride, and a $ 0.4 per mile cost was used for multi-passenger ride.

The 2016 Public Transportation Fact Book published by American Public Transportation Association (APTA) provides detailed statistics on public transit usage throughout the nation (*39*). Accordingly, the average transit trip length over the nation was 4.65 miles with an average fare of $1.14, which leads to an average cost of 24.5 cents/mile. This is very close to the average value reported by Victoria Transport Policy Institute in 2015 (21 cents/mile) (*36*).

Based on the results from the literature and considering the necessity to maintain reasonable trade-offs between the modes, the following base value for travel cost are used in this study:

- Private vehicle cost = 0.5*Distance + $ 0.8 (parking)
- Single-person ride cost = 0.5*Distance
- Multi-passenger ride cost = 0.4*Distance
- Transit cost = 0.25*Distance

Google map was used to estimate average travel time for vehicle trips and transit trips by distance segment. Six segments of distance were considered, from 5 miles to 30 miles with 5-mile increments. Parking time and wait time were also incorporated. The following values are used in this study:

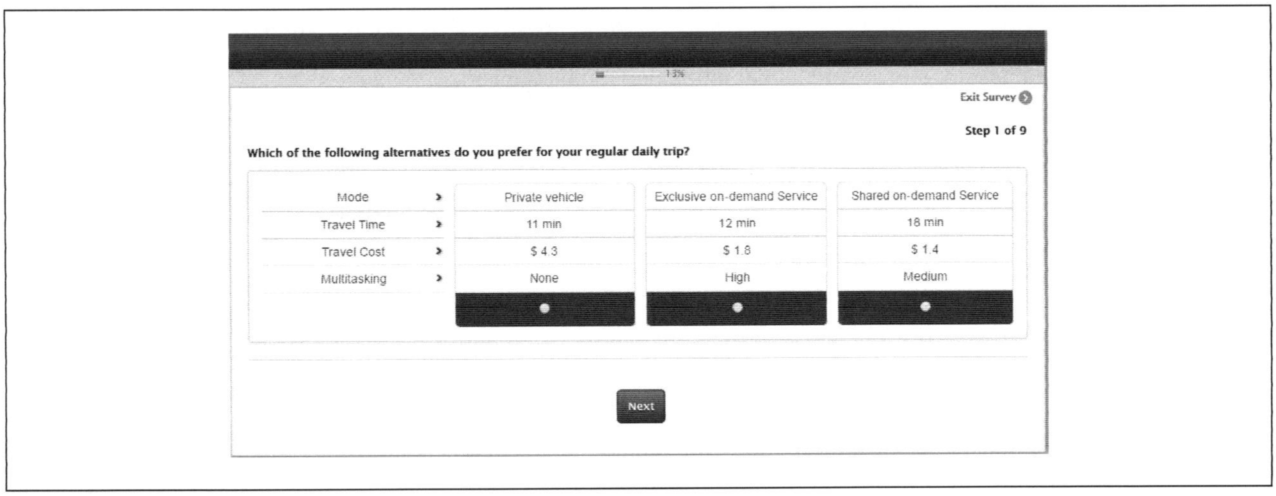

Figure 2. An example of a choice scenario.

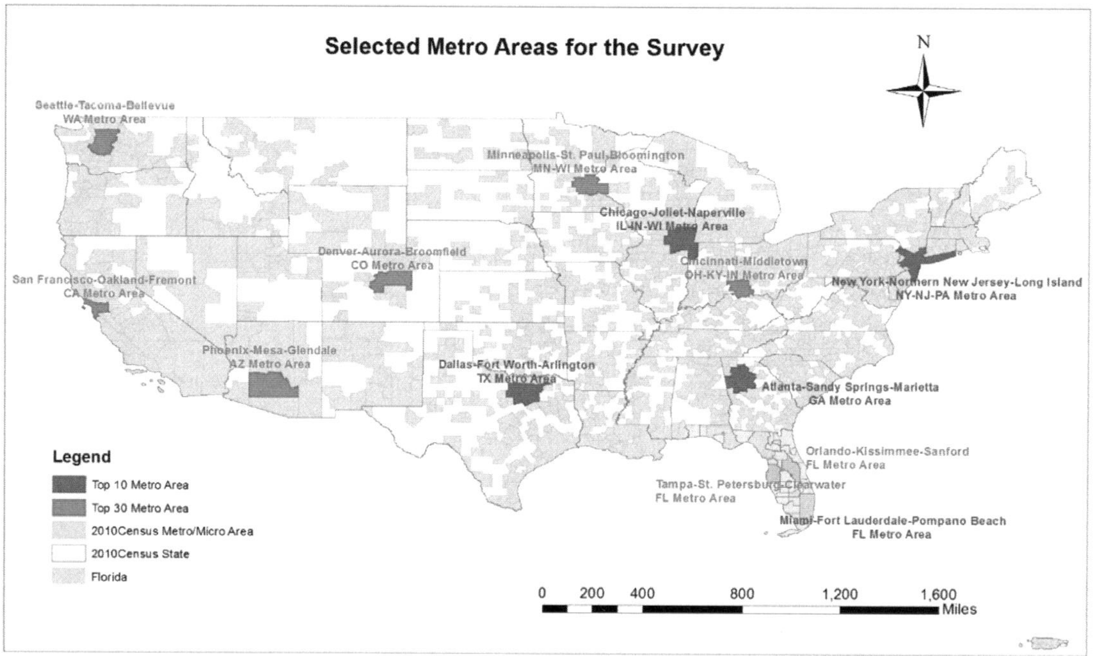

Figure 3. Selected metro areas for survey sampling.

- Private vehicle travel time = trip travel time + 5 min (parking & access/egress time)
- Single-person ride travel time = trip travel time + 2 min (wait time)
- Multi-passenger travel time = trip travel time + 8 min (wait & dead time)
- Transit travel time = trip travel time + 10 min (wait + access/egress time)

Full factorial design with manual adjustments to ensure reasonable representation of each mode (e.g., transit always has the lowest cost, shared ride always offers lower fare than single ride at the expenses of higher travel time, and so forth) yielded 21 choice sets for S1 and S2, and 24 choice sets for S3. The choice questions were divided into three blocks of seven or eight choice questions for each respondent. The scenarios were selected so that the attribute levels were balanced and covers various levels of trade-offs between the modes within each block.

Sampling

This survey targeted Census representative samples (in terms of age, gender, HH income, education, and ethnicity), with oversampling for age group 16–34. The survey focused on ten metropolitan areas throughout the nation plus three metro areas in Florida, as shown in Figure 3. The metro areas were selected based on the 2010 Census demographic profile for core based statistical areas (CBSA) (*40*), considering their sizes as well as geographic locations.

Survey Questionnaire

The survey questionnaire contains six major sections:

- Information about you and your family: collects basic socio-economic and demographic information on the survey participants and their households.
- Information about how you travel: acquires information on their current mode choice behavior, including how often a specific mode (including public transit, taxi, ride-sourcing, and car-sharing) was used, for what types of trips, and for how far.
- Information about how you make your transportation choices: focuses on the decision-making process to establish a general understanding of the underlying factors that lead to certain mode choice decisions.
- Information about a most frequent or regular trip you make: collects detailed trip information for a regular trip, which serves as the RP correspondence.

- Stated preference scenarios: as the core component of the survey, the respondents were presented with seven or eight SP questions, each question showing three alternatives defined by travel cost, travel time, and level of multitasking or availability of a driver.
- Information about your views toward mobility options: presents various questions measuring the respondents' attitudes and opinions toward different mobility options, including driving, shared services, and AV technologies.

Survey Results

The online survey was implemented through a survey firm between April and May 2017. The survey firm maintains a panel that covers a wide range of population groups across the country. The survey link was distributed to the panel. A sample monitoring mechanism was in place that assigned a quota to each demographic cohort based on gender, age, household income, ethnicity, and education according to the 2010 Census. The collected sample represents the population well in terms of gender and ethnicity. Besides the oversampling of young people aged between 16 and 34, the sample was slightly better educated (higher shares for those with college degrees or above), and less wealthy (less shares for those with income $100k or above) compared with the Census 2010 data, probably because of the focus on metropolitan areas and oversampling of young people. Detailed comparison between the sample and 2010 Census data is provided in Jin and Asgari (*27*).

The final cleaned dataset consists of 878 individuals, 478 from the three metropolitan areas in Florida and 400 respondents from the ten metro areas nationwide. This is because of the heavier focus on Florida per the funding agency's requirement.

How Americans Travel

The data showed that about 53% of respondents had ride-sourcing experience. More than half of the respondents seldom use transit or taxi, as shown in Figure 4. Looking closer to the non-private vehicle modes (the graph on the right), it is interesting to see that ride-sourcing has very comparable market size as transit. Car-sharing is not that far behind. This indicates rapid growth in shared mobility services, and implies great potential for frequent use on a monthly basis, although they have not yet been seen as a regular alternative for daily travel by most people.

Looking at the distance traveled by each mode, the data indicated that ride-sourcing had relatively higher share for trips of 6–10 miles, and less share for long trips (more than 20 miles). Regarding mode choice by trip purpose, as expected, drive alone was still the dominant mode for all purposes. It is worth noting that, ride-sourcing showed much higher potential for social trips than other purposes, given that 17.5% respondents had used it in the past three months for social trips whereas only 8.5% used it for commute and shopping trips.

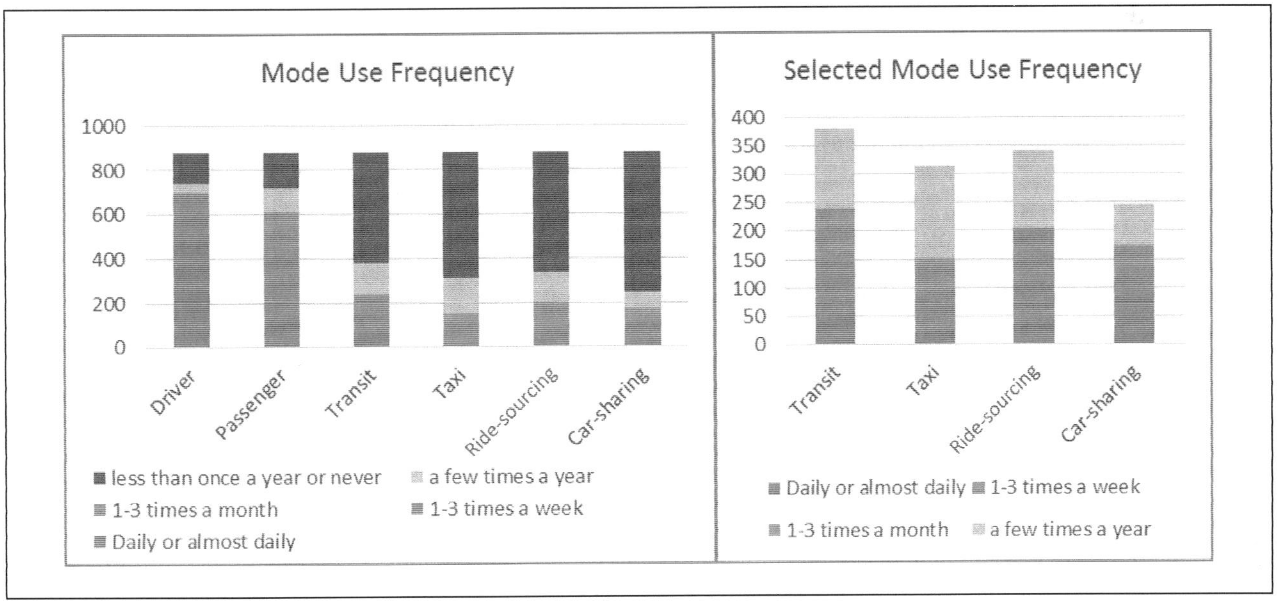

Figure 4. Frequency of usage for each mode.

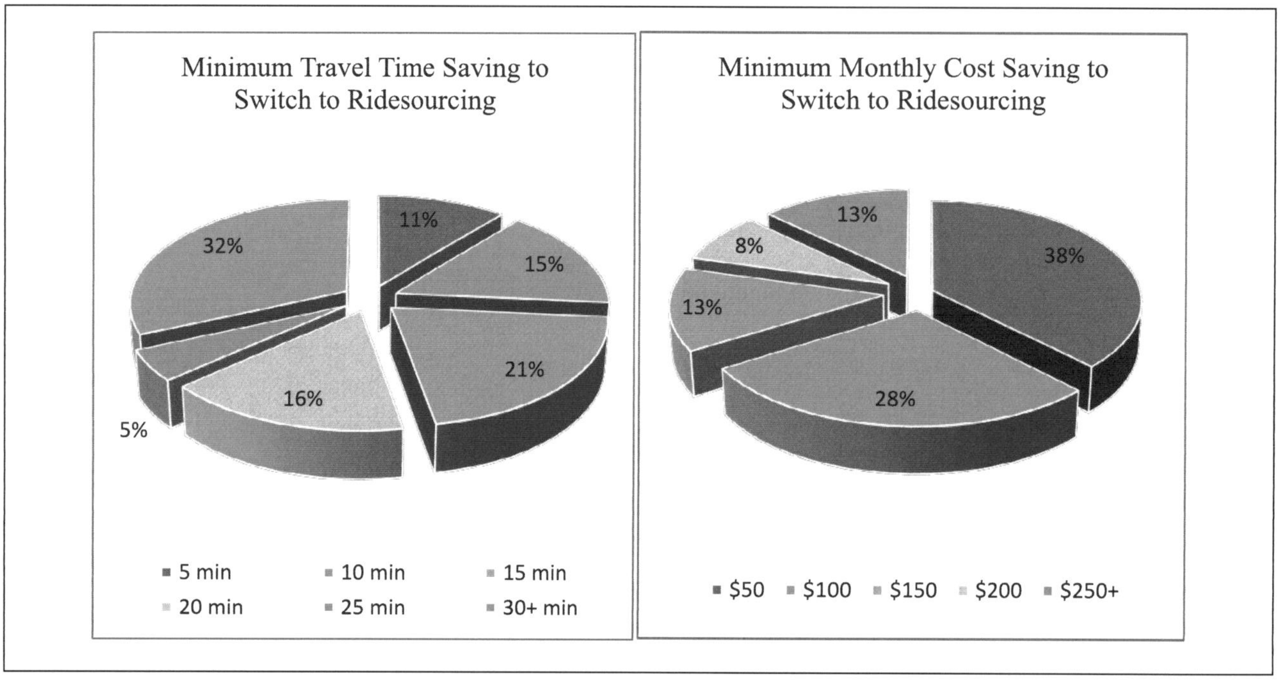

Figure 5. Sensitivity of respondents to mobility cost and travel time.

How Americans Make Transportation Choices

To investigate how mode choice may be influenced by the circumstance, the survey asked the respondents' choice of mode under different conditions, such as bad weather, bad traffic, tight budget, new to an area, or difficult to find parking, and so forth. The data showed that transit became a viable choice when traffic is bad, budget is tight, or there is no access to a vehicle, and especially when parking is expensive or difficult. It also shows that ride-sourcing would not be particularly attractive to respondents when the weather is bad, budget is tight, or they are running late or carrying a lot of stuff. On the other hand, ride-sourcing showed great potential for those with no access to vehicles, or when traveling to a new area, or when parking is expensive or difficult to find.

When asked about influential factors on mode choice decisions, convenience, cost, and reliability were voted as the top three that had the greatest impact. Not surprisingly, "environmental issues" is the factor perceived as least important by the respondents. Interestingly multitasking capabilities, expressed as cellphone/tablet usage during their trips, showed the lowest importance as well. This reveals that multitasking is not yet a decisive factor in individuals' mode choice decisions, although it has been viewed as an advantage of some modes, including ride-sourcing, taxi, and public transit.

In the aims to understand what it takes to make ride-sourcing attractive, the respondents were asked about the minimum savings in travel time and mobility costs that would persuade them to switch to ride-sourcing modes. The results are shown in Figure 5. It shows that around 38% of the respondents would consider switching to ride-sourcing if it could reduce their monthly mobility costs by $50, and this percentage would increase to 66% if the cost saving increases to $100 per month. An interpolation analysis derived a median cost saving of $72.4 per month. This means that 50% of Americans would switch to ride-sourcing if it saves them $72.4 in mobility cost per month. A similar analysis for travel time savings showed a median value of 15.9 min per trip. Considering that the average trip length for the sample is about 23.5 minutes, it calls for a significant reduction in travel time (about 67%) to result in a switch of mode to ride-sourcing.

How Americans View Emerging Mobility Options

To obtain a general understanding of Americans' attitudes and perceptions toward mobility options, the survey designed a section with questions in three categories: (1) general preferences in mobility, (2) views on ride-sourcing, and (3) perceptions of AV.

Sharing rides with strangers remained a big issue, as around 53.7% of the respondents either agreed or strongly agreed that they would hardly trust to travel with strangers, and around 79.1% stated that they felt more convenient traveling by themselves. Interestingly,

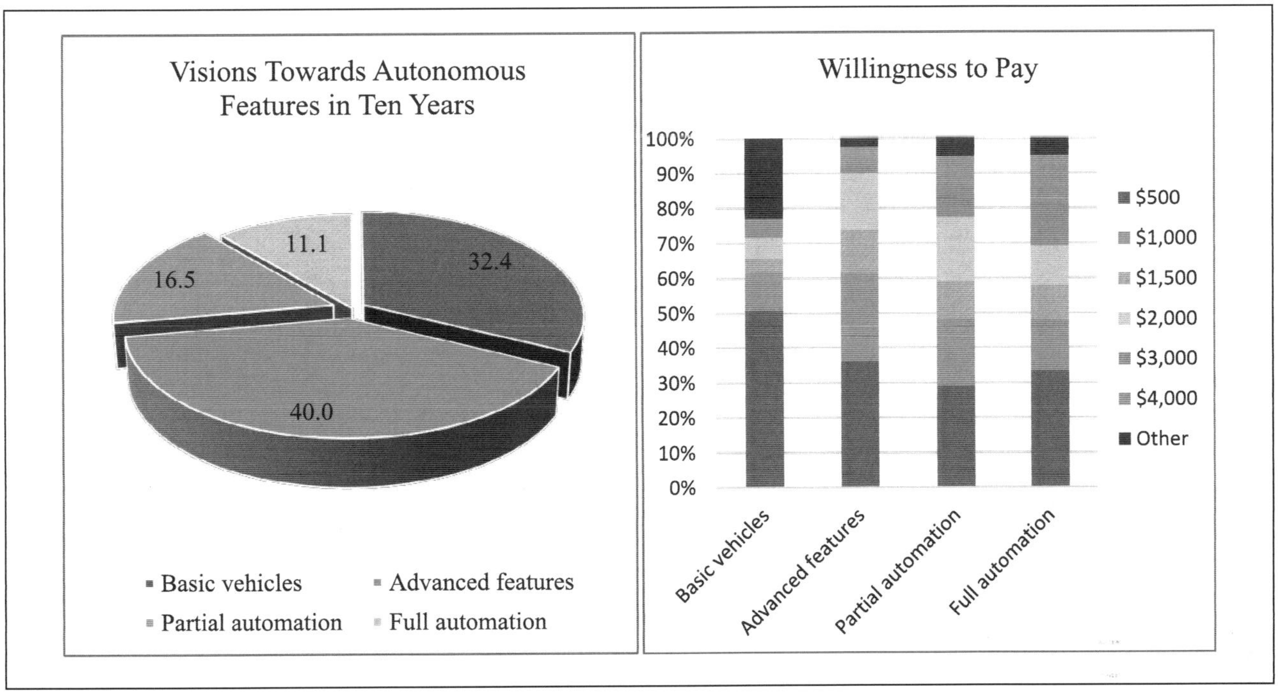

Figure 6. Willingness to pay for different levels of automation.

only around 45% of the respondents preferred to multitask during their commute trip if it was possible to do so, while about 28% preferred not to multitask.

Respondents were also asked about their visions of future enhancements in vehicle/mobility technology within the next 10 years, and their willingness to pay for these advanced features.

[*I see myself driving/riding in these vehicles in ten years.*

- *Basic vehicles (the driver is in complete and sole control of the primary vehicle functions)*
- *Advanced features (such as adaptive cruise control, lane centering, crash warning, etc.)*
- *Partial automation (allows the vehicle to take over only under certain conditions and returns control to the driver when conditions change)*
- *Full automation (allows the vehicle to take over all driving functions for an entire trip)*]

[*I'm willing to pay $ ___ more for these features*
a) 500 b) 1000 c) 1500 d) 2000 e) 3000 f) 4000 g) other______]

The results are shown in Figure 6. Around 32.4% of the respondents saw themselves driving basic vehicles with no advanced features. Some 40% of individuals expected to drive vehicles with some kind of advanced features such as crash warning or adaptive cruise control system, and 16.5% of respondents were anticipating partially automated vehicles (i.e., automation is active only under certain conditions), whereas only around 11.1% would drive fully automated cars in the next 10 years.

Less than 50% of the respondents were willing to pay $1,500 or more for partial automation and full automation features. The weighted average willingness to pay for advanced features, partially automated vehicles, and fully automated vehicles were about $1,052, $1,483, and $1,639, respectively.

In terms of major motivations to ride in AVs, improved safety, reduced driving stress, and better technology were the three most popular choices. Similarly, the most desired automated features selected by the respondents were safety enhancements expressed in terms of collision avoidance or collision severity reduction, and improvement in fuel efficiency, supported by 48%, and 44% of the participants, respectively.

SP Choices

As described earlier, respondents were allocated to different scenario types based on their current mode choice. The data contain 428 samples with 2,996 scenarios in the driver (S1) scenario type, 154 samples and 1,078 scenarios in the passenger (S2) scenario type, 95 samples and 760 scenarios in the transit-captive (S3) scenario type, and 582 samples and 4,656 scenarios in the visitor (S4) scenario type.

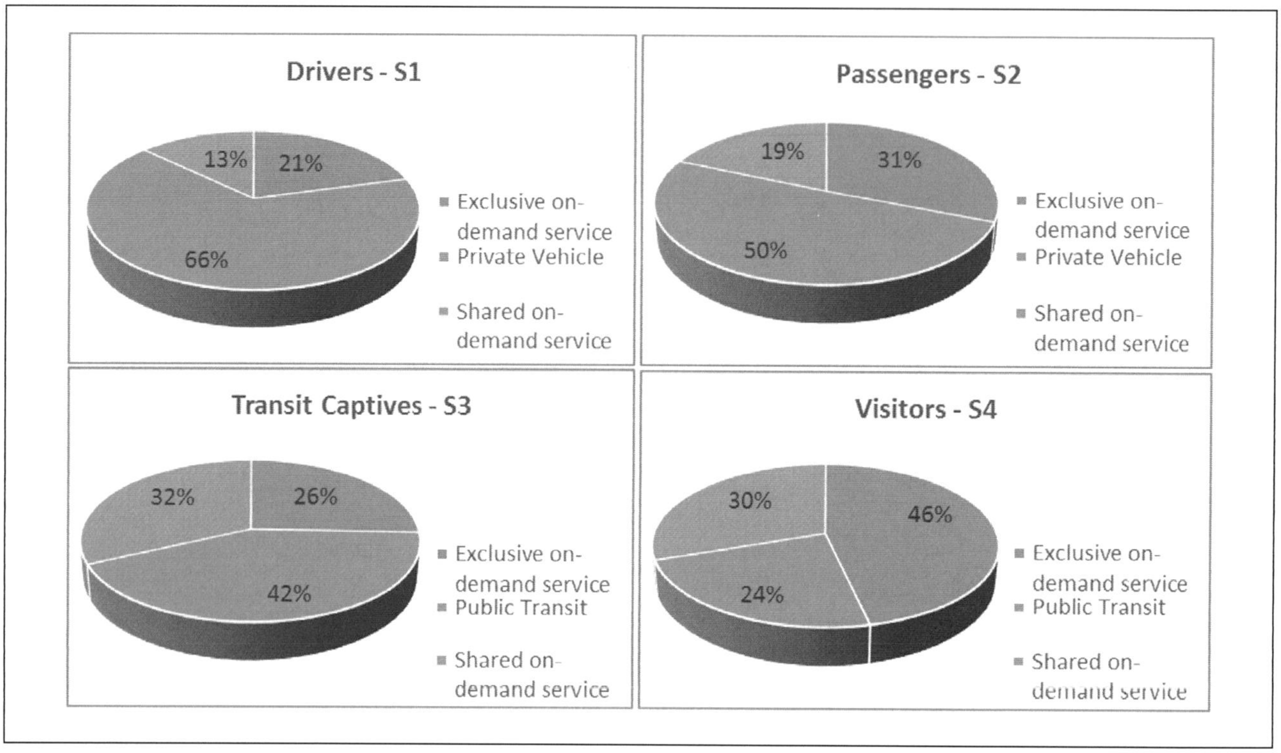

Figure 7. SP choices by scenario type.

Figure 7 presents the mode choice share for each scenario type. A comparison between the scenario types reveals some interesting findings. For the first two scenario types, nearly all respondents relied on private vehicle for their daily activities, 98% for drivers (S1) and 100% for passengers (S2) (by definition), based on the RP information. Figure 7 shows promising results for ride-sourcing modes, where 34% of drivers and 50% of passengers switched to on-demand modes given that they provide competitive services. The share between exclusive and shared rides was about the same for passengers (S2) and drivers (S1); a little more than one-third (38%) of the ride-sourcing trips was in shared rides.

When a private vehicle was not available (S3), public transit still showed the highest share for daily or regular arrangement, 72% for current condition, and 42% in the SP scenarios (S3 in Figure 7). This might indicate that on-demand services incorporating AV technologies, or shared automated vehicles (SAVs), with lower operating costs may be able to attract a significant portion of transit-captive users, but transit remains a viable option because of lower cost or other reasons. This may be confirmed by the notion that S3 showed higher propensity for shared services than exclusive services, which indicates that cost was a key factor in the mode choice decisions for these users, and/or they are not bothered by sharing services with others.

On the other hand, for visitors or those without access to a private vehicle in occasional conditions (S4), ride-sourcing services (76%) outperformed public transit (24%). Interestingly, the same share between exclusive and shared rides was observed here as in S1 and S2, about 39% of the ride-sourcing trips was in shared rides. It indicates that preferences on whether to share a ride with strangers do not vary much by the situation.

These findings are very important, given that the choice sets in transit scenario (S3) and visitor scenario (S4) were exactly the same, except that S3 was presented to transit users or those without access to a private vehicle on a regular basis, and S4 was designed for occasional situations when private vehicles are not available. It also should be noted that the same choice sets for driver (S1) and passenger (S2) were employed in terms of travel time and travel cost. Only the third attribute differs—level of multitasking for S1 and level of dependency (driver availability whether for ride-sourcing or household vehicles) for S2.

Conclusions

This paper presents preliminary findings from a nationwide survey conducted between April and May 2017, with the aim of understanding travelers' mode choice behavior in light of emerging mobility options, such as ride-sourcing and AV technologies. The survey targeted major metropolitan areas, and collected information on personal attributes, current mode choice patterns, influential factors and attitudes in mode choice decisions, as well as SP choices in different scenarios. Given the size limit, this paper focuses on descriptive analysis. Econometric modeling efforts will be presented in a separate paper.

The results suggest that ride-sourcing serves as a supplementary service to traditional mobility options. Currently, ride-sourcing has a comparable market size of transit in metropolitan areas. About 53% of the respondents had ride-sourcing experience, 37% of them used it once a month or more, and less than 10% used it daily. If the cost of ride-sourcing comes down significantly, it may become a viable mobility option for many travelers. The results showed that a monthly mobility cost savings of $72 may persuade 50% of the sample to switch to ride-sourcing modes. Similarly, a time savings of 16 minutes per trip may have the same effect.

In this regard, SP scenarios were designed to consider reduced operation cost of ride-sourcing services (integrated with AVs) and time savings in parking and access/egress. The results show a promising market propensity for on-demand services, especially when private vehicles are not available either on a long-term basis or under occasional conditions (out-of-town visitors, for example). Interestingly the choices between single ride and shared ride do not vary by situation. Most travelers prefer single ride than shared ride regardless of whether it is on a daily or occasional basis. However, for transit captives or those who use transit on a regular basis, shared rides showed higher potential than exclusive services, which may indicate that cost is a primary consideration in the mode choice decisions for these users.

This paper provides a comprehensive scan on current mode choice patterns and the factors and preferences that contributed to the decisions. The propensity of ride-sourcing under different conditions may provide some insights on designing on-demand services that target specific market segments. SP choice experiments also provided additional insight into travelers' mode choice behavior in view of the time and cost trade-offs among different mobility options. Future work will further examine the SP choices and investigate travelers' sensitivities toward time and cost, and how those may differ by personal attributes and attitudinal factors.

Acknowledgments

This work is funded by the research office of the Florida Department of Transportation.

Author Contributions

The authors confirm contribution to the paper as follows: study conception and design: Xia Jin, Hamidreza Asgari, Terrence Corkery; data collection: Xia Jin, Hamidreza Asgari; analysis and interpretation of results: Xia Jin, Hamidreza Asgari; draft manuscript preparation: Xia Jin, Hamidreza Asgari. All authors reviewed the results and approved the final version of the manuscript.

References

1. Litman, T. *Autonomous Vehicle Implementation Prediction, Implications for Transport Planning*. Victoria Transport Policy Institute, 2017.
2. Kim, K. H., D. H. Yook, Y. S. Ko, and D. H. Kim. *An Analysis of Expected Effects of the Autonomous Vehicles on Transport and Land Use in Korea*. Working Paper, NYU Marron Institute of Urban Management, 2015.
3. Bierstedt, J. A. Gooze, C. Gray, J. Peterman, L. Raykin, and J. Walters. *Effects of Next-Generation Vehicles on Travel Demand and Highway Capacity, FP Think*, 2014.
4. Lavasani, M., X. Jin, and Y. Du. Market Penetration Model for Autonomous Vehicles on the Basis of Earlier Technology Adoption Experience. *Transportation Research Record, Journal of the Transportation Research Board*, 2016. 2597: 67–74.
5. Anderson, J. M., K. Nidhi, K. D. Stanley, P. Sorensen, C. Samaras, and A. Oluwatobi. *Autonomous Vehicle Technology, A Guide for Policy Makers*. RAND Transportation, Space and Technology Program, 2014.
6. Pendyala, R. M., and C. R. Bhat. Activity-Travel Behavior Impacts of Driverless Cars. Presented at 93rd Annual Meeting of the Transportation Research Board, Washington, D.C., 2014.
7. Wallace., R., and G. Silberg. *Self-Driving Cars: The Next Revolution*. Center for Automotive Research, Transportation System Analysis Group, 2012.
8. Fagnant, D. J., and K. M. Kocleman. *Preparing a Nation for Autonomous Vehicles, Opportunities, Barriers and Recommendations*. Report Prepared for ENO Center for Transportation, 2013.
9. WallaceFreemark, Y. *Will Autonomous Cars Change the Role and Value of Public Transportation? The Transport Politic*, 2015. http://www.thetransportpolitic.com/2015/06/23/will-autonomous-cars-change-the-role-and-value-of-public-transportation/. Accessed on September 7, 2015.
10. Malokins, A., G. Circella, and P. L. Mokhtarian. How Do Activities Conducted while Commuting Influence Mode Choice? Testing Public Transportation Advantage and Autonomous Vehicle Scenarios. Presented at 94th Annual Meeting of the Transportation Research Board, Washington, D.C., 2015.
11. Levin, M. W., and S. D. Boyles. Effects of Autonomous Vehicle Ownership on Trip, Mode, and Route Choice. *Transportation Research Record: Journal of the Transportation Research Board*, 2015. 2493: 29–38.

12. KPMG. *The Clockspeed Dilemma, What Does It Mean for Automotive Innovation?* 2015.
13. Kornhauser, A., A. Chang, C. Clark, J. Gao, D. Korac, B. Lebowitz, and A. Swoboda. *Uncongested Mobility for All, New Jersey's Area-wide a Taxi System.* Operations Research and Financial Engineering, Princeton University, 2013.
14. Schoettle, B., and M. Sivak. *A Survey of Public Opinion About Autonomous and Self-driving Vehicles in the U.S., the U.K., and Australia.* Technical Report No. UMTRI-2014-21. 2014.
15. Schoettle, B., and M. Sivak. *Public Opinion about Self-Driving Vehicles in China, India, Japan, The U.S., The U.K., and Australia.* Report No. UMTRI-2014-30. The University of Michigan Transportation Research Institute, 2014.
16. Schoettle, B., and M. Sivak. *Potential Impact of Self-Driving Vehicles on Household Vehicle Demand and Usage.* Report No. UMTRI-2015-3. The University of Michigan, Transportation Research Institute, 2015.
17. Schoettle, B., and M. Sivak. *Motorists' Preferences for Different Levels of Vehicle Automation.* Report No. UMTRI-2015-22. The University of Michigan, Transportation Research Institute, 2015.
18. Menon, N. *Consumer Perception and Anticipated Adoption of Autonomous Vehicle Technology: Results from Multi-Population Surveys.* Graduate theses and dissertations. University of South Florida, 2015.
19. Duncan, M., N. Charness, T. Chapin, M. Horner, L. Stevens, A. Richard, D. Sounders, J. Crute, A. Riemondy, and D. Morgan. *Enhanced Mobility for Aging Populations Using Automated Vehicles.* Report No. 035048. Florida State University Department of Urban and Regional Planning, Draft Final report for The Florida Department of Transportation, 2015.
20. Shabanpour, R., N. Golshani, J. Auld, and A. Mohammadian. *Willingness-to-pay for Automated Vehicles Technology: A Random Parameters and random Thresholds HOPIT Model.* International Choice Modelling Conference 2017, Cape Town, South Africa, 2017.
21. Howard, D., and D. Dai. Public Perceptions on Self-driving Cars: The Case of Berkeley. Presented at 93rd Annual Meeting of the Transportation Research Board, Washington, D.C., 2014.
22. Cisco. *Consumers Desire More Automated Automobiles, According to Cisco Study*, 2013. [press release].
23. Seapine Software. *Study Finds 88 Percent of Adults would be Worried about Riding in a Driverless Car*, 2014. [press release].
24. Danise, A. *Women Say No Thanks to Driverless Cars, Survey Finds; Men Say Tell Me More*, 2015. http://www.nerdwallet.com/blog/insurance/2015/06/09/survey-consumer-fears-self-driving-cars/
25. Intel. *The Vote Is In: Citizens Support 'Smart Cities' with Driverless Cars, Public Service Drones and Surroundings the Sense Activities*, 2014. [press release].
26. Casley, S. V., A. S. Jardim, and A. M. Quartulli. *A Study of Public Acceptance of Autonomous Cars.* Bachelor of Science thesis. Worcester Polytechnic Institute, Worcester, MA, USA, 2013.
27. Jin, X., and H. Asgari. *Potential Implications of Automated Vehicle Technologies on Travel Behavior.* Final Report. Florida Department of Transportation, 2018.
28. Fagnant, D. J., and K. M. Kockelman. The Travel and Environmental Implications of Shared Autonomous Vehicles, Using Agent-Based Model Scenarios. *Transportation Research Part C*, No. 40, 2014, pp. 1–13.
29. Fagnant, D. J., K. M. Kockelman, and P. Bansal. Operations of a Shared Autonomous vehicle Fleet for the Austin, Texas Market. *Transportation Research Record, Journal of the Transportation Research Board*, 2015. 2536: 98–106.
30. Zhang, W., S. Guhathakurta, J. Fang, and G. Zhang. Exploring the Impact of Shared Autonomous Vehicles on Urban Parking Demand: An Agent-Based Simulation Approach. *Sustainable Cities and Society* No. 19, 2015, pp. 34–45.
31. Kreuger, R., T. H. Rashidi, and J. M. Rose Preferences for Shared Autonomous Vehicles. *Transportation Research Part C* No. 69, 2016, pp. 343–355.
32. The American Automobile Association. *Your Driving Costs*, 2016. http://exchange.aaa.com/automobiles-travel/automobiles/driving-costs/#.WBExMPkrJhE
33. Bert, J., B. Collie, M. Gerrits, and G. Xu. *What's Ahead for Car Sharing? The New Mobility and Its Impacts on Vehicle Sales.* The Boston Consulting Group, 2016.
34. Burns, L. D., W. C. Jordan, and B. A. Scarborough. *Transforming Personal Mobility.* The Earth Institute, Columbia University, 2013.
35. Corwin, S., J. Vitale, E. Kelly, and E. Cathles. *The Future of Mobility, How Transportation Technology and Social Trends Are Creating a New Business Ecosystem.* Deloitte University Press, 2015.
36. Litman, T. Evaluating Carsharing Benefits. Victoria Transport Policy Institute, 2015.
37. Uber. https://www.uber.com/. Accessed January 2017.
38. ZimRide. http://www.mssf.org/MSSF%20Zimride%20-FAQ.pdf. Accessed January 2017.
39. The American Public Transportation Association. *2016 Public Transportation Fact Book*, 2017. http://www.apta.com/resources/statistics/Documents/FactBook/2016-APTA-Fact-Book.pdf
40. Census. https://www.census.gov/programs-surveys/metro-micro.html. Accessed March 2017.

The Standing Committee on Traveler Behavior and Values (ADB10) peer-reviewed this paper (18-00445).

The opinions, findings and conclusions expressed in this publication are those of the authors and not necessarily those of the Florida Department of Transportation or the U.S. Department of Transportation.

Article

TRR
JOURNAL OF THE TRANSPORTATION RESEARCH BOARD

Transportation Research Record
2018, Vol. 2672(47) 23–34

DOI: 10.1177/0361198118787629
journals.sagepub.com/home/trr
SAGE

Tolerance and Indifference Bands in Regret–Rejoice Choice Models: Extension to Market Segmentation in the Context of Mode Choice Behavior

Sunghoon Jang[1], Soora Rasouli[2], and Harry Timmermans[2]

Abstract

Random regret minimization models (RRMs), based on seminal work in regret theory, have been introduced into transportation research as an alternative to expected/random utility models. With ample applications in diverse choice contexts, the RRMs have been extended to include the effect of "rejoice," the counterpart of the emotion of regret. The fundamental assumption of regret–rejoice models is that when the chosen alternative is inferior to non-chosen alternatives with respect to an attribute, individuals feel regret; otherwise, if the chosen alternative is superior to non-chosen alternatives, individuals rejoice. The regret and rejoice functions are assumed to be continuous in attribute differences. However, individuals may tolerate small attribute differences when judging regret and be indifferent to small differences when assessing rejoice. This paper therefore introduces tolerance and indifference bands in random regret–rejoice choice models, and compares the performance of these models against the performance of the original models. Furthermore, it is assumed that tolerance and indifference bands differ by trip purpose. Empirical results testify to the better performance of the models with the tolerance and indifference bands, and show that trip purpose is an important factor affecting tolerance and indifference bands.

Different disciplines have developed different concepts and modeling approaches to predict individual choice behavior. Dominant in economic theory has been the concept of (expected) utility, which states that individuals choose an alternative so as to maximize the utility they derive from the attributes of the choice alternatives. (Expected) utility theory (*1*) has been the foundation of many discrete choice models that have been applied in transportation research and many other application areas (*2*).

Although utility maximizing models have produced adequate results when forecasting individual choice behavior and aggregate market shares, several researchers have argued that in many cases observed choices do not reflect the maximization of utility. This has led to a quest in behavioral economics and other decision sciences for models that are based on concepts other than utility and decision rules different from utility-maximization to move away from models of purely rational behavior.

One of these alternative approaches, regret theory, is based on the contention that individuals experience regret when they realize the choice they made is not the best possible one (*3*, *4*). Consequent experiments have demonstrated that the negative emotion of regret significantly affects decision making processes. Chorus et al. introduced the random regret minimization model (RRM), which is inspired by, but different from, seminal regret theory, into transportation research (*5*). Regret avoidance was translated into the decision principle of regret minimization. Specifying the attribute regret as proportional to the attribute difference from the best alternative for that attribute, and assuming that overall regret equals the maximum of the sum of attribute regrets in the comparison of alternatives, adopting the independent and identically distributed (IID) assumption of the multinomial logit (MNL) model then yields the random regret minimization (RRM) model. Most applications, however, are based on an alternative specification, in which the double max operators are replaced with the

[1]Eindhoven University of Technology, Department of Urban Science and Systems, Urban Planning Group, Eindhoven, The Netherlands
[2]Eindhoven University of Technology, Urban Planning Group, Eindhoven, The Netherlands

Corresponding Author:
Address correspondence to Sunghoon Jang: s.jang@tue.nl

principle of all pairwise comparisons and the attribute difference function, truncated at zero, is approximated by a continuous function (*6*). Since its introduction, the RRM model has been applied to many diverse choice problems under certain (*7*–*14*) and uncertain (*15*, *16*) conditions.

Both conventional RRM and random utility maximization (RUM) models are based on the postulate that individuals are rational economic agents, implying that they discriminate even small differences between alternatives. Under various circumstances, this postulate may not be valid (*17*). Kahneman argued that cognitive processes are bounded by biological limitations that prevent individuals from making perfect (rational) choices (*18*). Rasouli and Timmermans argued that bounded rationality triggers individuals to use simplifying decision rules (*19*). Ignoring small attribute differences in judgmental processes is one example of simplifying decision rules.

There is a consistent body of knowledge, accumulated in a variety of disciplines across a diverse set of topics, suggesting that individuals are insensitive to small attribute (stimuli) differences, either because they cannot perceive the differences or because they are indifferent in that attribute range. For instance, Georgescu-Roegen argued that individuals perceive differences between two commodities only when attribute differences exceed some necessary minimum (*20*). Coombs et al. proposed the concept of thresholds of perception, called "just noticeable differences," to indicate that individuals do not perceive a difference if a change of a stimulus is below a certain threshold (*21*).

In discrete choice modeling, the principle of indifference has been operationalized by adding an indifference threshold to the deterministic utility difference of choice alternatives in calculating the probability of choosing an alternative under the assumption of utility maximizing behavior (*22*–*24*). If the deterministic utility difference between two alternatives is below a certain threshold, individuals are assumed to perceive or judge the two alternatives as being the same. However, it has not been acknowledged that incorporating indifference thresholds in RUM should satisfy two assumptions. First, the error terms should be independent of the deterministic utility value. Second, the error terms should be identically independently Gumbel distributed. If multiplicative error terms are assumed (*25*, *26*), or the error terms are not Gumbel distributed, the methodology is difficult to apply. Even if these conditions are met, because of the complex dependencies of indifference thresholds between utilities, it is very difficult to extend the model to many alternatives in the choice set.

In the context of regret, insensitivity can be interpreted as individuals being tolerant about small attribute differences. Regret is only felt if the chosen alternative is outperformed by some margin. Jang et al. incorporated attribute tolerance into deterministic regret functions (*27*, *28*). This specification can be easily applied to any type of error term. The tolerance based regret models outperformed their conventional counterparts for two data sets. This positive result stimulated exploration of the application of this idea to more complex choice models.

In this paper, tolerance thresholds are estimated in mixed regret–rejoice choice models. The models assume that individuals experience the negative emotion of regret for inferior attributes and rejoice at the superior attributes (*27*–*31*). Thus, it is necessary to extend the model and test whether the notion of thresholds applies to both negative and positive emotions which are triggered by attribute differences. Do individuals apply an indifference/tolerance band around the reference point of identical attribute values between pairs of alternatives (*32*–*35*)? Do people tolerate a small loss (regret) and are they indifferent to a small gain (rejoice)? Is the band asymmetric between regret and rejoice? In addition, market segmentation analysis based on trip purpose is conducted. When punctuality is required for a trip, such as a work-related trip, individuals may have quite a small tolerance for attribute difference. Otherwise, if they have time to spare, such as on a non-work-related trip, they may have larger tolerance.

The regret–rejoice model incorporating tolerance thresholds will be compared against the original regret–rejoice model using revealed choice data about mode choice. In addition to comparing the predictive performance of the models, *K*-fold validation tests will be conducted. Further, to examine the effect of trip purpose, market segmentation will be applied, and the thresholds between segments statistically compared. For further insights for transportation policy, elasticities will be compared in each case.

The remainder of this paper first summarizes the development from regret-based models to regret–rejoice-based models as they have been formulated in transportation research. Next, the proposed attribute indifference and tolerance models are outlined. Then, the estimation and validation results are discussed and the results of the market segmentation reported. The paper is completed by a discussion of the results and avenues of future research.

Regret–Rejoice Choice Model

The original RRM models (*5*, *6*) introduced into transportation research assume that individuals' choice only depends on the negative emotion (semi-compensatory decision rule), called regret. This assumption may be too extreme. Therefore, several recent studies have focused on the effect of positive emotions when the chosen alternative outperforms the non-chosen one, the concept of rejoice (*27*–*31*). With the inclusion of rejoice, the random regret–(rejoice) minimization models are based on a fully compensatory decision rule. In this study, to incorporate thresholds in the rejoice function in the same manner as in the regret function, the linear regret–rejoice model is used (*27*, *28*).

When chosen alternative i is compared with the non-chosen alternative j, regret–rejoice for individuals n can be defined as

$$RGJ_{ijn} = \sum_k \left(\varphi_{ijnk}^{RG} + \varphi_{ijnk}^{RJ} \right) \quad (1)$$

where φ_{ijnk}^{RG} is an attribute-level regret with respect to attribute k, defined as a negative emotion when people realize they made the wrong choice in the sense that the alternative they chose is outperformed by one or more foregone or non-chosen alternative(s). Therefore, it is formulated following the original RRM (*5*),

$$\varphi_{ijnk}^{RG} = \max\left[0, \beta_k^{RG}\left(x_{jnk} - x_{ink}\right)\right] \quad (2)$$

and φ_{ijnk}^{RJ} is an attribute-level rejoice, defined as a positive emotion when people realize they made the right choice in the sense that the alternative they choose outperforms one or more foregone or non-chosen alternative(s). Then, it is defined (*27*, *28*) as

$$\varphi_{ijnk}^{RJ} = \min\left[0, \beta_k^{RJ}\left(x_{jnk} - x_{ink}\right)\right] \quad (3)$$

Then, this regret–rejoice model can be defined for multi-alternative choice sets, either (1) against the best non-chosen alternative (Equation 4) or (2) against all non-chosen alternatives (Equation 5).

$$RGJ_{in} = \max_{j \neq i}\left(RGJ_{ijn}\right) \quad (4)$$

$$RGJ_{in} = \sum_{j \neq i}\left(RGJ_{ijn}\right) \quad (5)$$

Note that these regret–rejoice models represent flexible reference-dependent preferences with different taste weights for regret (β_k^{RG}) and rejoice (β_k^{RJ}). For more detail, see Jang et al. (*27*, *28*).

Total regret–rejoice is assumed to be stochastic, where $RRGJ_{in}$ is random regret-rejoice.

$$RRGJ_{in} = RGJ_{in} + \varepsilon_{in} \quad (6)$$

Based on the assumption that the negative error terms ($-\varepsilon_{in}$) are identically independently Gumbel distributed, where P is choice probability and C is choice set,

$$P_{in} = \frac{\exp\left(-RGJ_{in}\right)}{\sum_{j \in C} \exp\left(-RGJ_{jn}\right)} \quad (7)$$

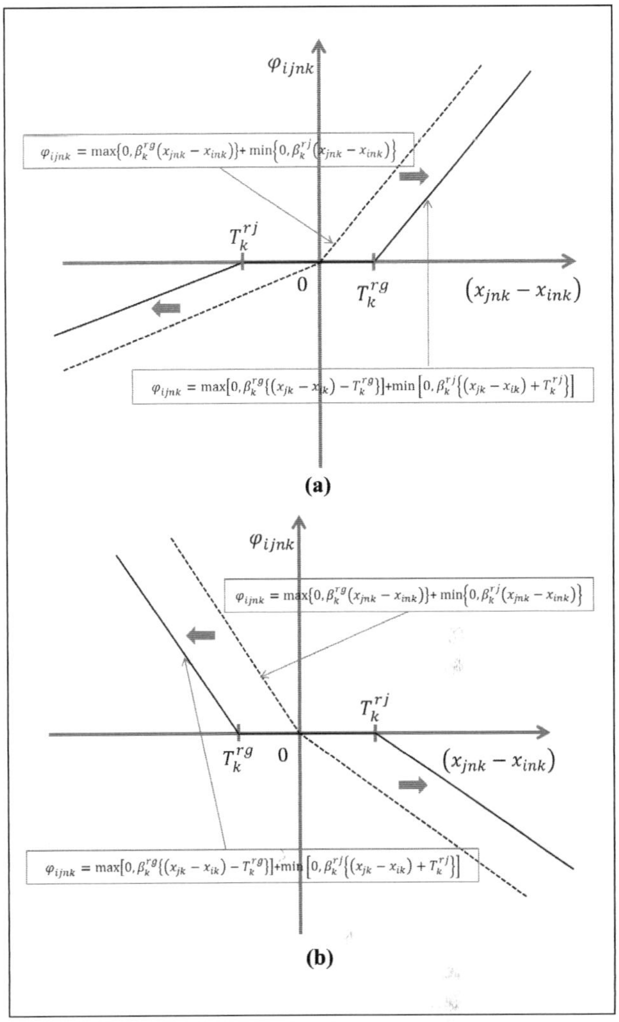

Figure 1. Regret–rejoice specification incorporating tolerance threshold, (*a*) when taste weight for regret (β_k^{rg}) and rejoice (β_k^{rj}) have positive values, and (*b*) when taste weight for regret (β_k^{rj}) and rejoice (β_k^{rj}) have negative values.

Tolerance Thresholds in Regret–Rejoice Models

How can the indifference/tolerance threshold be incorporated into the regret–rejoice model? In the regret domain, individuals feel regret when the attribute difference between the chosen and non-chosen alternative exceeds a certain tolerance threshold (Figure 1):

$$x_{jnk} - x_{ink} > \Delta_k^{RG} \mid \beta_k^{RG} > 0 \quad (8)$$

$$x_{jnk} - x_{ink} < \Delta_k^{RG} \mid \beta_k^{RG} < 0 \quad (9)$$

Similarly, individuals rejoice if

$$x_{jnk} - x_{ink} < \Delta_k^{RJ} | \beta_k^{RJ} > 0 \quad (10)$$

$$x_{jnk} - x_{ink} > \Delta_k^{RJ} | \beta_k^{RJ} < 0 \quad (11)$$

Therefore, the band defined by the indifference and tolerance thresholds equals

$$\Delta_k^{RJ} \leq x_{jnk} - x_{ink} \leq \Delta_k^{RG} | \beta_k^{RG} > 0, \beta_k^{RJ} > 0 \quad (12)$$

$$\Delta_k^{RG} \leq x_{jnk} - x_{ink} \leq \Delta_k^{RJ} | \beta_k^{RG} < 0, \beta_k^{RJ} < 0 \quad (13)$$

Within this band, individuals do not feel any negative (regret) or positive (rejoice) emotions for the chosen alternative from the comparison with non-chosen alternative(s). The thresholds can be incorporated as

$$\varphi_{ijnk}^{RGT} = \max\left[0, \beta_k^{RG}\left(\left(x_{jnk} - x_{ink}\right) - \Delta_{ink}^{RG}\right)\right] \quad (14)$$

$$\varphi_{ijnk}^{RJT} = \min\left[0, \beta_k^{RJ}\left(\left(x_{jnk} - x_{ink}\right) + \Delta_{ink}^{RJ}\right)\right] \quad (15)$$

$$RGJT_{ijn} = \sum_k \left(\varphi_{ijnk}^{RGT} + \varphi_{ijnk}^{RJT}\right) \quad (16)$$

$$RGJT_{in} = \max_{j \neq i}\left(RGJT_{ijn}\right) \quad (17)$$

$$RGJT_{in} = \sum_{j \neq i}\left(RGJT_{ijn}\right) \quad (18)$$

$$RRGJT_{in} = RGJT_{in} + \varepsilon_{in} \quad (19)$$

The thresholds may depend on the attribute value of the chosen alternative. If someone traveled for 10 min, and the travel time of the non-chosen route turned out to be 5 min, individuals might feel regret for the 5 min travel time difference. However, when someone traveled for 120 min, and the non-chosen route turned out to be 115 min, individuals might feel less or no regret for the 5 min less travel time. This is similar to the case in which the travel time of the chosen alternative is less: individuals may feel less or no rejoicing for the same travel time difference when the travel time for the chosen alternative is large enough. Therefore, it is assumed that the thresholds Δ_{ink} of chosen alternative i for attribute k for individual n are a percentage of the attribute value x_{ink} of the chosen alternative.

$$\Delta_k^{RG} = \delta_k^{RG} x_{ink} \quad (20)$$

$$\Delta_k^{RJ} = \delta_k^{RJ} x_{ink} \quad (21)$$

where δ_k^{RG} in Equation 20 is a threshold ratio for regret and δ_k^{RJ} in Equation 21 is a threshold ratio for rejoice. As shown in Figure 1, it is expected that the threshold parameters should have the same sign as the taste weights: Δ_k^{RG} should have the same sign as β_k^{RG}, and the sign of Δ_k^{RJ} should be the same as the sign for β_k^{RJ}.

$$\delta_k^{RG} = Y_k^{RG} \zeta_k^{RG} \quad (22)$$

$$\delta_k^{RJ} = Y_k^{RJ} \zeta_k^{RJ} \quad (23)$$

where Y_k^{RG}, Y_k^{RJ} are binary variables for regret and rejoice respectively: when $\beta_k^{RG} > 0$, $Y_k^{RG} = 1$, otherwise $Y_k^{RG} = -1$; similarly when $\beta_k^{RJ} > 0$, $Y_k^{RJ} = 1$, otherwise $Y_k^{RJ} = -1$. In addition, $\zeta_k^{RG}$and ζ_k^{RJ} are threshold parameters for regret and rejoice, to be estimated, and expected to be equal to or larger than zero.

The threshold parameters $\zeta_k^{RG}$and ζ_k^{RJ} can be deterministic or stochastic. As Jang et al. (*27*) found that stochastic tolerance thresholds for regret show better performance than deterministic thresholds, it is assumed that the threshold parameters are stochastic.

$$\zeta_k^{RG} = \bar{\zeta}_k^{RG} + \eta_{nk}^{RG} \quad (24)$$

$$\zeta_k^{RJ} = \bar{\zeta}_k^{RJ} + \eta_{nk}^{RJ} \quad (25)$$

where $\bar{\zeta}_k^{RG}$ and $\bar{\zeta}_k^{RJ}$ are the mean threshold values for regret and rejoice, and η_{nk}^{RG} and η_{nk}^{RJ} represent individual deviations following a certain distribution with zero mean and variance $\left(\sigma_\eta^{RG}\right)^2$ and $\left(\sigma_\eta^{RJ}\right)^2$, respectively.

Then, if the error terms (ε_{in}) are assumed to follow a mixture of the IID-Gumbel distribution and the distribution for the stochastic threshold as in the random coefficients model, the choice probability can be derived using maximum simulated likelihood estimation as

$$P_{in} = \iint L_{in}(\zeta^{RG}, \zeta^{RJ}) f(\zeta^{RG}|\theta^{RG}) f(\zeta^{RJ}|\theta^{RJ}) d\zeta^{RJ} d\zeta^{RG} \quad (26)$$

$$L_{in}(\zeta^{RG}, \zeta^{RJ}) = \frac{\exp(-R_{in}|\zeta^{RG}, \zeta^{RJ})}{\sum_{j \in C} \exp(-R_{jn}|\zeta^{RG}, \zeta^{RJ})} \quad (27)$$

where L is logit function. Thresholds (ζ^{RG}, ζ^{RJ}) are assumed to be distributed with density $f(\zeta^{RG}|\theta^{RG})$, $f(\zeta^{RJ}|\theta^{RJ})$, where θ^{RG}, θ^{RJ} respectively refer collectively to the threshold parameters of the distribution (generally mean and standard deviations of ζ) for regret and rejoice.

Because the computation in Equation 26 involves complex integrals with respect to distributions $f(\zeta^{RG})$, $f(\zeta^{RJ})$, it is difficult to solve it analytically. Therefore, the probabilities are approximated through simulation for any given value of ζ^{RG} and ζ^{RJ} (*36*):

1) Draw a value of ζ^{RG} and ζ^{RJ} from $f(\zeta^{RG})$ and $f(\zeta^{RJ})$ respectively, and label them ζ_r^{RG} and ζ_r^{RJ} with the subscript $r = 1$ referring to the first draw.

2) Calculate $\frac{\exp(-R_{in}|\zeta_r^{RG}, \zeta_r^{RJ})}{\sum_{j \in C} \exp(-R_{in}|\zeta_r^{RG}, \zeta_r^{RJ})}$ with this draw.

3) Repeat steps 1 and 2 R times, and average the results. This average is the simulated probability:

$$\hat{P}_{in} = \frac{1}{R}\sum_{r=1}^{R} L_{in}\left(\zeta_r^{RG}, \zeta_r^{RJ}\right) \quad (28)$$

where R is a certain number of draws estimating threshold parameters, and $\hat{P}_{in}$ is an unbiased estimator of P_{in}.

$$\ln SL^*(\beta;\zeta) = \sum_{n=1}^{N}\left[y_{in} ln\hat{P}_{in} + \cdots + y_{jn} ln\hat{P}_{jn}\right] \quad (29)$$

where SL^* is the simulated likelihood function and $\ln SL^*$ is the simulated log likelihood function. y_{in} is 1 if individual n chooses alternative i and 0 otherwise, and β, ζ are respectively the vector of regret weights and tolerance thresholds.

Empirical Evidence

To examine whether the inclusion of tolerance and indifference thresholds improves the predictive performance of the regret–rejoice model, a revealed preference data set about mode choice behavior in the Noord Brabant area of the Netherlands was used. Matlab R2015a was used to develop the code to estimate the various models. Likelihood values were calculated to compare the goodness-of-fit of the models. Because it was necessary to compare non-nested models with different numbers of estimated parameters, Ben-Akiva and Swait's test (*37*) for non-nested choice models was applied.

$$\overline{\rho_j}^2 = 1 - \frac{L(J) - K}{L(0)} \quad (30)$$

where $\overline{\rho_j}^2$ represents the fitness of model J, $L(J)$ is the log likelihood at convergence, and K is the number of estimated parameters.

$$\text{prob}\left(\left|\overline{\rho_2}^2 - \overline{\rho_1}^2\right| \geq Z\right) \leq \phi\left(-\sqrt{-2ZL(0) + (K_1 - K_2)}\right) \quad (31)$$

where Z is the difference between the fitness measures of the compared models and ϕ indicates the standard normal cumulative distribution function.

As Ben-Akiva and Swait's test (*37*) is based on Akaike's Information Criterion (AIC), which tends to prefer the larger model (by number of parameters) if the sample size is big enough, the Bayesian Information Criterion (BIC) (*38*) of each model was also calculated.

$$\text{BIC}_j = -2L(J) + K_J * \ln(N) \quad (32)$$

where N is the number of observations. By definition, the model with the lowest BIC is preferred.

In addition to the comparison of model performance, a K-fold cross-validation test was conducted for each case. The validation test requires splitting the data into K subsets, estimating the models using $K-1$ subsets, and testing the model using the remaining subset. This process is repeated K times because each subset should be used as a test set once. The 2-norms statistic, which quantifies the distance between predicted and observed choices (*39*) was used in this test. It can be expressed as

$$\text{2-norms} = \sqrt{D_{Alt.1}^2 + \cdots D_{Alt.i}^2} \quad (33)$$

where $D_{Alt.i}$ represents the absolute error between observed and predicted shares for alternative i. Note that if the predicted and observed market shares are the same, D is equal to zero. It increases if the difference between predicted and observed market share increases. Thus, by definition, a lower value of the 2-norms statistic indicates higher predictive power.

Moreover, elasticities were compared to examine how attribute difference thresholds affect choice probabilities. The elasticities compared were those of the original regret–rejoice choice model and the proposed variant incorporating tolerance thresholds. The elasticity of the regret–rejoice model indicates the relationship between a 1% change in the level of the attribute and the percentage change in the probability of choosing the alternative.

Further, it is assumed that attribute difference thresholds are affected by trip purpose. When the purpose requires punctuality (e.g., a commuting trip), they may be very sensitive to even small travel time differences, whereas for other trip purposes they may be more insensitive to small travel time differences. Therefore, a market segmentation by trip purpose was conducted, and the tolerance and indifference thresholds compared.

Data

For the first case study, the MON (Mobiliteit Onderzoek Nederlands—the Dutch National Travel Survey) data, collected in 2009, was used to estimate the regret models. A sample of representative residents in the Netherlands was randomly recruited to collect data about their daily travel. Data pertaining to the Province of Noord Brabant were extracted. The sample contains revealed preference data of a sample of 354 respondents with 1,158 trips. The extracted data include three mode choice alternatives (car, bike, walk) with two level-of-service variables. The estimated models thus predict the probability of choosing these transport modes as a function of these level-of-service variables. The travel time varies between 1 and 134 min, and travel distance ranges from 1 km to 91 km.

Estimation Results

Because of space limitations, the focus in this paper is limited to Equations 6 and 19. Table 1 shows the estimation results and the difference in predictive performance

Table 1. Summary of Estimation Results of Regret–Rejoice Models: Specification Incorporating Stochastic Threshold

	Regret–Rejoice model	
Mode choice	Original formulation	Stochastic threshold
Taste weight		
Travel time_regret	−0.0445	−0.0548
(t-value)	(−21.69)	(−17.59)
Travel distance_regret	−0.171	−0.247
(t-value)	(−15.50)	(−16.02)
Travel time_rejoice	−0.0327	−0.0364
(t-value)	(−4.31)	(−3.78)
Travel distance_ rejoice	−0.101	−0.129
(t-value)	(−3.90)	(−4.43)
Tolerance threshold		
Tolerance threshold _Travel time_regret_mean (t-value)		0.203 (3.94)
Tolerance threshold _Travel time_regret_standard deviation (t-value)		0.966 (3.27)
Tolerance threshold _Travel distance_regret_mean (t-value)		0.320 (4.09)
Tolerance threshold _Travel distance_regret_standard deviation (t-value)		1.145 (2.92)
Tolerance threshold _Travel time_rejoice_mean (t-value)		0.247 (5.26)
Tolerance threshold _Travel time_rejoice_standard deviation (t-value)		1.022 (3.31)
Tolerance threshold _Travel distance_rejoice_mean (t-value)		0.301 (4.55)
Tolerance threshold _Travel distance_rejoice_standard deviation (t-value)		1.488 (4.79)
Model fit		
R-squared	0.241	0.312
Adj. R-squared	0.238	0.302
Final log likelihood	−965.394	−875.392
BIC	1959.006	1835.437

of the original regret–rejoice model and the specification based on stochastic thresholds for both regret and rejoice. All parameters are statistically significant at the 95% significance level. Their signs are consistent with theoretical expectations.

Whereas the mean stochastic threshold is larger for rejoice than regret for travel time, it is smaller for rejoice than regret for travel distance. This means that individuals seem more sensitive to loss (regret) than gain (rejoice) for small travel time differences, but more sensitive to gain (rejoice) than loss (regret) for small travel distance differences.

Furthermore, the band for travel time ($-0.203{*}x_{int} \leq x_{jnt} - x_{int} \leq 0.247{*}x_{int}$) suggests that if an individual's chosen alternative has a travel time of 10 min, and the travel time of a non-chosen alternative is between 8 and 12.5 min, the individual perceives or judges the difference as irrelevant. Likewise, the band for travel distance ($-0.320{*}x_{ind} \leq x_{jnd} - x_{ind} \leq 0.301{*}x_{ind}$) indicates that if the travel distance of the chosen alternative is 10 km, and the distance of a non-chosen alternative is between 6.8 and 13.0 km, the difference is not perceived or judged as significant and both alternatives are considered equal.

Based on Ben-Akiva and Swait's test (*37*) for non-nested choice models, it is possible to conclude that the tolerance band significantly improves the predictive power of the regret–rejoice model. The probability that the original regret–rejoice model is superior to the

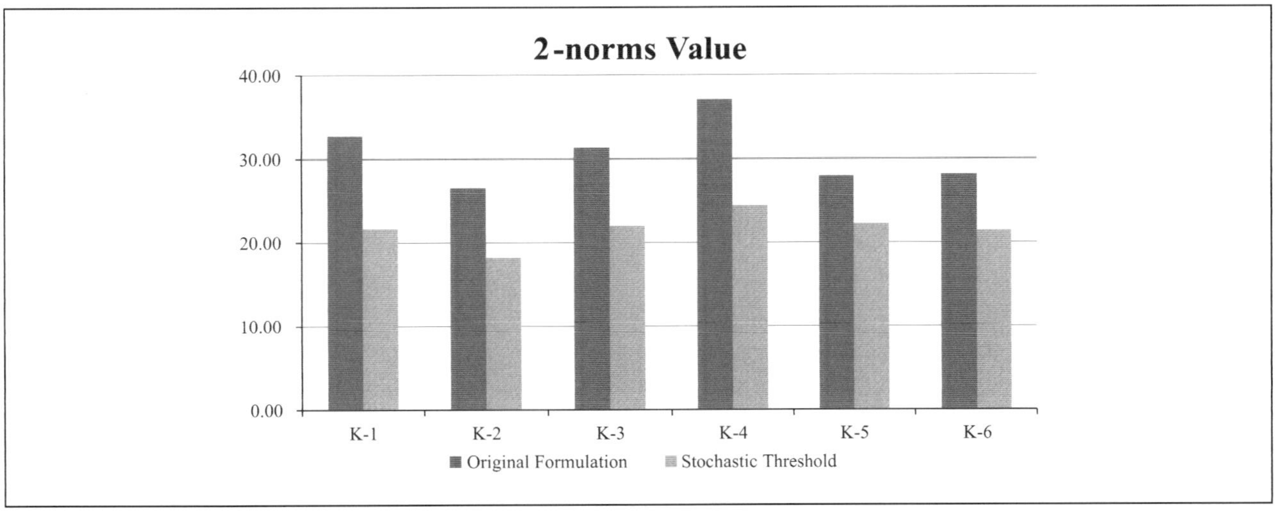

Figure 2. 2-norms values for *K*-fold cross-validation test: Comparison of original regret–rejoice formulation and its variants incorporating stochastic tolerance thresholds.

Table 2. Observed and Predicted Market Shares

Mode choice	Observed market share (%)	Predictive market share (%): Regret–rejoice model	
		Original formulation	Stochastic threshold
Alt. 1 (car)	40.59	38.50	39.03
Alt. 2 (bus)	32.64	36.12	34.06
Alt. 3 (walk)	26.77	25.38	26.91
RMSE	-	4.29	2.11

Note: RMSE = Root mean square error.

specification incorporating stochastic tolerance thresholds is zero ($p \leq (-12.444) \cong 0$). This result is also supported by BIC. The BIC of the threshold specification is 1835.437, which is lower than the BIC value of the original regret–rejoice model (1959.006).

Validation Results

The models were also validated using *K*-fold cross-validation. To that end, the data were randomly split into six different subsets. The models were estimated by pooling five of these subsets. The estimated model was then used to predict the observations of the remaining subset. By repeating this procedure six times, each subset was used for validating the model once.

Figure 2 shows the results of the cross-validation test using the value of the 2-norms statistic to compare the original regret–rejoice formulation and its variant incorporating stochastic thresholds. Results demonstrate that the estimated parameters do not differ much between the subsets used for model estimation. The model incorporating stochastic tolerance thresholds consistently shows better performance than its original counterpart in terms of the 2-norms statistic.

Market Shares

The comparison between observed and predicted market shares is shown in Table 2. It is clear that consideration of attribute difference tolerance makes the difference between the observed and predicted market smaller in the regret–rejoice model. Without consideration of the attribute difference tolerance, the market shares for alternatives 1 (car) and 3 (walk) are under-estimated, whereas the market share for alternative 2 (bus) is under-estimated. However, by considering the attribute difference tolerance, albeit the market share is still under-estimated for alternative 1 and over-estimated for alternative 2, the difference becomes smaller. It even becomes over-estimated for alternative 3.

Considering the root mean square error (RMSE) in Equation 34, by considering attribute difference

Table 3. Elasticities: Comparison between Original Regret–Rejoice Formulation and Variant that Includes a Stochastic Threshold

	Regret–Rejoice Model					
Mode choice	Original formulation			Stochastic threshold		
Attributes	Alt. 1 (Car)	Alt. 2 (Bike)	Alt. 3 (Walk)	Alt. 1 (Car)	Alt. 2 (Bike)	Alt. 3 (Walk)
Travel time	–0.66	–1.45	–2.87	–0.26	–0.64	–0.88
Travel distance	–0.91	–1.56	–3.01	–0.37	–0.75	–0.97

Table 4. Elasticities: Comparison between Market Segments

	Regret–Rejoice Model with Stochastic Threshold					
Mode choice	Work-related segment			Non-work-related segment		
Attributes	Alt. 1 (Car)	Alt. 2 (Bike)	Alt. 3 (Walk)	Alt. 1 (Car)	Alt. 2 (Bike)	Alt. 3 (Walk)
Travel time	–0.48	–1.14	–2.09	–0.11	–0.35	–0.51
Travel distance	–0.73	–1.21	–2.32	–0.20	–0.42	–0.55

tolerance, the value of RMSE decreases (original formulation: 4.29, incorporation thresholds: 2.11). By definition, this indicates that the consideration of attribute difference tolerance generates market shares closer to the observed ones.

$$\text{RMSE} = \sqrt{\sum_j (\text{Observed market share\_Predictive market share})^2} \quad (34)$$

Elasticity

Analysis of the effect of attribute difference tolerance to changes in the travel demand of each mode was carried out. Because of the space limitation, it is not possible to include details of the derivation of elasticity. For the detailed derivation of elasticity with tolerance thresholds, readers can refer to Jang et al. (*27*).

The elasticities of the original regret–rejoice formulation and the regret–rejoice model with thresholds are shown in Table 3. The stochastic thresholds decrease the elasticities of the regret–rejoice model in every case. This is because it is assumed that individuals feel regret or rejoice for small attribute differences in the original regret–rejoice model, but not in the model with the stochastic thresholds. For example, whereas a 1% increase in travel distance decreases the probability of choosing alternative 3 (walk) by 3.01% in the original regret–rejoice model, it only decreases it by 0.97% if the stochastic tolerance threshold is incorporated. These results provide relevant insights for transportation policy. For instance, policy makers believe that if they impose a penalty on car traffic (increasing travel time by car), the market share of public transport will increase, even for small increases in travel time. However, these expectations are ill-founded, as the existence of thresholds implies that the market share of public transportation will increase much less than expected, or may even not change at all.

Table 4 shows the elasticities of each segment when considering attribute difference tolerance in regret–rejoice models. The elasticities of work-related segments are higher than those of non-work-related segments in every case. This means that when individuals travel with work-related purpose, they are more sensitive to changes in attribute values. This implies that when travel time by car increases, many travelers with work-related purpose will use public transportation. However, if the increase is small, most travelers with a non-work-related purpose will stick to car travel.

Market Segmentation

Even when stochastic tolerance thresholds are used, individuals' tolerance may be quite different for different travel purposes. In the context of mode choice behavior, travelers' tolerance for attribute difference may depend on their purpose of travel: Individuals may be more sensitive to small travel time differences during the work commute than during non-work travel. Therefore, analysis of attribute indifference tolerance with market segmentation will provide more insights into traveler behavior.

To examine whether thresholds vary by trip purpose, this section proposes market segmentation,

Table 5. Summary of Estimation Results of Market Segmentation: Regret–Rejoice Specification Incorporating Stochastic Threshold

Mode choice	Regret–Rejoice with Stochastic Threshold: Work-related segment	Non-work-related segment
Taste weight		
Travel time_regret	−0.0692	−0.0363
(*t*-value)	(−22.27)	(−15.34)
Travel distance_regret	−0.284	−0.178
(*t*-value)	(−16.63)	(−12.76)
Travel time_rejoice	−0.0477	−0.0231
(*t*-value)	(−5.59)	(−3.86)
Travel distance_rejoice	−0.175	−0.084
(*t*-value)	(−3.94)	(−6.85)
Tolerance threshold		
Tolerance threshold _Travel time_regret_mean (*t*-value)	0.061 (2.85)	0.325 (5.07)
Tolerance threshold _Travel time_regret_standard deviation (*t*-value)	0.294 (3.66)	0.587 (3.21)
Tolerance threshold _Travel distance_regret_mean (*t*-value)	0.202 (4.77)	0.373 (4.59)
Tolerance threshold _Travel distance_regret_standard deviation (*t*-value)	0.498 (2.50)	0.629 (3.31)
Tolerance threshold _Travel time_rejoice_mean (*t*-value)	0.128 (3.64)	0.356 (2.73)
Tolerance threshold _Travel time_ rejoice _standard deviation (*t*-value)	0.272 (2.91)	0.804 (2.86)
Tolerance threshold _Travel distance_ rejoice _mean (*t*-value)	0.194 (2.26)	0.339 (2.45)
Tolerance threshold _Travel distance_ rejoice_standard deviation (*t*-value)	0.547 (1.81)	1.013 (1.27)
Model fit		
Number of observations	605	553
R-squared	0.307	0.353
Adj. *R*-squared	0.289	0.334
Final log likelihood	−460.331	−392.875

differentiating between work and non-work trip purposes. The work-related segment includes business and education with a total of 605 trips, equaling 52.25% of all trips. The non-work-related segment includes other purposes such as recreation, shopping, and touring with 553 trips (47.75%).

Table 5 shows the estimation results of this market segmentation. All parameters are significant at the 95% level, except for the standard deviation of the tolerance threshold for travel distance in both segments. First, to validate the market segmentation, a log likelihood ratio test (*40*) was conducted. The null hypothesis is that there are no differences between parameter values, which would mean that the market segmentation is not valid.

$$\chi^2 = -2 * \left[L_N(\hat{\beta}) - \sum_{g=1}^{G} L_{N_g}\left(\widehat{\beta^g}\right) \right] \quad (35)$$

where $L_N(\hat{\beta})$ is the final log likelihood value for the full data set (−875.392, from Table 1), and $L_{N_g}\left(\widehat{\beta^g}\right)$ is the final log likelihood value for each segment (−460.331 and −392.875, from Table 5).

Therefore, $\chi^2 = -2*[-875.392-(-460.331-392.875)] = 44.372$. Considering $\chi^2_{0.05} = 21.026$ with 12 degrees of freedom, the null hypothesis is rejected. Therefore, the proposed market segmentation is statistically valid.

The absolute values of the taste weights for regret and rejoice are always larger for the work-related segment. This means that individuals are more sensitive to travel time and travel distance differences when their purpose is work-related. To check whether the parameters from each segment are statistically different, an asymptotic *t*-test (*40*) was conducted:

$$t = \frac{\hat{\beta}_k^1 - \hat{\beta}_k^2}{\sqrt{var(\hat{\beta}_k^1) + var(\hat{\beta}_k^2)}} \quad (36)$$

where $\hat{\beta}_k^1$ is the estimated parameter for attribute k for the first segment, $\hat{\beta}_k^2$ is the estimated parameter for attribute k for the second segment, $\text{var}(\hat{\beta}_k^1)$ is the variance of the $\hat{\beta}_k^1$ and $\text{var}(\hat{\beta}_k^2)$ is the variance of the $\hat{\beta}_k^2$. The result for the weight of travel time for regret is $t = \left[\frac{(-0.0692)-(-0.0363)}{\sqrt{(0.003107)^2 + (0.011032)^2}} = -8.42\right]$.

Thus, the null hypothesis is rejected. Therefore, it can be concluded that the taste weight of travel time for regret is statistically different between the segments at the 95% confidence level. Likewise, the result of the asymptotic *t*-test for travel distance for regret, travel time for rejoice, and travel distance for rejoice are −4.81, −2.36, and −1.97. The null hypothesis is rejected at the 95% confidence level in every case, indicating that all taste weights for travel time and travel distance are statistically different between the segments for both regret and rejoice.

In the work-related segment, travelers are more sensitive to small travel time differences than travel distance differences. Particularly, they are quite sensitive to small travel time differences in the regret domain: The mean stochastic tolerance threshold is only 0.061. Consequently, when individuals travel for a work-related purpose, they want to avoid any delay in arrival times. Therefore, when their chosen mode is slightly inferior to the non-chosen mode, they immediately feel regret. Conversely, when the chosen mode is slightly superior to the non-chosen mode, they immediately rejoice, but the indifference threshold is larger than the tolerance threshold: The mean stochastic indifference threshold is about 0.128. For travel distance, the threshold is more similar for both regret and rejoice (it is slightly larger for regret): The mean stochastic tolerance threshold is 0.202 for regret and the stochastic indifference threshold is 0.194 for rejoice.

In the non-work-related segment, the thresholds are larger than for the work-related segment, and relatively similar across attributes for both regret and rejoice: All mean stochastic thresholds are around 0.35. This means that when individuals travel for a non-work-related purpose, they exhibit higher indifference and tolerance for attribute differences, and the thresholds do not differ much for loss (regret) and gain (rejoice). Furthermore, the standard deviation of the threshold is always higher for non-work-related trips, implying that when individuals travel for a non-work activity, their tolerance is more heterogeneous than in the case of a work-related purpose. Compared with the whole data set, the heterogeneity (standard deviation) of tolerance for each attribute is smaller in each segment in all cases. This is reasonable because similar respondents are assigned in each group, and therefore their tolerance tends to be more similar.

Conclusions and Discussion

The regret–rejoice models that have been introduced in transportation research have in common the assumption that both the regret and rejoice functions are continuous in attribute differences. However, in contexts other than travel choice, there is evidence that individuals may be insensitive to small attribute differences. Related to the very nature of the concepts of regret and rejoice, individuals may tolerate or be indifferent to small attribute differences between the chosen alternative and one or more non-chosen alternatives. In this study, therefore, tolerance and indifference thresholds are incorporated in regret–rejoice choice models. Then, the study explores two main issues using empirical data: 1) How individuals' tolerance is different between loss (regret) and gain (rejoice), or symmetric; and 2) How individuals' tolerance differs by market segmentation.

Empirical results based on revealed mode choice data support the hypothesis that individuals are insensitive to small attribute differences in both the regret and rejoice domains, at least for this dataset. The sensitivity, however, depends on the attribute and is asymmetric between regret and rejoice. Moreover, indifference and tolerance thresholds vary by trip purpose. When individuals have a trip purpose related to work, they are less tolerant of small attribute differences, especially when they feel regret for travel time. However, when they travel for a non-work-related purpose, they are less sensitive to attribute differences, and the sensitivity is different across attributes for both regret and rejoice. Further elasticity analysis implies that the effect of transportation policy may become smaller when considering the attribute difference tolerance. In addition, the effect would be quite different at the individual level depending on travel purpose.

The suggested approach, which is based on the inclusion of attribute thresholds in the pairwise comparison of choice alternatives, solves the typical limitations of models which define a threshold at the level of overall utility.

The approach is scalable to choice problems involving large choice sets. Still, some interesting and relevant research questions remain for future work. The current model assumes that the thresholds are proportional to the attribute differences. However, previous research by the authors has provided evidence that regret models based on perceptual differences, and which account for attribute variation, outperform conventional regret models. Therefore, it would be potentially valuable to more critically examine the specification of indifference and tolerance thresholds. Would a specification in perceptual space be better than the current specification in attribute space? Should the specification be made a function of (perceived) attribute variation? The authors plan to explore these issues in future research.

Author Contributions

The authors confirm contribution to the paper as follows: study conception and design: S. Jang, S. Rasouli, H. J. P. Timmermans; data collection: H. J. P. Timmermans; analysis and interpretation of results: S. Jang, S. Rasouli, H. J. P. Timmermans; draft manuscript preparation: S. Jang. All authors reviewed the results and approved the final version of the manuscript.

References

1. Von-Neumann, J., and O. Morgenstern. *Theory of Games and Economic Behavior*. Princeton University Press, Princeton, NJ, 1944.
2. McFadden, D. Conditional Logit Analysis of Qualitative Choice Behaviour. In *Frontiers in Econometrics* (P. Zarembka, ed.), Academic Press, NY, 1974, pp. 105–142.
3. Loomes, G., and R. Sugden. Regret-Theory: An Alternative Theory of Rational Choice under Uncertainty. *The Economic Journal*, Vol. 92(368), 1982, pp. 805–824.
4. Bell, D. E. Regret in Decision Making under Uncertainty. *Operation Research*, Vol. 30, No. 2, 1982, pp. 961–981.
5. Chorus, C. G., T. A. Arentze, and H. J. P. Timmermans. A Random Regret Minimization Model of Travel Choice. *Transportation Research Part B: Methodological*, Vol. 42, No. 1, 2008, pp. 1–18.
6. Chorus, C. G. A New Model of Random Regret Minimization. *European Journal of Transport and Infrastructure Research*, Vol. 10, No. 2, 2010, pp. 181–196.
7. Chorus, C. G., T. A. Arentze, and H. J. P. Timmermans. Spatial Choice: A Matter of Utility or Regret? *Environment Planning B: Urban Analytics and City Science*, Vol. 36, No. 3, 2009, pp. 538–551.
8. Chorus, C. G., and G. C. De Zong. Modeling Experienced Accessibility for Utility-Maximizers and Regret-Minimizers. *Journal of Transport Geography*, Vol. 19, No. 6, 2011, pp. 1155–1162.
9. Kaplan, S., and C. G. Prato. The Application of the Random Regret Minimization Model to Drivers' Choice of Crash Avoidance Maneuvers. *Transportation Research Part F: Traffic Psychology and Behaviour*, Vol. 15, No. 6, 2012, pp. 699–709.
10. Beck, M. B., C. G. Chorus, J. M. Rose, and D. A. Hensher. Vehicle Purchasing Behavior of Individuals and Groups. *Journal of Transport Economics and Policy*, Vol. 47, No. 3, pp. 475–492.
11. Hensher, D. A., W. H. Green, and C. G. Chorus. Random Regret Minimization or Random Utility Maximization: An Exploratory Analysis in the Context of Automobile Fuel Choice. *Journal of Advanced Transportation*, Vol. 47, No. 7, 2013, pp. 667–678.
12. Boeri, M., R. Scarpa, and C. G. Chorus. Stated Choices and Benefit Estimates in the Context of Traffic Calming Schemes: Utility Maximization, Regret Minimization, or Both? *Transportation Research Part A: Policy and Practice*, Vol. 61, 2014, pp. 121–135.
13. Jang, S., S. Rasouli., and H. J. P. Timmermans. Incorporating Psycho-Physical Mapping into Random Regret Choice Models: Model Specifications and Empirical Performance Assessments, *Transportation*, Vol. 44, No. 5, 2017, pp. 999–1019.
14. Jang, S., S. Rasouli., and H. J. P. Timmermans. Accounting for Cognitive Effort in Random Regret-Only Models: Effect of Attribute Variation and Choice Set Size. Accepted for publication, *Environment Planning B: Urban Analytics and City Science*. DOI: 10.1177/0265813516688687.
15. De Ramos, G., W. Daamen, and S. P. Hoogendoorn. Expected Utility Theory, Prospect Theory, and Regret Theory Compared for Prediction of Route Choice Behavior. *Transportation Research Record: Journal of the Transportation Research Board*, 2011. 2230: 19–28.
16. Chorus, C. G. Acquisition of Ex-Post Travel Information: A Matter of Balancing Regrets. *Transportation Science*, Vol. 48, No. 2, 2014, pp. 243–255.
17. Simon, H. A. A Behavioral Model of Rational Choice. *The Quarterly Journal of Economics*, Vol. 69, No. 1, 1955, pp. 99–118.
18. Kahneman, D. Maps of Bounded Rationality: Psychology for Behavioral Economics. *The American Economic Review*, Vol. 93, No. 5, 2003, pp. 1449–1475.
19. Rasouli, S., and H. J. P. Timmermans. Models of Bounded Rationality under Certainty. In *Bounded Rational Choice Behaviour: Applications in Transport* (S. Rasouli, and H. J. P. Timmermans, eds.), Emerald Group Publishing, Bingley, 2015, pp. 1–30.
20. Georgescu-Roegen, N. Threshold in Choice and the Theory of Demand. *Econometrica*, Vol. 26, No. 1, 1958, pp. 157–168.
21. Coombs, C. H., R. M. Dawes, and A. Tversky. *Mathematical Psychology: An Elementary Introduction*. Princeton-Hall, Englewood Cliffs, NJ, 1970.
22. Krishnan, K. S. Incorporating Thresholds of Indifference in Probabilistic Choice Models. *Management Science*, Vol. 23, No. 11, 1977, pp. 1224–1233.
23. Cantillo, V., J. Amaya, and J. de D. Ortuzar. Thresholds and Indifference in Stated Choice Surveys. *Transportation Research Part B: Methodological*, Vol. 44, No. 6, 2010, pp. 753–763.
24. Obermeyer, A., M. Treiber, and C. Evangelinos. On the Identification of Thresholds in Travel Choice Modelling. *Journal of Choice Modeling*, Vol. 17, 2015, pp. 1–9.

25. Castillo, E., J. M. Menéndez, P. Jiménez, and A. Rivas. Closed Form Expressions for Choice Probabilities in the Weibull Case. *Transportation Research Part B: Methodological*, Vol. 42, No. 4, 2008, pp. 373–380.
26. Fosgerau, M., and M. Bierlaire. Discrete Choice Models with Multiplicative Error Terms. *Transportation Research Part B: Methodological*, Vol. 43, No. 5, 2009, pp. 494–505.
27. Jang, S., S. Rasouli., and H. J. P. Timmermans. Incorporating Principles of Psychophysics into Discrete Choice Models - Part 1. Issues and Model Formulations. *Transportation Research Part B: Methodological*, under review.
28. Jang, S., S. Rasouli., and H. J. P. Timmermans. Incorporating Principles of Psychophysics into Discrete Choice Models - Part 2. Empirical Applications. *Transportation Research Part B: Methodological*, under review.
29. Chorus, C. G. A Generalized Random Regret Minimization Model. *Transportation Research Part B: Methodological*, Vol. 68, 2014, pp. 224–238.
30. Leong, W., and D. A. Hensher. Contrasts of Relative Advantage Maximization with Random Utility Maximisation and Regret Minimisation. *Journal of Transport Economics and Policy*, Vol. 49, No. 1, 2015, pp. 167–186.
31. van Cranenburgh, S., C. A. Guevara, and C. G. Chorus. New Insights on Random Regret Minimization Models. *Transportation Research Part A: Policy and Practice*, Vol. 74, 2015, pp. 91–109.
32. Gupta, S., and L. G. Cooper. The Discounting of Discounts and Promotion Thresholds. *Journal of Consumer Research*, Vol. 19, 1992, pp. 401–411.
33. Kalyanaram, G., and J. D. C. Little. An Empirical Analysis of Latitude of Price Acceptance in Consumer Packaged Goods. *Journal of Consumer Research*, Vol. 21, 1994, pp. 408–418.
34. Helson, H. *Adaptation-Level Theory*. Harper and Row Inc, New York, 1964.
35. Woodruff, R. B., E. R. Cadotte, and R. L. Jenkins. Modeling Consumer Satisfaction Processes Using Experience-Based Norms. *Journal of Marketing Research*, Vol. 20, No. 3, 1983, pp. 296–304.
36. Train, K. E. *Discrete Choice Analysis with Simulation*, 2nd ed. Cambridge University Press, 2009.
37. Ben-Akiva, M. E., and J. D. Swait. The Akaike Likelihood Ratio Index. *Transportation Science*, Vol. 20, 1986, pp. 133–136.
38. Schwarz, G. Estimating the Dimension of a Model. *The Annals of Statistics*, Vol. 6, No. 2, 1978, pp. 461–464.
39. Cherchi, E., and C. Cirillo. Validation and Forecasts in Models Estimated from Multiday Travel Survey. *Transportation Research Record: Journal of the Transportation Research Board*, 2175. 2010: 57–64.
40. Ben-Akiva, M., and S. Lerman. *Discrete Choice Analysis: Theory and Application to Travel Demand.* MIT Press, Cambridge, MA, 1985.

The Standing Committee on Traveler Behavior and Values (ADB10) peer-reviewed this paper (18-00733).

Article

TRR
JOURNAL OF THE TRANSPORTATION RESEARCH BOARD

Transportation Research Record
2018, Vol. 2672(47) 35–45

DOI: 10.1177/0361198118773556
journals.sagepub.com/home/trr
SAGE

Machine Learning Travel Mode Choices: Comparing the Performance of an Extreme Gradient Boosting Model with a Multinomial Logit Model

Fangru Wang[1] and Catherine L. Ross[2]

Abstract

The multinomial logit (MNL) model and its variations have been dominating the travel mode choice modeling field for decades. Advantages of the MNL model include its elegant closed-form mathematical structure and its interpretable model estimation results based on random utility theory, while its main limitation is the strict statistical assumptions. Recent computational advancement has allowed easier application of machine learning models to travel behavior analysis, though research in this field is not thorough or conclusive. In this paper, we explore the application of the extreme gradient boosting (XGB) model to travel mode choice modeling and compare the result with an MNL model, using the Delaware Valley 2012 regional household travel survey data. The XGB model is an ensemble method based on the decision-tree algorithm and it has recently received a great deal of attention and use because of its high machine learning performance. The modeling and predicting results of the XGB model and the MNL model are compared by examining their multi-class predictive errors. We found that the XGB model has overall higher prediction accuracy than the MNL model especially when the dataset is not extremely unbalanced. The MNL model has great explanatory power and it also displays strong consistency between training and testing errors. Multiple trip characteristics, socio-demographic traits, and built-environment variables are found to be significantly associated with people's mode choices in the region, but mode-specific travel time is found to be the most determinant factor for mode choice.

Travel mode choice modeling is one of the most studied areas in travel behavior research and is a critical step in the travel demand forecasting process. Many different factors can influence travel mode choices, including the level of service of certain travel modes, travelers' socio-demographic traits, and the built environment characteristics associated with each trip. In recent times we have seen the emergence of new travel modes like ride-sourcing and autonomous vehicles. These and other emerging transportation technologies and the availability of new travel modes will bring about changes in travel behavior including mode choice. Moreover, the emerging transport technologies may come with new data and information sources that may result in changes in the methods we use to analyze travel behavior and forecast travel mode choices.

The multinomial logit (MNL) model and its variations, such as the nested logit model, have been dominating the travel mode choice modeling field for decades. Advantages of the MNL model include its elegant closed-form mathematical structure and its interpretable model estimation result based on random utility theory. Its main limitation is the strict statistical assumptions, such as the independence of irrelevant alternatives (IIA), which often require very careful model specification and well-formatted data structures. Another limitation of the MNL model is that when the dataset is very unbalanced, meaning that different classes are represented very unequally, the estimation of the MNL model may be biased, resulting in particularly higher prediction error for classes with smaller shares (*1*). Recent exploration of applying machine learning models to travel mode choice modeling has shown its merits and most of the existing studies that compare machine learning models with statistical models suggest machine learning models could achieve similar or even higher prediction accuracy. The main strength of most machine learning models is that they

[1]School of City and Regional Planning, Georgia Institute of Technology, Atlanta, GA
[2]School of City and Regional Planning and Civil and Environmental Enginering, Georgia Institute of Technology, Atlanta, GA

Corresponding Author:
Address correspondence to Fangru Wang: fangru@gatech.edu

do not have strict statistical assumptions and can therefore be applied to different data structures with great flexibility. Some machine learning techniques, such as ensemble methods, are found to be able to improve prediction accuracy of weak learners, such as low-depth decision tree models. In this exploratory research, we employ an extreme gradient boosting model and an MNL model to predict travel mode choices in the Delaware Valley (DV) region, and examine the results of the two models by comparing their average multi-class prediction errors. The extreme gradient boosting (XGB) model is a tree-based ensemble method that has recently enhanced its great reputation by consistently winning machine learning competitions on Kaggle (https://www.kaggle.com/). In this paper, we discuss the predictive results of the two models and compare and document the model implementation processes and implications for future travel mode choice modeling.

The rest of the paper is divided into four sections. Literature Review summarizes what factors have been identified as associated with travel behavior and reviews past studies that applied machine learning to travel mode choice modeling. The Data and Methodology section explains how the two models are implemented and describes data sources. Results compare the prediction accuracy of the two models and provide interpretation of model estimation results. The Conclusions section evaluates the two models, summarizes the findings and points to potential efforts to improve travel mode choice modeling.

Literature Review

Extensive studies have been conducted to explore the factors associated with people's travel mode choices and what modeling structures should be applied to mode choice modeling. This literature review provides a glimpse of the numerous studies and summarizes recent findings.

Travel Mode Choice Modeling Techniques

Modeling travel mode choices is never an easy task as trip-makers' choice of travel modes is intertwined with many different factors. Travel mode choice has been modeled extensively with random utility maximization theory over the decades. One of the most commonly applied models is the MNL model. Stopher (*2*) and McFadden (*3*) are the earliest contributors to applications of the MNL model to travel behavior modeling. The detailed model structure and applications were thoroughly discussed in the book by Ben-Akiva & Lerman (*4*). The MNL model captures the underlying mode choice process with utility maximization assumptions that travelers are rational decision makers who are fully informed and are able to choose the mode that has the largest utility for them. Though there are limitations in using utility maximization to represent the mode choice process (*5*), MNL models are effective in quantifying the effects of trip characteristics on people's mode choice. The MNL model has a closed form mathematical estimation that is computable. Another merit of the MNL model is that it can always replicate the shares of different classes.

The recent advancement in computation power has brought interest in applying machine learning techniques to modeling travel mode choices. Karlaftis & Vlahogianni (*6*) comprehensively review the differences and similarities of using statistical methods versus neural network models in transportation research. They concluded the merit of the neural network model is its flexibility in dealing with complex datasets and its great predictive power, while the most significant challenge is the lack of explanatory power compared with conventional statistical models. Similar to the merits of neural network models, most machine learning models do not have strict statistical assumptions behind the model estimation and are thus more flexible. Some machine learning techniques, such as bagging and ensemble methods, are often found to perform well with unbalanced datasets meaning that different classes account for very unequal shares.

There are not many existing studies that have applied machine learning models to predicting travel mode choices. Among the limited number of such studies, all of them find that machine learning models can achieve at least similar prediction accuracy compared with conventional statistical models. Several studies find that the machine learning methods are substantially better (*7–11*). Machine learning models and statistic models do not have comparable interpretation power, as most machine learning models do not allow meaningful interpretation. The commonly used machine learning methods in modeling travel mode choice include decision trees (*7, 12–15*), neural networks (*8, 9, 11, 13, 15, 16*), support vector machines (*7, 8, 11*), and random forests (*10*). Some models' predictive powers are improved by combining other machine learning concepts such as fuzzy sets and ensemble methods (*8, 13, 17*). Only three of the existing studies on machine learning travel mode choices use data from the United States, as far as the authors can determine. The lack of understanding about the application of machine learning methods to travel mode choice modeling generally and specifically in the United States calls for more research and exploration.

Factors Associated with Travel Mode Choices

The factors associated with travel mode choices can be roughly categorized into three types: trip characteristics, personal/household factors, and neighborhood variables. Trip characteristics, often measured by level of service and trip-specific factors, can directly influence people's mode choices. Level of service is often measured by travel time and travel cost and they vary by travel modes (*18–21*). Frank et al. (*22*) found that travel time is the most important factor that influences people's choice of travel modes in the Puget Sound region, among many other built environment and

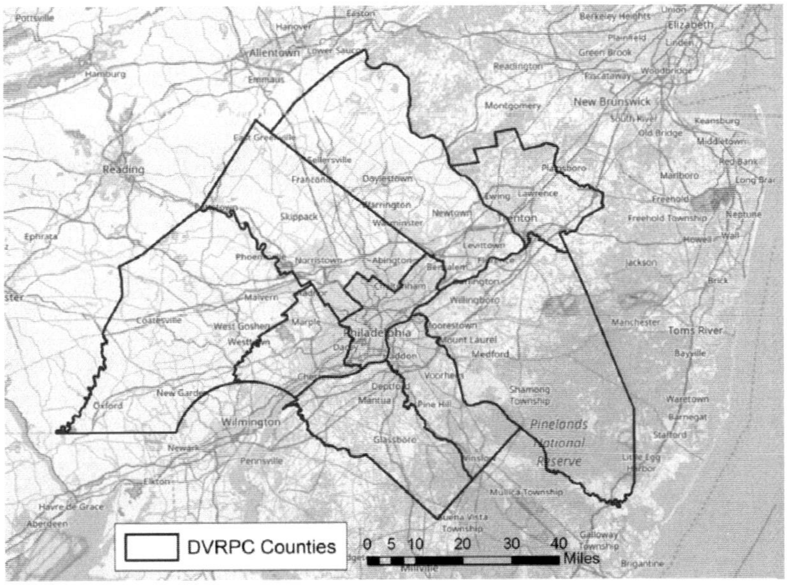

Figure 1. Study area of the DVRPC region.

socio-demographic variables. The departure time of a trip is also found to be related to people's mode choice (*23*) and it is correlated with the travel time of certain modes. Trip purpose is another important factor that directly affects people's mode choice. Most of the existing mode choice studies focus on commuting trips or home-based non-work trips and some others focus on special trip purposes or traveler groups, such as students' travel (*24–26*), trips to the airport (*27*), shopping trips (*28*), etc.

Personal and household variables reflect the socio-economic and demographic characteristics of travelers. Income, vehicle ownership, household size, number of workers, age, and gender, are found to be closely related to a traveler's mode choice. Income and vehicle ownership are directly determinant on people's choice of driving. Household size and number of workers also have effect on the mode choices of the household members and work-related trips. Special population groups, like the elderly, low-income, women, and disabled people, often have special needs limiting their travel options. The elderly are often found to have distinguishable activity and mode choice patterns (*29–32*). Similar to the elderly whose mobility is impaired by aging, low-income populations mobility is constrained by economic disadvantages and limited travel options in areas where they can afford to live (*33, 34*). Female travelers are also found to present distinct travel behavior compared with men, though the findings are mixed. Cervero, (*35*) found that female travelers were more dependent on cars, probably because women often undertake both working outside their homes and house work that requires chained trips between work, shops, and child-care centers. Conversely, it is found that women have a greater willingness to reduce car use and a potentially stronger preference for public transportation (*36*).

There are numerous studies of the relationship between built environment factors and travel behavior. Ewing & Cervero (*37*) developed a comprehensive review of studies of the relationship between travel behavior and the built environment and categorized the built environment factors into five "D" categories, including density, diversity, design, destination accessibility, and distance to transit. Policy and planning have been leveraged on these positive relationships between these built environment factors and people's mode choices to promote compact development. Land use is an indispensable component of the built environment and many studies have identified the association between land use measures and mode choice. Land use related metrics include the dominant land use type, land use mix, and the distance to certain land use types such as retail, and so forth (*22, 35, 38–40*).

Data And Methodology

Data Preparation

Our major data source is the Delaware Valley Regional Planning Commission (DVRPC) household travel survey data collected in the Delaware Valley region in 2012. The Delaware Valley region consists of nine counties in Pennsylvania and New Jersey (Figure 1). The dataset contains 81,940 unique trip

Table 1. Variables Included in the Two Models, and Their Names, Explanation, and Data Sources

Variable name	Explanation	Data source
Travel mode	Travel mode (dependent variable): 1= car; 2 = walk; 3 = bike; 4 = fixed route transit	DVRPC 2012 travel survey data
Trip variables		
Travel time	Travel time by mode in minutes	DVRPC 2012 travel survey data and Google Maps Distance Matrix API data for the unchosen mode.
Morning peak	The trip is happening in the morning peak hour (7 am–9 am)	DVRPC 2012 travel survey data
Evening peak	The trip is happening in the evening peak hour (4 pm–6 pm)	
Late night	The trip is happening in late night (9 pm–4 am)	
Activity duration	Activity duration in minutes	
Tour HBW	Dummy variable: the tour purpose is home-based work	
Tour HBO	Dummy variable: the tour purpose is home-based other	
Tour stops	Number of stops in the tour	
HH member	Number of household members in the trip	
Personal/household variables		
License	The traveler has a driver's license	DVRPC 2012 travel survey data
Low income	The traveler is low income (annual income < $30,000)	
High income	The traveler is high income (annual income > $100,000)	
Young	The traveler is younger than 16	
Elderly	The traveler is elderly than 65	
Disability	The traveler has disability	
Female	The traveler is female	
Household size	The traveler's household size	
Employed	The traveler is employed	
Student	The traveler is a student	
Parker must pay	Employee must pay for workplace parking out of pocket	
Transit subsidy	Employer offers to subsidize/pay for part of transit fare	
Number of vehicles per capita	Number of vehicles per capita in the household	
Neighborhood data (calculated at census tract level for the DVRPC region)		
O/D: Population density	Population density (persons per square mile) at the origin or destination	ACS 5-year estimate 2011–2015
O/D: employment density	Employment density (jobs per square mile) at the origin or destination	LEHD 2012 data
O/D: employment entropy	Employment diversity (3-category entropy*) at the origin or destination	LEHD 2012 data
O/D: job-housing balance	Employment population balance (entropy metric) at the origin or destination	Calculated using the ACS and LEHD data above
O/D: bus stop density	Bus stop density (# bus stops per square km) at the origin or destination	Calculated using GTFS data
O/D: rail station density	Rail station density (# subway density per square km) at the origin or destination	Calculated using GTFS data
O/D: land use entropy	Land use entropy index (calculated as residential and commercial two-category entropy)	DVRPC land use GIS data

*Employment entropy is calculated as an entropy index considering three types of employment: retail, service, and finance.

records, but more than 20 thousand of them do not include information about travel mode and about 6 thousand do not have information about income levels. After the data cleaning process and removal of the trips with missing values in the dependent and independent variables we want to model, there are 51,910 trips made by car (driver or passenger), biking, walking, or transit that we use for the mode choice analysis.

Table 1 summarizes the variables included in the model-development process, their explanation, and data sources. The dependent variable is mode choice indicating whether the trip is made by car (driving or passenger), biking, walking, or transit. The independent variables include three types of factors suggested by the literature, including trip characteristics, personal/household features, and neighborhood-level variables. Most of the variables are constructed using publicly available datasets, as shown in the table, except that the travel times of alternative transportation modes are requested using the Google Maps Distance Matrix API. The DVRPC household travel survey data contain the reported travel time and modeled travel time for every trip, but it only has the travel time for the "chosen" travel modes. To implement discrete choice models, we also need to know the travel time of the "unchosen" modes to form a complete choice set for each individual. The Google Maps Distance Matrix API provides estimated travel time by driving, biking, walking, and transit at different times of day. We use it to request the estimated travel time of all the "unchosen" cases by specifying the departure time of each trip and the trip origin/destination (centroids of traffic analysis zones (TAZs) in the DVRPC region). For the chosen travel modes, we employed the modeled travel time, which are travel times between recorded departure TAZ and arrival TAZ modeled using DVRPC's travel demand model (TIM 2.2) based on mode choice. We chose to use the modeled travel time rather the reported travel time because reported travel times are often known to be error-prone and also because both the "modeled" travel times and the travel times we acquired from Google API are estimated using TAZ centroids, which make it more consistent between the chosen and unchosen modes. We include 51,910 unique trips from the DVRPC household travel survey data, but our final data panel included 207,640 observations as each unique trip has four entries for the four different travel modes.

Models and Implementation

We employ the XGB model and the MNL model to analyze travel mode choices in the DVRPC region and compare the predictive powers of the models and their implications. The XGB model is a tree-based ensemble method, so it allows understanding of the importance of the independent variables in determining the dependent variable. Details of model implementation are documented in the following sections.

Multinomial Logit Model. The MNL model is based on random utility theory that assumes the utility of choosing a certain travel mode is a random variable that travelers always want to maximize. We implemented the model with maximum likelihood estimation using the "mnlogit" package in R. Except that the variable of travel time (by mode) is set as a generic variable, all other variables are included in the model as individual-specific variables as their values vary by traveler. We selected a set of independent variables when developing the MNL model using the whole dataset, mainly according to three criteria: first, a variable's sign needs to be consistent with existing theory; second, the variable is statistically significant; and third, groups of variables were selected by developing chi-squared tests to compare whether adding a group of variables improves the goodness of fit of the model. The MNL model prediction was implemented by conducting a Monte Carlo simulation process.

Extreme Gradient Boosting Model. The Extreme gradient boosting (XGB) Model is a tree-based ensemble method that can be used for both regression and classification. The XGB model was first proposed by Friedman (*41*). It is an ensemble method that is built upon iteratively growing low-depth decision trees based on the idea of additive training. To put it simply, in an XGB model, each low-depth decision tree is built to minimize a defined loss function, but each time the estimation puts more weight on the cases that are wrongly predicted by previously developed trees. The final model result is collectively determined by the results of all the developed trees. We implemented the XGB model using the "XGboost" package in R. In addition to estimating the model based on the independent variables we included, training the XGB model relies on tuning some hyperparameters for optimizing the model's performance. "Hyperparameter" is a term that can be used interchangeably with parameters, but in machine learning, it more often refers to the parameters that need to be set by the modeler or need to be tuned before the final model is learned based on independent and dependent variables. Hyperparameters directly influence the performance of a model and are often tuned using cross-validation techniques in machine learning. Seven hyperparameters need to be specified in the XGB model, including:

(1) "nrounds": the maximum number of iterations (similar to the number of trees to grow in our case);
(2) "eta": it controls the learning rate;
(3) "max_depth": it controls the depth of the tree;
(4) "min_child_weight": it controls the minimum value of sum of instance weight of a node;
(5) "subsample": it controls the number of samples supplied to a tree;
(6) "colsample_bytree": it controls the number of variables supplied to a tree;
(7) "max_delta_step": it controls regularization and is used to avoid overfitting.

The hyperparameters of "nrounds", "eta", and "max_depth" directly influence whether the model will overfit or underfit the data. To optimize the performance of the XGB model and also to avoid the overfitting problem. We first tuned the parameters of "nrounds" and "eta" together using grid-search and found that "eta" = 0.3 always produces the best model performance for this dataset. The larger the value of "nrounds", the more iteration the model will run, which will more likely result in the overfitting problem. We found that the hyperparameter of "nrounds" of 50 can achieve great model predicting performance yet does not generate big gaps between training and testing errors, a sign that the model is overfitting, so we set nrounds to 50. Then we used 10-fold cross-validation to tune all other parameters except "max_delta_step" in every run. The "max_delta_step" was set as 2 as it improves the performance of predicting rare case mode choices (e.g., trips made by biking). We set the maximum value that the hyperparameter "max_depth" can take to 10 to avoid overfitting, and specific values of "max_depth" were automatically set using cross-validation in every run. During the parameter tuning processes, we set the loss function as the multiclass classification error rate that is calculated as number of cases wrongly predicted divided by the number of all cases. To make the performance of the XGB model and the MNL model comparable, we included almost the same set of independent variables in both models, except that travel distance is included in the XGB model but not in the MNL model due to the significant correlation between travel time and travel distance.

Results

Descriptive Statistics

Four travel modes are included in the mode choice modeling analysis for the DVRPC region, including car (driving or passenger), biking, walking, and transit. The total number of unique trips included in our analysis after removing missing and outlier values is 51,910, among which 43,196 are made by car, 511 are made by bike, 5,467 are made by walking, and 2,736 are made by transit. One commonly encountered challenge of modeling travel mode choices is the issue of unbalanced datasets. In this case, the majority of trips are made by car while only about 1% of trips are made by bike and only about 5% are made by transit. The dataset is very unbalanced and may cause the MNL and XGB models to perform bad especially when predicting mode choices with smaller shares, such as bike and transit. Table 2 presents the descriptive statistics of all the variables used in the analysis.

Comparing Prediction Accuracy of the MNL Model and the XGB Model

Each of the two models were run 100 times to compare their average prediction accuracy and robustness to data changes. For each run, the dataset was randomly split into a training subset (75% of the data) and a testing subset (25% of the data) and the training errors and testing errors are averaged for the 100 times run and summarized in Table 3. To compare the predictive power of the MNL model with the XBG model, we examined the average total errors of the two models and also compared their prediction error for each of the four travel modes. The total error is calculated as the number of trips that are predicted to have the wrong mode choice out of the total number of trips. The mode-specific prediction error is calculated as the number of trips wrongly predicted out of the total number of trips made by that mode.

In addition to running the MNL model and the XGB model with the complete datasets, we also tested how the two models perform when biking, which has the smallest mode share, is removed from the data. Therefore, we also run each of the two models 100 times for a subset of the data that only contains the travel modes of car, walking, and transit. The results of the two models run with complete datasets and subsets of the data are shown in Table 3 respectively. The predicted mode shares of the two models are also presented in Table 3.

As Table 3 shows, for the complete choice set of four alternatives, both models show good overall prediction accuracy while the XBG model has lower errors than the MNL model. The XGB model has a total training error of 3.5% and a total testing error of 15.5%, while the MNL model has a total training error of 18.2% and a total testing error of 18.3%. This means that the XGB model has an overall prediction accuracy of 94.5% while the MNL model's overall prediction accuracy is 92.7%. When the alternative of biking is removed from the dataset, both models perform even better, but the XGB model performs significantly better compared with the result of the complete dataset. When biking is removed, The XGB model has a total training error of 3.7% and a total testing error of 10.5%, while the MNL model has a total training error of 17.8% and a total testing error of 17.9%.

In the case of mode-specific prediction accuracy, both models did similarly badly in predicting mode choices of biking, which is not surprising, as biking has much lower mode share, only about 1%, compare with other travel modes in the DVRPC region. Nevertheless, the XGB model still has a much better performance when predicting the choice of walking while it has lower performance of predicting the choice of transit, compared with the MNL model. The XGB model's walk-specific testing error is only 28.7% but the MNL model has a walk-specific testing error as high as 58.6%. The XGB model's transit-specific testing error is 80.7%, while the MNL model achieves a transit-specific testing error of 66.7%. Interestingly, when biking is removed from the model, the predicting errors of the XGB model regarding all the three alternatives significantly improve and are all significantly lower than that of the MNL model.

The MNL model can always replicate the shares of all alternatives and it is why the predicted mode shares of the MNL model are very similar to the observed ones, shown in Table 3. The XGB model also performs well in predicted mode shares, and the difference between the observed mode shares and the predicted ones are all smaller than 2%. In summary, the XGB model can lower the overall prediction error compared to the MNL model. Neither model performs well when dealing with extremely unbalanced data (when biking is included), but the XGB model performs significantly better in overall mode choices and all three specific modes when biking is removed.

Explanatory Results

The result of the MNL model estimated using the whole dataset is shown in Table 4 for evaluating the models' validity. Overall, the model has an excellent goodness of fit, as the adjusted Rho-squared (McFadden R-squared) of this model is

Table 2. Mode Share and Descriptive Statistics

Travel mode share	Car	Bike	Walk	Transit	Total
Number of unique trips	43,196 (83.20%)	511 (1.00%)	5,467 (10.5%)	2,736 (5.30%)	51,910 (100.00%)
Continuous variables					
Variable name	Mean	S.D.	Minimum	Maximum	
Travel time	126.82	260.63	0	1772.22	
Tour stops	0.7	1.2	0	19	
Activity duration	124.8	173.2	1	1215	
HH member	0.4	0.7	0	6	
HH_WORK	1.4	1	0	6	
Vehicle per capita	0.8	0.4	0	8	
O/D: population density	23,760	66,285	0	506,607.20	
O/D: employment density	170,082	1,052,925	6.4	9,515,692	
O/D: bus stop density	184	693.1	0	4,790	
O/D: job-housing balance	0.7	0.2	0	1	
O/D: rail station density	12	64.6	0	501	
O/D: land use entropy	0.7	0.2	0	1	
Dummy variables					
Variable name	Number of 1		Number of 0		
Tour HBW	18,687	36%	33,223	64%	
Tour HBO	29,588	57%	22,322	43%	
Morning peak	10,901	21%	41,009	79%	
Evening peak	12,458	24%	39,452	76%	
Late night	2,076	4%	49,834	96%	
Household size	8,305	16%	43,605	84%	
Low income	6,748	13%	45,162	87%	
High income	20,244	39%	31,666	61%	
Female	28,550	55%	23,360	45%	
Young	6,229	12%	45,681	88%	
Elderly	11,939	23%	39,971	77%	
License	44,123	85%	7,787	15%	
Disability	1,557	3%	50,353	97%	
Parker must pay	4,671	9%	47,239	91%	
Transit subsidy	2,076	4%	49,834	96%	
Student	8,824	17%	43,086	83%	

about 0.71, indicating that about 71% of the information contained in the data is explained by this model. This model uses the car as the base mode for estimation and the signs of all the variables included in this model are consistent with theory.

Several variables of trip characteristics are significantly associated with mode choices. Compared with the choice of car, "home-based other" trips are negatively associated with biking and transit, but are positively associated with walking. Biking trips are more likely to be in morning peak hours and less likely to occur in evening peak hours compared with trips by car. Walking trips are less likely to be in evening peak hours and late night compared with car trips, probably for safety concerns. People prefer transit to car during morning peak hours, probably wanting to avoid roadway congestion. Biking trips are associated with longer activities at trip origins and transit trips are associated with shorter activities.

Multiple personal and household characteristics are found to be statistically significant in the model. Travelers from larger households are negatively associated with all three alternative travel modes, biking, walking, and transit, compared with car. The number of vehicles per capita and availability of a driver's license are also negatively associated with the three alternative modes. Low-income travelers are more likely to choose walking and transit, while high-income travelers are less likely to use transit. Female travelers are less likely to choose the three alternative travel modes, which is consistent with the finding from Cervero (*35*) that females are more dependent on the car. Children, elderly people, and travelers with disabilities are also found to be more dependent on cars according to the model result. Parking costs reduces the likelihood to choose driving and transit subsidies encourage people to choose travel modes other than the car.

Table 3. Average Training Errors, Testing Errors, and Predicted Mode Shares of 100 Model Runs

	MNL Model - 4 Alternative				XGB Model - 4 Alternative			
	Training error		Testing error		Training error		Testing error	
	Mean	Variance	Mean	Variance	Mean	Variance	Mean	Variance
Total	18.2%	0.000004	18.3%	0.000007	3.5%	0.001651	15.5%	0.000148
Car	8.8%	0.000003	8.8%	0.000009	1.8%	0.000548	8.6%	0.000056
Bike	94.6%	0.000139	94.9%	0.000291	19.4%	0.059651	95.2%	0.000636
Walking	58.1%	0.000054	58.6%	0.000163	5.5%	0.006854	28.7%	0.001171
Transit	66.5%	0.000102	66.7%	0.000370	1.8%	0.000548	80.7%	0.000650

	MNL Model - 3 Alternative				XGB Model - 3 Alternative			
	Training error		Testing error		Training error		Testing error	
	Mean	Variance	Mean	Variance	Mean	Variance	Mean	Variance
Total	17.8%	0.000003	17.9%	0.000008	3.7%	0.000485	10.5%	0.000065
Car	8.3%	0.000002	8.3%	0.000007	0.6%	0.000047	3.5%	0.000005
Walking	58.5%	0.000063	58.5%	0.000153	8.5%	0.005302	27.8%	0.000495
Transit	67.2%	0.000092	67.4%	0.000302	0.6%	0.000047	65.9%	0.001284

Predicted mode shares	MNL Model 4 Alternative		MNL Model 3 Alternative		XGB Model 4 Alternative		XGB Model 3 Alternative	
	Training	Testing	Training	Testing	Training	Testing	Training	Testing
Total	100.0%	100.0%	99.4%	95.3%	100.0%	100.0%	98.4%	95.6%
Car	83.3%	83.0%	82.9%	83.5%	84.0%	84.5%	84.2%	83.5%
Bike	1.0%	1.0%	1.0%	0.3%				
Walking	10.6%	10.5%	10.4%	9.0%	10.6%	10.7%	9.8%	9.1%
Transit	5.1%	5.5%	5.2%	2.5%	5.4%	4.8%	4.4%	3.0%

Table 4. MNL Model Result

Generic variable	Est.	S.E	Sig.						
Travel time	−0.0047	0.0001	***						
Individual-specific variables	Biking			Walking			Transit		
	Est.	S.E	Sig.	Est.	S.E	Sig.	Est.	S.E	Sig.
(Intercept)	−0.74	0.25	**	1.30	0.10	***	1.39	0.12	***
Tour HBO	−0.26	0.11	*	0.07	0.04	.	−0.71	0.06	***
Morning peak	0.38	0.12	**	0.06	0.04		0.39	0.06	***
Evening peak	−0.02	0.12		−0.24	0.04	***	0.09	0.06	
Late night	0.11	0.23		−0.27	0.09	**	−0.12	0.13	
Activity duration	0.0007	0.0003	**	0.0001	0.0001		−0.0006	0.0001	***
Household size	−0.50	0.04	***	−0.42	0.02	***	−0.41	0.02	***
Vehicle per capita	−3.03	0.14	***	−1.96	0.06	***	−2.27	0.07	***
Low income	−0.23	0.15		0.30	0.05	***	0.68	0.06	***
High income	−0.19	0.11	.	0.03	0.04		−0.27	0.06	***
Female	−0.87	0.10	***	−0.18	0.03	***	−0.22	0.05	***
Young	0.16	0.23		−0.31	0.08	***	−1.05	0.11	***
Elderly	−0.89	0.16	***	−0.62	0.05	***	−0.39	0.07	***
License	−0.46	0.19	*	−1.25	0.07	***	−1.84	0.08	***
Disability	−0.96	0.36	**	−0.62	0.10	***	0.03	0.11	
Parker must pay	1.17	0.11	***	0.78	0.05	***	1.10	0.06	***
Transit subsidy	1.28	0.13	***	0.80	0.07	***	0.96	0.08	***
O: population density	6.8E-06	4.9E-07	***	8.4E-06	2.7E-07	***	7.8E-06	2.9E-07	***
D: population density	8.2E-06	5.0E-07	***	9.8E-06	2.9E-07	***	9.9E-06	3.1E-07	***

Note: Sig. = significance codes; O = origin of a trip; D = destination of a trip; '***' = 0 – 0.001; '**' = 0.001 – 0.01; '*'= 0.01 – 0.05; '.'= 0.01 – 0.1; ' '= 0.1 – 1. Pseudo R-squared (McFadden R-squared) = 0.701.

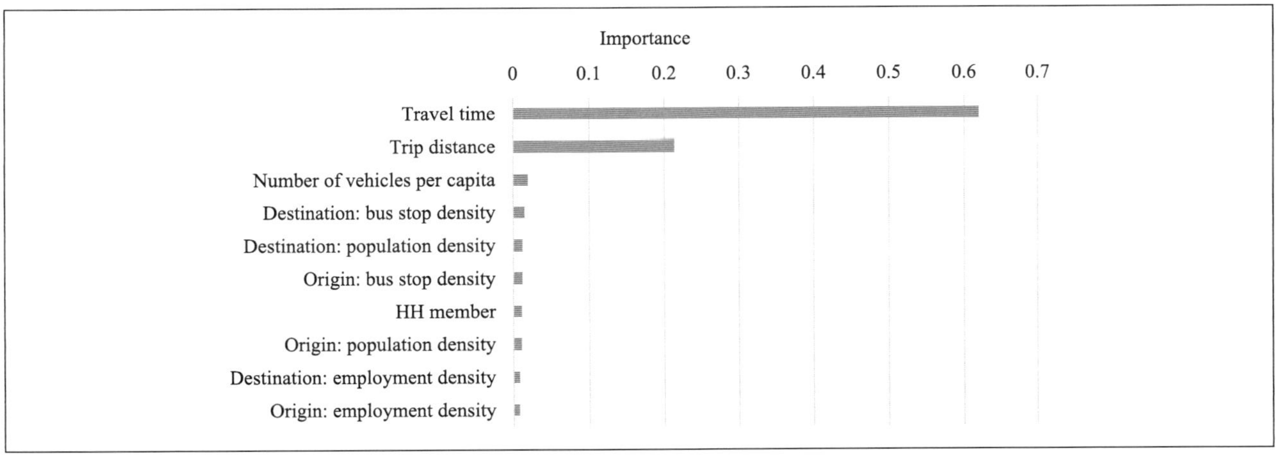

Figure 2. Top 10 most important independent variables of the XGB model.

Many of the built environment variables listed in Table 1 are highly correlated, such as the density measures and the entropy measures, so only variables of population density at both trip origin and destination are included in the final model. Population densities at both ends of a trip are positively associated with biking, walking, and transit, though the direction of causation is unclear.

As illustrated previously, the XGB model is a tree-based ensemble method that allows measuring the "importance" of a variable in forming the final rule-based classifications. The importance measures used by the XGB model are calculated by integrating three importance-related measures, including "gain", "cover", and "frequency". "Gain" measures the improvement in accuracy brought by a feature to the branches it is on; "cover" measures the relative quantity of observations concerned by a feature; and "frequency" counts the number of times a feature is used in all generated trees (*42*). Figure 2 plots the top 10 most important independent variables of the XGB model. Travel time is the most important variable influencing people's mode choices and it has an importance measure that is far larger than other independent variables. Trip distance is the second important variable and travel time and trip distance together account for more than 80% of the importance that all independent variables have. Vehicles per capita, number of household members in the trip, and multiple built environment variables are also associated with travel mode choices.

Conclusions

Travel mode selection is intertwined with many different factors and may have noticeable changes as technological advances and new travel modes and data sources are available. Modeling travel mode choice is a critical step in travel demand forecasting and may also face challenges and opportunities as new travel modes and data sources emerge. In this paper, we apply a popular machine learning model to predict travel mode choices in the Delaware Valley region and compare the results with an MNL model by examining their multi-class prediction errors. The paper is among the limited number of studies that explore machine learning's application to travel mode choice modeling and has included a relatively comprehensive list of independent variables that are ready for practical use.

Both the XGB and the MNL models show high prediction accuracy for travel mode choices in the Delaware Valley region, but both models perform poorly for predicting the choice of biking that has an extremely small share of 1% in the dataset. Nevertheless, when biking is removed from the dataset, the performance of the XGB model exceeds that of the MNL model substantially. Interestingly, after biking is removed from the dataset, the XGB model substantially exceeds the performance of the MNL model not only in predicting the choices of all three modes together, but also by significantly improving the accuracy for predicting every individual mode.

The main advantage of the MNL model is its great interpretability and it maintains high consistency between the training and testing errors. This is a result of the fact that the MNL model can easily avoid the issue of overfitting. The MNL model directly contributes to understanding the relationships between people's travel mode choices and other factors, and is useful for variable selection and deriving policy implications.

Regarding the effort of developing and implementing the models, the XGB and the MNL models have different strengths and challenges. The advantage of the XGB model is that it has very little limitation on the data structure and model specification. For example, travel distance is included in the XGB model and in Figure 2 it is shown that travel distance is an important independent variable in estimating mode choice, but it cannot be included in the MNL model because of the correlation between travel time and distance. While the XGB model requires effort in tuning some hyperparameters to optimize the model's performance, the whole model fitting process requires less attention and effort compared with the MNL model that requires very careful model specification and testing to examine whether the model assumptions hold. In contrast, the MNL model can easily avoid the overfitting issue. Especially for a very unbalanced dataset, the XGB model may not suffer from overfitting issues for all choices combined, but may have overfitting issues when predicting the choice with small shares. For example, in this analysis, the hyperparameters of the XGB models are tuned by minimizing the multi-class predicting error and the overfitting issue is controlled at the whole dataset level. However, since biking only accounts for 1% of the data, the tuned hyperparameters will likely result in overfitting for predicting biking choices. Therefore, machine learning techniques that can handle choice-specific overfitting control will be very helpful in this situation.

Future mode choice modeling efforts should consider using machine learning techniques or integrating some machine learning techniques into conventional statistical modeling. The results of this study demonstrate the advantages of the XGB and the MNL models. The XGB model has higher prediction accuracy than the MNL model. The MNL model allows intuitive interpretation that machine learning methods cannot surpass. An easy way to combine the advantages of the two types of models is to use the MNL model as a before-hand variable selection and interpretation tool, while using machine learning models and techniques to perform or improve the mode choice forecasting. Unbalanced data are a notorious problem in machine learning and the analysis of this research may suggest that when the smallest share of a dataset is larger than 5% the XGB model can have significantly better performance compared with the MNL model. Of course, this may not hold true for other datasets, but avoiding extremely unbalanced datasets by collecting more samples for small share modes or applying resampling techniques may be useful for improving models' performance in predicting travel mode choices. Machine learning models that perform well with unbalanced datasets are useful in this case.

The study has its limitations. The travel mode choices are modeled at the trip level without considering trip chaining effects or tour-level factors. How to incorporate both trip-level and tour-level considerations into travel mode choice analysis has been explored with statistical models, but has not been researched for machine learning models. This might be the next step for facilitating real-world applications of machine learning models to travel mode choice modeling. Another limitation is that the models are only applied to the Delaware Valley region, so the generalizability of the findings will need to be confirmed by future research. Nevertheless, this paper reveals the great potential of using machine learning methods for improving travel mode choice prediction accuracy and it provides an early contribution to documenting the implementation of machine learning models and techniques.

Author Contributions

The authors confirm contribution to the paper as follows: study conception and design: Fangru Wang and Catherine L. Ross; data collection: Fangru Wang; analysis and interpretation of results: Fangru Wang; draft manuscript preparation: Fangru Wang and Catherine L. Ross. All authors reviewed the results and approved the final version of the manuscript.

References

1. King, G., and L. Zeng. Logistic Regression in Rare Events Data. *Political Analysis*, Vol. 9, No. 2, 2001, pp. 137–163.
2. Stopher, P. A Multinomial Extension of the Binary Logit Model for Choice of Mode of Travel. *Northwestern University*, 1969. (Unpublished)
3. McFadden, D. The Measurement of Urban Travel Demand. *Journal of Public Economics*, Vol. 3, No. 4, 1974, pp. 303–328.
4. Ben-Akiva, M., and S. R. Lerman. *Discrete Choice Analysis: Theory and Application to Travel Demand*. MIT Press, Cambridge, MA, 1985.
5. Gärling, T. Behavioural Assumptions Overlooked in Travel Choice Modelling. In *Travel Behaviour Research: Updating the State of Play* (J. Oriuzar, D. Hensher, and S. Jara-Diaz, eds.), Elsevier Science Ltd, Oxford, 1998, pp. 3–18.
6. Karlaftis, M. G., and E. I. Vlahogianni. Statistical Methods versus Neural Networks in Transportation Research: Differences, Similarities and Some Insights. *Transportation Research Part C: Emerging Technologies*, Vol. 19, No. 3, 2011, pp. 387–399.
7. Biagioni, J. P., P. M. Szczurek, P. C. Nelson, and A. Mohammadian. Tour-Based Mode Choice Modeling: Using an Ensemble of (Un-) Conditional Data-Mining Classifiers. Presented at 88th Annual Meeting of the Transportation Research Board, Washington, D.C., 2008.
8. Omrani, H., O. Charif, P. Gerber, A. Awasthi, and P. Trigano. Prediction of Individual Travel Mode with Evidential Neural Network Model. *Transportation Research Record: Journal of the Transportation Research Board*, 2013. 2399: 1–8.
9. Rao, P. S., P. K. Sikdar, K. K. Rao, and S. L. Dhingra. Another Insight into Artificial Neural Networks through Behavioural Analysis of Access Mode Choice. *Computers, Environment and Urban Systems*, Vol. 22, No. 5, 1998, pp. 485–496.
10. Sekhar, C. R., Minal, and E. Madhu. Mode Choice Analysis Using Random Forrest Decision Trees. *Transportation Research Procedia*, Vol. 17, 2016, pp. 644–652.
11. Zhang, Y., and Y. Xie. Travel Mode Choice Modeling with Support Vector Machines. *Transportation Research Record: Journal of the Transportation Research Board*, 2008. 2076: 141–150.
12. Celikoglu, H. B. Application of Radial Basis Function and Generalized Regression Neural Networks in Non-Linear Utility Function Specification for Travel Mode Choice Modelling. *Mathematical and Computer Modelling*, Vol. 44, No. 7, 2006, pp. 640–658.
13. Shukla, N., J. Ma, R. Wickramasuriya, N. N. Huynh, and P. Perez. Tour-Based Travel Mode Choice Estimation Based on Machine Learning and Fuzzy Techniques. In *International Symposium for Next Generation Infrastructure (ISNGI 2014)* (T. Dolan and B. S. Collins, eds.), University College London, UK, 2015, pp. 215–220.
14. Wets, G., K. Vanhoof, T. Arentze, and H. Timmermans. Identifying Decision Structures Underlying Activity Patterns:

An Exploration of Machine learning Algorithms. *Transportation Research Record: Journal of the Transportation Research Board*, 2000. 1718: 1–9.

15. Xie, C., J. Lu, and E. Parkany. Work Travel Mode Choice Modeling with Machine learning: Decision Trees and Neural Networks. *Transportation Research Record: Journal of the Transportation Research Board*, 2003. 1854: 50–61.
16. Hensher, D. A., and T. T. Ton. A Comparison of the Predictive Potential of Artificial Neural Networks and Nested Logit Models for Commuter Mode Choice. *Transportation Research Part E: Logistics and Transportation Review*, Vol. 36, No. 3, 2000, pp. 155–172.
17. Vythoulkas, P. C., and H. N. Koutsopoulos. Modeling Discrete Choice Behavior Using Concepts from Fuzzy Set Theory, Approximate Reasoning and Neural Networks. *Transportation Research Part C: Emerging Technologies*, Vol. 11, No. 1, 2003, pp. 51–73.
18. Bhat, C. R. Work Travel Mode Choice and Number of Non-Work Commute Stops. *Transportation Research Part B: Methodological*, Vol. 31, No. 1, 1997, pp. 41–54.
19. Bowman, J. L., and M. E. Ben-Akiva. Activity-Based Disaggregate Travel Demand Model System with Activity Schedules. *Transportation Research Part A: Policy and Practice*, Vol. 35, No. 1, 2001, pp. 1–28.
20. Cervero, R., and M. Duncan. Walking, Bicycling, and Urban Landscapes: Evidence from the San Francisco Bay Area. *American Journal of Public Health*, Vol. 93, No. 9, 2003, pp. 1478–1483.
21. Dissanayake, D., and T. Morikawa. Investigating Household Vehicle Ownership, Mode Choice and Trip Sharing Decisions Using a Combined Revealed Preference/stated Preference Nested Logit Model: Case Study in Bangkok Metropolitan Region. *Journal of Transport Geography*, Vol. 18, No. 3, 2010, pp. 402–410.
22. Frank, L., M. Bradley, S. Kavage, J. Chapman, and T. K. Lawton. Urban Form, Travel Time, and Cost Relationships with Tour Complexity and Mode Choice. *Transportation*, Vol. 35, No. 1, 2008, pp. 37–54.
23. Ye, X., R. M. Pendyala, and G. Gottardi. An Exploration of the Relationship between Mode Choice and Complexity of Trip Chaining Patterns. *Transportation Research Part B: Methodological*, Vol. 41, No. 1, 2007, pp. 96–113.
24. Ewing, R., W. Schroeer, and W. Greene. School Location and Student Travel Analysis of Factors Affecting Mode Choice. *Transportation Research Record: Journal of the Transportation Research Board*, 2004. 1895: 55–63.
25. Klöckner, C. A., and T. Friedrichsmeier. A Multi-Level Approach to Travel Mode Choice – How Person Characteristics and Situation Specific Aspects Determine Car Use in a Student Sample. *Transportation Research Part F: Traffic Psychology and Behaviour*, Vol. 14, No. 4, 2011, pp. 261–277.
26. McDonald, N. C. Critical Factors for Active Transportation to School Among Low-Income and Minority Students: Evidence from the 2001 National Household Travel Survey. *American Journal of Preventive Medicine*, Vol. 34, No. 4, 2008, pp. 341–344.
27. Gupta, S., P. Vovsha, and R. Donnelly. Air Passenger Preferences for Choice of Airport and Ground Access Mode in the New York City Metropolitan Region. *Transportation Research Record: Journal of the Transportation Research Board*, 2008. 2042: 3–11.
28. Bhat, C. R. Analysis of Travel Mode and Departure Time Choice for Urban Shopping Trips. *Transportation Research Part B: Methodological*, Vol. 32, No. 6, 1998, pp. 361–371.
29. Georggi, N. L., and R. M. Pendyala. *An Analysis of Long-Distance Travel Behavior of the Elderly and the Low-Income*. University of South Florida, Tampa, FL, 2000.
30. Golob, T. F. A Simultaneous Model of Household Activity Participation and Trip Chain Generation. *Transportation Research Part B: Methodological*, Vol. 34, No. 5, 2000, pp. 355–376.
31. Kim, S., and G. Ulfarsson. Travel Mode Choice of the Elderly: Effects of Personal, Household, Neighborhood, and Trip Characteristics. *Transportation Research Record: Journal of the Transportation Research Board*, 2004. 1894: 117–126.
32. Kuppam, A. R., and R. M. Pendyala. A Structural Equations Analysis of Commuters' Activity and Travel Patterns. *Transportation*, Vol. 28, No. 1, 2001, pp. 33–54.
33. Giuliano, G. Low Income, Public Transit, and Mobility. *Transportation Research Record: Journal of the Transportation Research Board*, 2005. 1927: 63–70.
34. Murakami, E., and J. Young. *Daily Travel by Persons with Low Income*. US Federal Highway Administration, Washington, D.C., 1997.
35. Cervero, R. Built Environments and Mode Choice: Toward a Normative Framework. *Transportation Research Part D: Transport and Environment*, Vol. 7, No. 4, 2002, pp. 265–284.
36. Matthies, E., S. Kuhn, and C. A. Klöckner. Travel Mode Choice of Women The Result of Limitation, Ecological Norm, or Weak Habit? *Environment and Behavior*, Vol. 34, No. 2, 2002, pp. 163–177.
37. Ewing, R., and R. Cervero. Travel and the Built Environment. *Journal of the American Planning Association*, Vol. 76, No. 3, 2010, pp. 265–294.
38. Pinjari, A. R., R. M. Pendyala, C. R. Bhat, and P. A. Waddell. Modeling the Choice Continuum: An Integrated Model of Residential Location, Auto Ownership, Bicycle Ownership, and Commute Tour Mode Choice Decisions. *Transportation*, Vol. 38, No. 6, 2011, p. 933.
39. Rajamani, J., C. Bhat, S. Handy, G. Knaap, and Y. Song. Assessing Impact of Urban Form Measures on Nonwork Trip Mode Choice after Controlling for Demographic and Level-of-Service Effects. *Transportation Research Record: Journal of the Transportation Research Board*, 2003. 1831: 158–165.
40. Zhang, M. The Role of Land Use in Travel Mode Choice: Evidence from Boston and Hong Kong. *Journal of the American Planning Association*, Vol. 70, No. 3, 2004, pp. 344–360.
41. Friedman, J. H. Greedy Function Approximation: A Gradient Boosting Machine. *Annals of Statistics*, Vol. 29, No. 5, 2001, pp. 1189–1232.
42. Xgboost 0.6 Documentation. *Understand Your Dataset with XGBoost*. http://xgboost.readthedocs.io/en/latest/R-package/discoverYourData.html#feature-importance. Accessed July 20, 2017.

The Standing Committee on Traveler Behavior and Values (ADB10) peer-reviewed this paper (18-01087).

Article

TRR
JOURNAL OF THE TRANSPORTATION RESEARCH BOARD

Transportation Research Record
2018, Vol. 2672(47) 46–56

DOI: 10.1177/0361198118773557
journals.sagepub.com/home/trr
SAGE

Are Activity Patterns Stable or Variable? Analysis of Three-Year Panel Data

Tim Hilgert[1], Sascha von Behren[1], Christine Eisenmann[1], and Peter Vortisch[1]

Abstract

Routines and mandatory activities, such as work and school, shape the activity patterns of individuals and strongly influence travel demand. Knowledge about stability and variability of these routines could strengthen travel demand modelling and forecasting. A longitudinal perspective is required to investigate these aspects. In this study, the activity patterns of a sample of people is compared for one week in two successive years. It is analyzed whether the activity patterns of a given person vary from year to year, to what degree, and how this variability and stability can be measured. It is considered whether socio-demographic factors and life events determine stability in weekly activity patterns. The study is based on the representative panel survey, German Mobility Panel. The weekly activity patterns of the same respondents in different years is assessed, using two methods to measure stability and variability. The survey respondents are clustered into three groups according to the degree of variability in their activity patterns. A logistic regression model is also used to identify socio-economic and demographic covariates for similarity in weekly activity patterns. Results show that about one-third of the sample had the same or very similar weekly activity patterns in the two years examined. A person's occupation status is a good predictor for the variability of activity patterns. Moreover, persons undergoing a change in occupation status are quite likely to show a greater variability in their activity patterns.

Routines and mandatory activities, such as work and school, shape the activity patterns of individuals and strongly influence their travel behavior. To investigate the stability or variability of these routines, a longitudinal time perspective is required. Understanding stability and variability in travel behavior and activity execution is, in turn, essential for developing rational planning policies and forecasting travel demand (*1–3*). Therefore, various research methods have been applied to address this issue. When describing behavior at different points in time, one can choose from several terms. These include "habits and routines," "rhythmic patterns," "stability," and "variability" (*1*, *2*, *4–6*).

In this study, the activity patterns on an intra-personal basis are analyzed in different weeks of different years. The term stability is used to describe a constant behavior and variability to describe a changing behavior. Since variability is the opposite and thus not stable, the term "stability" is also used to refer to the measurement of both. Scope is not limited to day-to-day differences in behavior but behavior investigated for a whole week in two different years. Furthermore, the focus is on activity patterns found in the two survey weeks. The combination of longer time periods and the focus on activity patterns is an under-explored research field. By analyzing behavior from year to year, it can be explored whether changes in activity patterns result from life events, such as changes in occupational status or residence. The research questions are: Do the activities of the same individual differ from year to year? If so, how much, and how can stability in activity patterns be measured? Which socio-demographic factors and life events determine stability in weekly activity patterns?

The study draws on data from the German Mobility Panel (MOP), a representative panel survey of everyday travel in Germany. The weekly activity patterns of the same respondents who participated in the panel survey in different years between 2005 and 2015 is assessed. Activity patterns from trip diaries of the MOP are identified and two methods introduced to measure the stability

[1]Institute for Transport Studies, Karlsruhe Institute of Technology (KIT), Karlsruhe, Germany

Corresponding Author:
Address correspondence to Tim Hilgert: tim.hilgert@kit.edu

of these patterns. Based on these measures, survey participants are then divided into three similarity groups according to the stability of activity patterns between weeks.

The paper is structured as follows: First, the existing literature on behavior stability and on the distance measuring methods used in behavior analysis is reviewed. Second, the paper describes in more detail the methodology sketched briefly above. Third, results are presented for the different activity patterns and distance measures. Finally, a logistic regression model is used to identify socio-economic and demographic covariates for similarity in weekly activity patterns.

Literature Review

The idea of measuring similarity has its origins in biology, in which two DNA strings or protein sequences must often be compared. When using distance measures, one generally calculates the similarity between two sequences. For alphabetical strings, two standard distance measures are widely used in research: the Hamming distance (*7*) and the Damerau-Levenshtein distance (*8*). Whereas both distance measures quantify similarities between two strings, the Hamming distance is more appropriate for sequences of equal length (*8*).

Over time, the basic idea of similarity measurement was refined and adapted for different research fields, such as computer science and transportation. The resulting methods are often referred to as sequence alignment methods. (For an extended overview, see Hosangadi (*8*) and Joh et al. (*9*).) The measurement of similarity has become increasingly important in transportation research. For example, distance measures are used to analyze mode choice, activity patterns, activity departure times, or some combination thereof (*2–4*, *10–18*). Joh et al. (*9*) divided the methods used in travel behavior research into three groups. The first examines the composition of elements in two or more activity patterns to find differences. Here, the corresponding characters of two patterns are compared (*19*). The second considers element composition and sequence, but compares each element separately. The third consists of multidimensional sequence alignment that considers element composition and the sequence of the activities through simultaneous comparison (*9*). Wilson (*11*) introduced the use of sequence alignment to assess the stability of daily activities by translating activity patterns into a sequence of characters. This led to further developments in multidimensional sequence alignment methods with advanced applications and complex sequence structures (*9*).

Some research deals with classification studies, which identify interpersonal similarities by comparing activity patterns within a certain period (*10–12*). Other researchers (*3*, *9*) have examined an individual's activity patterns from day to day within the same week and found significant variability in travel behavior. Most of these studies do not capture variability over long periods, such as weeks or years, however, since they are based on cross-sectional studies using travel diaries limited to a maximum of two days. Only a few studies use longitudinal surveys to capture variability. Cherchi & Cirillo (*1*) and Schlich (*3*), for example, use the 6-week Mobidrive survey to compare the corresponding weekdays of different weeks in the same year. Bargeman (*10*) investigates variability in tourism travel behavior over several years by means of vacation histories. Cherchi & Cirillo (*1*) define tours with associated activities and modes. Based on these tours, logit models were developed to identify the factors impacting day-to-day variability in mode choice within a week (*1*). Raux et al. (*2*) investigate temporal rhythms in activity patterns and estimate influences on those variations.

In reviewing previous literature, however, no evidence is found of an intrapersonal day-to-day or week-to-week comparison between two years of panel data. Moreover, only a few studies investigate the relevant parameters that influence stability in activity patterns or mode choice. Further research on year-to-year stability in travel behavior is clearly needed, since longer periods can better capture the changes in life situation that typically influence behavior. In this paper, panel data is used to help close this research gap by studying the stability of activity patterns across two weeks located in different years.

Data

The study draws on travel survey data from the German Mobility Panel. This section gives an overview of the panel and introduces the study sample.

The German Mobility Panel

The German Mobility Panel (MOP) is a German National Household Travel Survey (NHTS) that has been conducted each year since 1994. It is carried out on behalf of, and funded by, the German Federal Ministry of Transport and Digital Infrastructure. The market research firm KANTAR TNS is responsible for the field work (i.e., recruitment and data collection) and the Institute for Transport Studies of the Karlsruhe Institute of Technology is in charge of the survey's design and scientific supervision (*20*, *21*). The MOP survey is conducted each autumn and, since the aim is to track everyday travel, the survey weeks are chosen to avoid school and bank holidays. The participants provide a complete trip diary for an entire week. This contains information about all their trips during a whole week, that is, distances traveled, means of transportation, trip purposes, and start and end times. Information about the socio-demographics of the participants, the availability of cars and bicycles, and the possession of transit passes is also collected. Participants also indicate whether the trip diary period was typical or non-typical, for example, whether they were ill or on holiday.

The sample size is 1,000 to 1,700 households containing 2,000 to 2,700 persons (aged ten years and older). The MOP is a rotating panel survey, and participants are asked to take part for three consecutive years. Each year, a portion of the households is dropped from the subsequent wave and replaced with new households.

Study Sample

The sample used for analyses is based on data collected between 2005 and 2015. This data contains 7,685 persons from 5,089 households that participated at least once during these years. They reported over 311,000 trips. Since the study focuses on year-to-year differences in activity patterns, only people that participated for at least two years in a row were of interest to the study. Those survey participants reporting non-typical periods such as illness or vacation were also removed from the sample since these diaries may have captured atypical behavior instead of behavioral change—the goal of the study. The final study sample contained 2,573 participants from 1,983 households. To review changes between two years, 3,491 transitions were ultimately used;2,573 between participants' first and second year participation and 918 between their second and third year participation.

Methods

To compare activity patterns from different years, first it is shown how activity schedules are generated from trip diary data. Next, it is explained how patterns are derived from those schedules and the measures to estimate differences between them are introduced. Finally, the logistic regression method used to further interpret the data is shown.

Data Transformation

To compare the activity patterns of people, activity data is needed. Since the MOP survey contains a trip diary, the activities (i.e., what people do between trips) can be inferred indirectly from the dataset only. Table 1 shows a one-day example of the transformation from trip data (left side—result of the survey) to activity data (right side—result of inference). For this transformation, it is assumed that people start their trip diary at home on the beginning of the first day. The analyses differentiate between the following activities: work, education, leisure, shopping, transport (pick-up, drop-off), and being at home. Detailed information about the activities a person performs at home was not queried in the MOP survey.

To facilitate the use of the chosen distance measures, the information about activity schedules was reduced to one main activity purpose per day. This main activity purpose was based on the construct of tours. Each day was divided into tours that both start and end at home. For example, the sample day in Table 1 consists of two tours. The purpose of any given tour was identified by the longest activity of the tour (e.g., work, for the first tour on the sample day). The main activity purpose per day was then determined by the following rule: If a day contained a work or education tour—that is, a mandatory activity—then the main activity purpose of that day would be work or education, respectively. Otherwise, the main activity purpose of the day was given by the purpose of the longest tour of that day. More information about the construction of activity schedules and the transformation process can be found in Hilgert et al. (*22*).

The reduction to one main purpose per day is a strict simplification of the activity patterns. It does not consider other aspects, such as tour concatenation, timing, or other activities during the week. When initially discussing this research idea, an appropriate dimension was

Table 1. One-day Example: Trip Diary and Inferred Activity Diary

Trip diary			Inferred activity diary				
Start	End	Purpose	Start	End	Purpose	Tour number	Tour purpose
			midnight	8:00 a.m.	Home	-	-
8:00 a.m.	8:30 a.m.	Work	8:30 a.m.	12:15 p.m.	Work	1	Work
12:15 p.m.	12:20 p.m.	Leisure	12:20 p.m.	1:10 p.m.	Leisure	1	Work
1:10 p.m.	1:15 p.m.	Work	1:15 p.m.	4:30 p.m.	Work	1	Work
4:30 p.m.	5:00 p.m.	Home	5:00 p.m.	7:00 p.m.	Home	-	-
7:00 p.m.	7:15 p.m.	Shopping	7:15 p.m.	7:45 p.m.	Shopping	2	Shopping
7:45 p.m.	8:00 p.m.	Home	8:00 p.m.	11:59 p.m.	Home	-	-

sought for comparing activity patterns. Investigating all activities for the entire week would require analyzing patterns of different lengths. This, in turn, would require the application of different methods and increase the interpretation complexity. Moreover, there might be some activities during the week that are less important for the year-to-year comparison and the assessment of behavioral changes. In the end, the simplification described above was chosen to reduce the complexity of the approach and to allow the use of unidimensional methods (see next sections) that are more practical to handle and straightforward to interpret.

Distance Measures

As evidenced in the literature review, many different methods have been applied to investigate stability in travel behavior between two points of time. The level of stability depends strongly on the chosen measurement type and the number of attributes the measurement covers (*3*). With an increasing number of measurement attributes, the similarity of the strings usually decreases. Complex methods, such as multidimensional sequence alignment, are not useful for all research questions, because of the sharp increases in interpretation complexity. However, they are suitable for long, multidimensional sequences. When studying simple, short sequences, approaches such as the Hamming distance are also appropriate. Several studies in travel behavior apply these methods (partly as adapted versions) to compare strings (*3, 10, 14, 23*). Here, two methods were chosen that permit straightforward interpretation and are appropriate for the research question. Using these methods, the frequencies and order of activities taking place within a week are compared.

Pairwise Edit Distance. The first distance measure implemented compares the corresponding days of two weeks from two different years. Here, weekly activity schedules were reduced to the information about main activity purposes. These, were called activity patterns. Then, each day of the first year was compared, pairwise, with the corresponding day of the second year. This distance is also known as the *Hamming distance* (*7*). The comparison is based on the days of the week (e.g., Monday in year 1 and Monday in year 2). If a given day has a different main activity purpose than in the previous year, a value of 1 is added to the distance. Thus, the maximum of this measure is 7 (all days have different purposes) and the minimum is 0 (all days have the same purposes) and the result can only be an integer. The Pairwise Edit Distance (PED) is calculated as

$$\mathrm{PED} = \sum_{i=1}^{7} d(i) \qquad (1)$$

$$d(i) = \begin{cases} 0, & \mathrm{MAP}_{i,1} = \mathrm{MAP}_{i,2} \\ 1, & \mathrm{MAP}_{i,1} <> \mathrm{MAP}_{i,2} \end{cases} \qquad (2)$$

where
i = weekday number,
$\mathrm{MAP}_{i,1}$ = main activity purpose of day i in year 1, and
$\mathrm{MAP}_{i,2}$ = main activity purpose of day i in year 2.

Main Activity Distance. The second distance measure implemented also compares the main activity purposes of each day between different years. Unlike the PED, however, it does not take into account the sequence of the days. Instead, this measure compares the number of days for each activity purpose and counts differences between the two weeks. Differences in the purposes work, education, and home are weighted with 1.5; other purposes are weighted with 1. This weighting emphasizes the importance of mandatory activities, such as work and education. Home was also weighted with 1.5 to indicate whether a person stayed at home or undertook an activity on a given day. This results in a maximum of 21 for this measure (e.g., 7 working days in year one and 7 home days in year two) and a minimum of 0 (i.e., the distribution of activity purposes remains unchanged from year one to year two); only .0 and .5 values are feasible. The Main Activity Distance (MAD) is calculated as

$$MAD = 1.5^{*}\sum_{x}\left|\mathrm{NOD}_{x,1} - \mathrm{NOD}_{x,2}\right| + 1.0^{*}\sum_{y}\left|\mathrm{NOD}_{y,1} - \mathrm{NOD}_{y,2}\right| \qquad (3)$$

for $x \in \{\text{work, education, home}\}$ and $y \in \{\text{leisure, shopping, transport}\}$
where
$\mathrm{NOD}_{\mathrm{purpose},1}$ = number of days with activity purpose in year 1, and
$\mathrm{NOD}_{\mathrm{purpose},2}$ = number of days with activity purpose in year 2.

Distance Measures Example. To illustrate the differences between the distance measures, the following example is offered (see Table 2). Here, the individual worked from Monday to Friday in the first year. The weekend was dominated by leisure activities. In the second year, Wednesday was also dominated by leisure activities. Counting the number of different days (PED) yields 1. Counting the number of days per purpose (MAD) yields 2.5, since the "missing" working day is weighted by 1.5.

Table 2. Distance Measures Example

Activity pattern information							
Weekday	Monday	Tuesday	Wednesday	Thursday	Friday	Saturday	Sunday
$MAP_{weekday,1}$	Work	Work	Work	Work	Work	Leisure	Leisure
$MAP_{weekday,2}$	Work	Work	Leisure	Work	Work	Leisure	Leisure
PED							
Weekday	Monday	Tuesday	Wednesday	Thursday	Friday	Saturday	Sunday
d(i)	0	0	1	0	0	0	0
PED	0 + 0 + 1 + 0 + 0 + 0 + 0 = 1.0						
MAD							
Purpose	Work	Education	Home	Leisure	Shopping	Transport	
$NOD_{purpose,1}$	5	0	0	2	0	0	
$NOD_{purpose,2}$	4	0	0	3	0	0	
Weighted difference	1.5 * 1	1.5 * 0	1.5 * 0	1.0 * 1	1.0 * 0	1.0 * 0	
MAD	1.5 + 0 + 0 + 1 + 0 + 0 = 2.5						

Grouping Participants

To evaluate differences between years, the PED and the MAD measures were combined to cluster participants into groups. Different groups of participants were tested in conjunction with the logistic regression model. Finally, participants were distinguished primarily according to the number of different days between the years (PED). Persons exhibiting larger differences were then further distinguished using the MAD measure. In the end, three groups of participants were defined according to their activity patterns:

1. PED <= 1: Participants having no or only small differences in their activity patterns.
   n = 1,238 observations (35.5%)
2. PED > 1 AND MAD <= 6: Participants having bigger differences in their activity patterns but only small differences concerning the activity types over the whole week.
   n = 1,642 observations (47.0%)
3. PED > 1 AND MAD > 6: Participants having bigger differences in their activity patterns and big differences concerning the activity types over the whole week.
   n = 611 observations (17.5%)

Logistic Regression Model

To determine the relevant parameters that influence whether activity patterns between two years remain stable or not, a logistic regression model was used. Incorporating different variables into the model, tests were run to see which variables best predicted the "choice" of grouping. Thus, the model revealed parameters relevant for stable or instable behavior, as given by the distance measures PED and MAD. For the model, two different types of variables were used. The first type describes the participants' situation in the second year (state variables). These variables include the socio-demographic characteristics (e.g., education level, household income, employment status) and the availability of mobility options (e.g., a car). The second type indicates differences between the years (change variables). For each transition between two years, it was determined whether variables such as education level or household income had changed. These changes were indicated using binary variables. If the variable had an ordinal scale, the size of the change would also be investigated. For the regression model estimate, the software tool SAS was used and the grouping of the participants as the dependent variable. Participant Group 1 was used as reference category.

Results

This section first describes the patterns for different population groups, which should deliver a first impression of relevant combinations among different people. It also shows the results of the implemented measures (PED and MAD) in a cross-table. Different state and change variables are then investigated and, finally, the results of the logistic regression model are shown.

Activity Patterns

To gain a better understanding of how often the various activity patterns (i.e., seven-digit chains of main daily activities during one week) occur among the different

Table 3. Activity Patterns by Frequency and Employment Status

Full-time workers		Part-time workers		Workers in vocational programs	
Activity pattern	Frequency	Activity pattern	Frequency	Activity pattern	Frequency
WWWWWLL	31.8%	WWWWWLL	14.7%	EEEEELL	24.0%
WWWWWSL	15.4%	WWWWWSL	7.4%	WWWWWLL	8.3%
WWWWWWL	6.1%	WWWWWWL	3.0%	EEEEESL	7.4%
WWWWWWW	2.2%	WWWWWWW	2.3%	EEEEELH	2.8%
WWWWWLH	2.2%	WWWWLLL	2.0%	WWWWLLL	2.8%
Other	42.3%	other	70.6%	other	54.7%
Students at school or university		**Pensioners**			
Activity pattern	Frequency	Activity pattern	Frequency		
EEEEELL	44.4%	LLLLLLL	8.8%		
EEEEESL	8.0%	LSLLLLL	2.7%		
EEEEELH	4.7%	LLLLLSL	2.6%		
EEEELLL	2.3%	LLLLSLL	2.6%		
EEEEEHL	2.2%	SLLLLLL	2.4%		
Other	38.4%	other	81.0%		

Note: W = work; E = education; L = leisure; S = shopping; H = home; T = transport.

population groups, their frequency was compiled as a function of employment status (see Table 3). For each status, the five most common activity patterns are shown. Patterns occurring more rarely are summarized as "other."

Most activity patterns during weekdays (i.e., the first five digits of the chains) were characterized by obligatory activities, such as work or school. The most frequent activity pattern of both full-time and part-time workers was WWWWWLL. That is, Monday to Friday the main activities were working, whereas leisure activities dominated Saturday and Sunday. This activity pattern was more common for full-time workers (31.8%) than for part-time workers (14.7%), which suggests that the activity patterns of full-time workers are more homogeneous than those of part-time workers.

The most common activity pattern of workers in vocational programs (24.0%) and students (44.4%) was EEEEELL. In Germany, vocational education consists of two parts: theory learned in public vocational schools (activity purpose: education) and training embedded in a real-life work environment (activity purpose: work) (*24*). This dual scheme is evidenced by this group's second most frequent activity pattern: WWWWWLL. For workers and students, leisure activities only dominated on the weekend. In contrast, the most frequent activity pattern of pensioners was LLLLLLL (8.8%), indicating that leisure was also likely to be the main activity on weekdays.

The highest share of "other" activity patterns (81.0%) among pensioners shows that their seven-day activity chains were more flexible than those of other employment groups. However, one must also take into consideration that pensioners generally had fewer activities than other groups (*21*). The activities they performed (e.g., shopping, leisure) were less often subject to weekly rhythms and their patterns were less often determined by mandatory activities. The lowest shares of "other" activity patterns were found among full-time workers (42.3%) and students (38.4%), a result that indicates a higher similarity in the activity patterns for members of these two groups and that might be expected, given the constraints of their work and study obligations.

Distance Measures

To illustrate the linkage of the similarity measures PED and MAD, Table 4 shows a combined frequency distribution. In total, 477 persons (13.7% of the sample) have results of 0 for both measures. In other words, their activity patterns were identical in the two years examined. If the main activities differ on only one day (e.g., WWWWWLL in year 1 and WWWWWSL in year 2), this person is still assigned to similarity group 1. The PED results for members of similarity group 2 (47.0% of the sample) and 3 (17.5%) range from 2 to 7. In other words, the main activities of at least two days differ from one year to the next. Similarity groups 2 and 3 are further distinguished using MAD. Group 2 contains people with lower MAD values (0 to 6) and group 3, people with higher MAD values (>6). Low MAD values (group 2) indicate similar frequencies for the main activity types. Here, it may be that the main activity types are merely allocated to different days of the week. The PED–MAD combination 3–2 is most common in similarity group 2, with 163 cases. One example of this activity pattern combination is WWWSWLL and WWSWWSL. Here, one

Table 4. Combined Frequency Distribution of PED and MAD

	PED							
MAD	0	1	2	3	4	5	6	7
0	477	0	125	8	41	3	1	0
2	similarity group 1	432	48	163	37	25	6	0
2.5		315	107	77	34	22	5	1
3		14	11	4	1	2	1	0
4			134	29	81	17	4	0
4.5			130	63	44	20	7	1
5			124	54	26	16	9	1
5.5		similarity group 2	22	13	7	2	1	1
6			6	69	18	19	2	0
6.5				24	21	22	4	3
7				26	20	12	3	2
7.5				37	29	15	7	1
8				11	20	6	7	0
8.5				7	12	11	3	0
9				3	5	5	4	1
9.5					10	2	4	0
10					28	16	11	5
10.5					3	10	2	2
11					5	5	2	0
11.5				similarity group 3	8	6	4	1
12					1	5	9	2
12.5						38	19	4
13						3	2	1
13.5						5	4	0
14						3	1	0
14.5						4	5	1
15						12	12	2
15.5							2	0
16							1	0
16.5							2	0
17							6	1
17.5							9	8
18								1
18.5								0
19								0
19.5								2
20								1
20.5								0
21								0

leisure day is changed to a shopping day and the activity types of Wednesday and Thursday are switched. Since group 3 contains people with MAD values greater than 6, there are also larger differences in activity frequencies. There are 38 cases (6.2%) with the PED–MAD combination 5–12.5, as exemplified by the activity pattern combination **WWWWW**LL and **LLLSS**LL.

State Variables

Variables of MOP data were categorized into state and change variables (see Table 5). State variables describe the participant in the second year, whereas change variables characterize a change between two years. Table 5 shows three state variables—occupation status, education level, and household type—along with their characteristic values, which sum to 100%. A majority (54.9%) of full-time workers landed in similarity group 1. Only students were more stable (57.4% in group 1). Part-time workers exhibited greater variability, with a plurality (46.4%) landing in similarity group 2. A plurality of temporarily unemployed persons and workers in vocational programs landed in similarity group 3. Persons with a pre-graduation education level—most of them are students in school—exhibited high stability, with almost 70% in group 1. Persons with secondary school and no vocational training had a higher percentage in group 2 (57.9%) than persons with other education levels. Single households were mostly in similarity group 2 and family households were to a large part in similarity group 1. Regarding their main

Table 5. Descriptive Results—State and Change Variables

State variables	Description	*n*	Group 1	Group 2	Group 3
Whole sample		3,491	35.5%	47.0%	17.5%
Occupation status	Full-time worker	1,140	54.9%	33.4%	11.7%
	Part-time worker	573	31.8%	46.4%	21.8%
	Temporarily unemployed	81	16.0%	40.7%	43.2%
	Student	350	57.4%	23.4%	19.1%
	Worker in vocational program	65	35.4%	26.2%	38.5%
	Homemaker	125	9.6%	67.2%	23.2%
	Pensioner	1,129	15.0%	68.0%	17.0%
Education level	Secondary school, no vocational education	76	21.1%	57.9%	21.1%
	Secondary school, vocational education	643	32.3%	51.5%	16.2%
	General certificate of secondary education	942	32.8%	48.0%	19.2%
	University-entrance diploma, university (of applied science) diploma	1,525	33.2%	47.9%	19.0%
	Pre-graduation	236	69.9%	23.7%	6.4%
Household type	Single household	747	30.7%	51.8%	17.5%
	Family household with children	1,013	48.6%	34.6%	16.9%
	Other household types	1,731	29.9%	52.3%	17.9%

Change variables (n = 4,038 for each variable)	Changes between years per similarity group: No			Yes			Changes between years	
	Group 1	Group 2	Group 3	Group 1	Group 2	Group 3	No	Yes
Impaired mobility	35.9%	46.6%	17.5%	24.5%	58.0%	17.5%	95.9%	4.1%
Education level	35.6%	47.0%	17.3%	33.8%	47.0%	19.2%	90.6%	9.4%
Occupation status	38.0%	48.7%	13.3%	18.2%	35.5%	46.4%	87.4%	12.6%
Job changes	36.0%	47.9%	16.1%	26.0%	32.7%	41.3%	94.4%	5.6%
Transit accessibility	32.2%	49.9%	17.9%	48.6%	35.4%	16.0%	80.1%	19.9%
Driving license possession	35.3%	47.2%	17.4%	41.3%	38.7%	20.0%	97.9%	2.1%
Car availability	35.1%	47.7%	17.2%	37.7%	42.6%	19.6%	87.0%	13.0%
Transit pass possession	35.3%	47.6%	17.1%	38.1%	37.6%	24.3%	94.6%	5.4%
Rail card possession	35.3%	47.1%	17.6%	39.0%	44.9%	16.2%	96.1%	3.9%
Residence (location)	35.5%	47.3%	17.2%	34.2%	35.5%	30.3%	97.8%	2.2%
Household type	36.1%	47.5%	16.4%	30.6%	43.3%	26.1%	88.5%	11.5%
Household size	35.2%	47.6%	17.2%	38.5%	40.6%	20.9%	92.0%	8.0%
Number of children under 10	34.9%	47.7%	17.4%	46.0%	34.9%	19.0%	94.6%	5.4%
Income class	34.0%	49.9%	16.1%	37.8%	42.4%	19.8%	62.1%	37.9%
Number of cars	35.3%	47.8%	17.0%	37.4%	39.8%	22.8%	90.6%	9.4%
Second home	35.6%	46.5%	17.9%	34.6%	49.8%	15.5%	83.8%	16.2%

activities, family households presumably exhibited more stable behavior than single households because of the more systematic routines needed to organize children households.

Change Variables

The change variables in Table 5 describe a possible change (e.g., of occupation status) between the two years. The frequency of these changes occurring among all individuals in the sample and for each similarity group are displayed. The impaired mobility variable signifies changes in physical mobility from one year to the next. Here, only a few changes (4.1%) are recorded in the sample. People whose physical mobility does change are most likely to be in similarity group 2 (58.0%). More changes are visible in the education level variable, which indicates a graduation event between the first and second years. In 9.4% of the sample, there was a change in education level, and most of these people are found in similarity group 2. A change in occupation status makes it more likely for people to be in similarity group 3, showing the important influence of this variable. In 19.9% of survey participants, there was a change in transit accessibility (i.e., the accessibility of work place or school by public transit). This might be correlated with changes in job, occupation, or residence location. Most people with transit accessibility changes still belong to similarity group 1, implying that such a change does not lead to significantly altered activity patterns. Changes in

Table 6. Results—Logistic Regression

No.	Variable	Wald chi-square	Parameter estimates	
			Similarity group 2	Similarity group 3
0	Intercept		1.364***	−0.198*
1	person is full-time worker	354.339***	−1.94***	−1.712***
2	person is part-time worker	62.256***	−1.026***	−0.746***
3	person is student	165.301***	−2.134***	−0.956***
4	person is worker in vocational program	27.652***	−1.766***	−0.814**
5	person holds a general qualification for university entrance, high school, or university degree	15.766***	0.278***	0.387***
6	person lives in family household with children	8.359**	−0.266***	−0.251**
7	occupation status has changed	213.634***	0.286*	1.837***

Note: ***,**,* = significance at 1%-, 5%-, 10%-level; ns = not significant.
LL (0) = −3,835; LL (constant) = −3,586; LL (full model) = −3,174.
McFadden pseudo R^2 (based on constant LL) = 0.12.
n = 3,491.

residence or in the possession of driving license, transit pass, or rail card take place relatively seldom (in each case, less than 6% of the sample). There were many shifts in income class in the sample (37.9%). Most of these shifts were in similarity groups 1 and 2. Changes in household type are also considered in the analysis (e.g., changing to a family household when a child arrives through birth or adoption). The variable household size captured changes in the number of people in the household. The second home variable was included, since it might also explain why travel patterns change from year to year. In 16.2% of the sample, there were changes in second home ownership, and most of these people belonged to similarity group 2.

Logistic Regression Model

The logistic regression model helps illustrate how various socio-demographic characteristics influence the probability of a person belonging to one of the three similarity groups. Table 6 shows estimates and significance levels for the parameters included in the model. All socio-demographic characteristics (state and change variables) have been defined as binary variables. The reference category was similarity group 1.

The Wald Chi Square results show that all variables included in the regression model were significant. The values of the Wald Chi Square results indicate that occupation has the highest explanatory power for both the state (occupation status) and change (change of occupation status) variables. For the occupation state variables, the reference consists of all people without employment, that is, pensioners, unemployed people, and homemakers. People of all occupation groups (variables 1 to 4) are less likely to be in similarity group 2 or 3 than in the reference category, since all estimates are negative. Survey participants with higher education levels tended to be more flexible than those with lower education levels. Possibly, they had more activity options on the weekend, resulting in both higher PED and higher MAD values. Interestingly, households with children were more stable than the reference households (households with 2 + adults but no children; single-households). Living with children typically requires more plans and agreements in everyday life, and once these routines have been established, families might be reluctant to give them up.

Considering the change variables, it was seen that a change in occupation status, such as moving from education to full-time work or from full-time work to retirement, explained differences in similarity groups quite well. Participants with a change in occupation status were more likely to be in group 3, since a new status is often associated with different activities. These results are in line with the descriptive findings.

The other change and state variables tested such as car ownership, spatial type, income, gender, the existence of children in the household (category 1), change of residence, employment status, and household size turned out not be significant in the regression model.

Conclusions

Routines and mandatory activities, such as working, determine individual activity patterns and strongly influence travel demand. However, the variability or stability in people's activity patterns from one year to the next has not been previously researched. Knowledge about the year-to-year stability of activity patterns and information about how changes in life circumstances affect this stability can make an important contribution to rational transport policies and facilitate the modeling and forecasting of travel demand.

To analyze activity patterns, two distance measures were used—PED and MAD. These measures compare weekly activity patterns. The activity information for a given week was distilled down to the main activity purposes for each day of that week. Using survey responses from the MOP, those participants who had responded multiple times and evaluated their activity pattern changes between two successive years of participation were selected for the sample. First the activity patterns were analyzed descriptively. About one-third of the sample participants had the same or very similar main activity patterns in the two weeks examined. The group identified as workers had the highest propensity of similar activity patterns. Around 50% had patterns consisting of five days of work and two days dominated by leisure and shopping. Pensioners, in contrast, had more flexible activity patterns. Since their behavior is not so restricted by mandatory activities, such as working or attending school, a wider variety of activity patterns becomes possible. Results reveal the variable "occupation status" (e.g., worker, student or pensioner) to be the most important for describing similarity groups. About 55.1% of full-time workers were in similarity group 1, but only 14.3% of pensioners. In addition, the change variable "occupation status has changed" had a high explanatory power for activity pattern variability: people whose occupation status changed were more likely to belong to similarity group 2 or 3.

Nevertheless, the study has several limitations. First, the definition of similarity groups is very subjective. Since the degree of freedom in defining a similarity group is very high, future research should explore how changing this definition affects the results. Second, the small number of observations involving life events might have obscured the significance of change variables in the logistic regression. An extension of the data with new survey waves would improve the levels of significance. Third, further research is needed on the definition of activity patterns for year-to-year comparison. Here, only the main daily activities for the study week were considered. A refinement of the approach would be to consider all activities during the week. However, this would expand the complexity of the activity patterns and possibly require the use of other distance measures. Another interesting project would be to combine activity pattern measures with land use aspects or mode choice decisions, an endeavor that would also tend to increase the complexity of the analysis.

Acknowledgments

This publication was written within the framework of the Profilregion Mobilitätssysteme Karlsruhe, which is funded by the Ministry of Science, Research and the Arts in Baden-Württemberg.

Author Contributions

The authors confirm contribution to the paper as follows: study conception and design: T. Hilgert, S. von Behren, C. Eisenmann, P. Vortisch; data collection: T. Hilgert, S. von Behren, C. Eisenmann; analysis and interpretation of results: T. Hilgert, S. von Behren, C. Eisenmann, P. Vortisch; draft manuscript preparation: T. Hilgert. All authors reviewed the results and approved the final version of the manuscript.

References

1. Cherchi, E., and C. Cirillo. Understanding Variability, Habit and the Effect of Long Period Activity Plan in Modal Choices: A Day to Day, Week to Week Analysis on Panel Data. *Transportation*, Vol. 41, 2014, pp. 1245–1262.
2. Raux, C., T.-Y. Ma, and E. Cornelis. Variability in Daily Activity-travel Patterns: The Case of a One-week Travel Diary. *European Transport Research Review*, Vol. 8, No. 4, 2016, p. 26.
3. Schlich, R. Analysing Intrapersonal Variability of Travel Behaviour Using the Sequence Alignment Method. *Proc., AET European Transport Conference*, Cambridge, UK, 2001. http://webarchiv.ethz.ch/ivt/vpl/publications/reports/ab80.pdf
4. Schönfelder, S. *Urban Rhythms - Modelling the Rhythms of Individual Travel Behaviour*. PhD thesis. Institute of Transport Planning and Systems, Swiss Federal Institute of Technology, Zurich, 2006.
5. Bayarma, A., R. Kitamura, and Y. Susilo. Recurrence of Daily Travel Patterns: Stochastic Process Approach to Multiday Travel Behavior. *Transportation Research Record: Journal of the Transportation Research Board*, 2007. 2021: 55–63.
6. Buliung, R. N., M. J. Roorda, and T. K. Remmel. Exploring Spatial Variety in Patterns of Activity-travel Behaviour: Initial Results from the Toronto Travel-Activity Panel Survey (TTAPS). *Transportation*, Vol. 35, No. 6, 2008, pp. 697–722.
7. Hamming, R. W. Error Detecting and Error Correcting Codes. *Bell System Technical Journal*, Vol. 29, No. 2, 1950, pp. 147–160.
8. Hosangadi, S. *Distance Measures for Sequences*, 2012, pp. 1–16.
9. Joh, C.-H., T. A. Arentze, and H. J. P. Timmermans. Multidimensional Sequence Alignment Methods for Activity-Travel Pattern Analysis. A Comparison of Dynamic Programming and Genetic Algorithms. *Geographical Analysis*, Vol. 33, No. 3, 2001, pp. 247–270.
10. Bargeman, B., C.-H. Joh, and H. Timmermans. Vacation Behavior Using a Sequence Alignment Method. *Annals of Tourism Research*, Vol. 29, No. 2, 2002, pp. 320–337.
11. Wilson, C. Analysis of Travel Behavior Using Sequence Alignment Methods. *Transportation Research Record: Journal of the Transportation Research Board*, 1998. 1645: 52–59.
12. Saneinejad, S., and M. Roorda. Application of Sequence Alignment Methods in Clustering and Analysis of Routine

Weekly Activity Schedules. *Transportation Letters: The International Journal of Transportation Research*, Vol. 1, No. 3, 2009, pp. 197–211.
13. Kwan, M.-P., N. Xiao, and G. Ding. Assessing Activity Pattern Similarity with Multidimensional Sequence Alignment based on a Multiobjective Optimization Evolutionary Algorithm. *Geographical Analysis*, Vol. 46, 2014, pp. 297–320.
14. Joh, C.-H., T. Arentze, and H. Timmermans. Pattern Recognition in Complex Activity Travel Pattern : Comparison of Euclidean Distance, Signal-processing Theoretical, and Multidimensional Sequence Alignment Methods. *Transportation Research Record: Journal of the Transportation Research Board*, 2001. 1752: 16–22.
15. Jones, P., and M. Clarke. The Significance and Measurement of Variability in Travel Behaviour. *Transportation*, Vol. 15, 1988, pp. 65–87.
16. Huff, J. O., and S. Hanson. Repetition and Variability in Urban Travel. *Geographical Analysis*, Vol. 18, No. 2, 1986, pp. 97–114.
17. Hanson, S., and O.J. Huff. Systematic Variability in Repetitious Travel. *Transportation*, Vol. 15, 1988, pp. 111–135.
18. Pas, E. I., and F. S. Koppelman. An Examination of the Determinants of Day-to-day Variability in Individuals' Urban Travel Behavior. *Transportation*, Vol. 14, 1987, pp. 3–20.
19. Pas, E. I. A Fexible and Integrated Methodology for Analytical Classififaction of Daily Travel-Activity Behavior. *Transportation Science*, Vol. 17, No. 4, pp. 405–429.
20. Weiss, C., B. Chlond, C. Minster, C. Jödden, and P. Vortisch. Assessing the Effects of a Mixed-Mode Design in a Longitudinal Household Travel Survey. Presented at 96th Annual Meeting of the Transportation Research Board, Washington, D.C., 2017.
21. Weiss, C., B. Chlond, S. von Behren, T. Hilgert, and P. Vortisch. *Deutsches Mobilitätspanel (MOP) - Wissenschaftliche Begleitung und Auswertungen Bericht 2015/2016*: Alltagsmobilität und Fahrleistung [German Mobility Panel - Scientific Supervision and Analyses Report 2015/2016: Everyday Travel and Car Use], Karlsruhe, 2016.
22. Hilgert, T., M. Heilig, M. Kagerbauer, and P. Vortisch. Modeling Week Activity Schedules for Travel Demand Models. *Transportation Research Record: Journal of the Transportation Research Board*, 2017. 2666: 69–77.
23. Streit, T., C.-E. Allier, C. Weiss, B. Chlond, and P. Vortisch. Changes in Variability and Flexibility of Individual Travel in Germany – Trends and Drivers. Presented at 94th Annual Meeting of the Transportation Research Board, Washington, D.C., 2015.
24. Federal Ministry of Education and Research. *The German Vocational Training System*. https://www.bmbf.de/en/the-german-vocational-training-system-2129.html. Accessed July 13, 2017.

The Standing Committee on Traveler Behavior and Values (ADB10) peer-reviewed this paper (18-01379).

Article

TRR
JOURNAL OF THE TRANSPORTATION RESEARCH BOARD

Transportation Research Record
2018, Vol. 2672(47) 57–67

DOI: 10.1177/0361198118787080
journals.sagepub.com/home/trr
SAGE

Coordinated Daily Activity Pattern Type in an Applied Activity-Based Travel Model System: Specifics of Modeling Large Households and Modality Preferences

Gaurav Vyas[1], Peter Vovsha[1], Danny Givon[2], and Yehoshua Birotker[2]

Abstract

This paper describes a modified approach to modeling an individual daily activity-travel pattern (DAP) coordinated at the household level. The model was primarily introduced to handle large households that are typical for the city of Jerusalem. However, the developed method proved useful in adding more behavioral aspects to the model. The study introduced daily modality and added the emphasis on modeling it for all household members simultaneously. It is of special practical value for Jerusalem since such population sectors as ultra-Orthodox Jewish and Arabs are characterized by a large share of persons who have an entire day of travel implemented in a non-motorized fashion. In addition to daily modality, this paper presents a simple but useful approach to understand at-home time-use. A binary choice model was formulated for the main reason of being inactive (working-at-home or other).

The paper also discusses possible extensions of the current approach as well as alternative approaches to frame DAP type choice in the context of a complete activity-based travel model (ABM). In particular, this model can be easily extended in order to better address the joint nature of the choices using the Gibbs sampler. The paper discusses how DAP type choice can be framed in many ways and with different levels of detail subject to the ultimate model system design.

Over the past decade, activity-based travel models (ABMs) have become the standard for travel demand models in practice for many major cities in the U.S. and around the world. The activity participation component in these ABMs is the centerpiece of the travel model that largely defines the set of activities for each individual. The original concept of a coherent individual daily pattern of activities and trips as a replacement for independently calculated trip rates by trip purpose pertinent to trip-based four-step models (*1*, *2*) served as the historical springboard for applied ABMs in practice. Although there is a wide spectrum of ABM system designs in practice and (even more) in research, the first step for modeling day-level travel in most ABMs is to identify the daily activity-travel pattern (DAP) types for all household members. Individual DAP type comprises a variety of patterns that share the most principal features. DAP type is the leading travel sub-model in the ABM chain. Subsequently other sub-models add details of the activities and trips for each individual conditional on the chosen principal DAP type.

One of the limitations of the original concept of individual DAP was that the model was designed to be applied for each household member independently and without an explicit accounting for intra-household interactions. Several attempts have been made since to account for intra-household interactions by considering only the main dimensions of the DAP type, which opened a way to model impacts of household members on each other.

In most ABMs in practice, individual DAP types are classified into three main categories: 1 = mandatory travel pattern, 2 = non-mandatory travel pattern, and 3 = at-home pattern (*3*). A mandatory travel pattern is one in which a person makes at least one work, university, or school trip during the day. A non-mandatory travel pattern involves at least one non-mandatory trip, with no mandatory trips taken. The at-home pattern describes a person who does not participate in out-of-home activities and makes no trips. The identification of individual DAP types represents one of the crucial demand modeling steps since a DAP type largely shapes the entire

[1]WSP USA Inc., New York, NY
[2]Jerusalem Transportation Masterplan Team (JTMT), Jerusalem, Israel

Corresponding Author:
Address correspondence to Gaurav Vyas: Gaurav.vyas@wsp.com

travel pattern. For example, a full-time worker with a non-mandatory pattern would behave in a very similar way to a non-worker due to the absence of the work activity on the travel day.

In one of the earlier versions of DAP type, Vovsha et al (*3*) accounted for intra-household interactions through sequential modeling of household members according to the assumed priority. This way DAP types chosen by some of the prioritized household members affected DAP types chosen by the other household members. This research showed that there is a great deal of correlation between DAP types of different household members, although the method itself can be criticized for a somewhat arbitrary established order of processing of the household members.

Bradley and Vovsha illustrated the importance of modeling DAP types for all household members simultaneously (*4*). They introduced intra-household interactions and formulated a simultaneous choice model for coordinated modeling of DAP type. The model was structured as a trinary choice applied to all household members jointly. The utility function of this joint model incorporated strong joint participation "bonuses" when different household members choose the same pattern type. This approach has been utilized in many ABMs in practice, developed for such cities in the U.S. as Columbus, OH; Atlanta, GA; San Diego, CA; Philadelphia, PA; and others. However, a principal drawback of this approach is an inevitable "explosion" in the number of alternatives of the joint choice with either growing household size or adding details to the DAP type itself. The most frequently applied manageable structure for five household members includes $3^5 = 243$ alternatives. If another household member is added in order to model six household members jointly, the number of alternatives would grow to $3^6 = 729$. If another DAP type is specified, for example, the at-home pattern is segmented between working from home and staying at home for a different reason, that would make the number of DAP types four instead of three, and the resulted dimensionality for a joint choice of DAP type for five household members would grow to $4^5 = 1{,}024$.

Lemp utilized a latent class approach to include intra-household interaction in DAP modeling (*5*). The author proposed an individual-level latent class DAP model, where class membership is defined at the household level. The latent classes were specified to capture intra-household interactions. This model is essentially a two-step approach where intra-household interactions are included in the first step (household class membership) but then followed by a set of individual trinary choice models for each household member. While this approach resolves the dimensionality explosion issue, it only implicitly accounts for the observed intra-household interactions and may result in a quite complex household classification when the number of household members grows.

In a similar vein, but using an explicit segmentation rather than latent class approach, Vuk et al (*6*) introduced the interesting characteristic of family in-home quality time (when all household members spend at least a certain amount of time at home together) as a household classification principle that subsequently affects choices of individual DAPs. However, this extension does not address the fundamental issue of a combinatorial explosion when DAP types for different household members are modeled jointly.

Bhat et al (*7*) formulated a parsimonious approach for household level activity pattern generation. The model was formulated as a multiple discrete-continuous extreme value (MDCEV) model in which household members allocate time to different activities, including joint activities given the total time constraint. This approach eliminates the need for sequential models and the enumeration of all possible combinations of activity participation by household members. While this model was able to efficiently model five-member households being applied as part of the ABM developed for Southern California, it required a significant simplification of the model structure in order to handle bigger households, due to an increase in the number of alternatives with the growing household size. It can be said that, while this structure resolves the issues of a combinatorial explosion with respect to the number of activities for each person, it still "explodes" with respect to the joint activity participation where each subset of household members has to be enumerated explicitly for each joint activity. While the suggested simplifications could be acceptable in practical terms for most cities in the U.S. where large households are rare, for a city such as Jerusalem where the average household size is 3.5 and many households have more than six members, a model formulation that can handle large households and still account for intra-household interactions is welcome.

It should be noted that the core model structure discussed in the paper, that is an individual daily activity pattern type, represents only one approach to constructing an activity-based modeling system that is largely pertinent to ABMs in practice due to simplicity and readily available data. There are many other approaches to activity-based modeling of travel demand that are, perhaps, theoretically richer, and in terms of modeling more flexible. Some of them do not even require a model of individual daily activity type per se.

The rest of the paper is structured as follows. The next section explains motivation for the current research and its main focuses, followed by discussions on the available data, model structure, estimation results, model implementation, conclusions, and future research directions.

Motivation for Current Research and Application for Jerusalem

The motivation to explore a different approach for DAP type modeling for the city of Jerusalem was manifold.

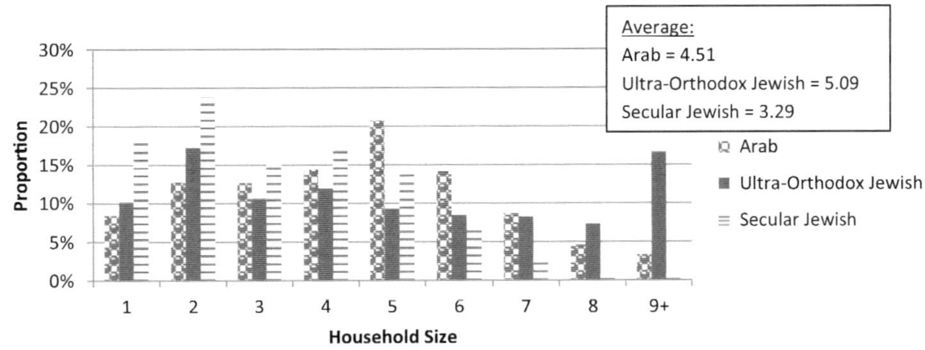

Figure 1. Household size distribution by population sector.

First, the existing approaches were unable to effectively handle larger households. As an example, the coordinated trinary choice approach for DAP modeling, prevalent in several ABMs in practice, can handle up to five household members simultaneously, which is slightly more than the average household size in Jerusalem (see Figure 1). For such population sectors in Jerusalem as ultra-Orthodox Jewish people and Arabs, there is a substantial percentage of households of size six or larger. Thus, there is a significant number of households with more members than a standard DAP type model can practically handle. For example, for a household with eight members, the trinary choice approach will have 9,840 alternatives ($\sum_{n=1}^{N} 3^n$, *where n = household size, N = maximum household size*), and the number of alternatives increases exponentially with an increase in the maximum household size. Thus, there is a need to explore alternate ways of DAP modeling for cities with larger household sizes, such as Jerusalem. An intermediate solution applied in many ABMs in practice involved an iterative process for large households where the DAP model would be applied for five prioritized household members jointly and then the other household members would be modeled sequentially and conditional on the DAPs chosen for the prioritized household members. This is of course a simplified solution that is justified in practical terms for ABMs developed for U.S. cities where households with six or more persons constitute less than 5%.

Secondly, several researchers have introduced an important notion of individual modality in a broader sense than just mode choice for the given trip or tour. Modality is defined as general mode choice preferences that can be modeled as a mid-term individual person attribute that does not change on a daily basis (*8*). Dugundji and Walker (*9*) associated modality with impacts of neighborhood and social networks. The TCRP Report H-37, 2013 (*10*) provides a comprehensive analysis of sources of modality in the context of transit awareness and use in which the explanatory variables included many individual and environmental characteristics. Modality choice is closely intertwined with the choice of individual mobility attributes (such as car ownership or transit pass) (*11*). The difference between modality as essentially a lifestyle component and mobility attributes is that modality is frequently formulated in latent-class terms such as a "transit-oriented" or "transit-averse" person in terms of the mental preferences rather than physical attributes. Modality is important from the econometric perspective for accounting for the fact that mode choices made by the same person and household for different trips are not independent but rather strongly correlated through sharing the same inherent mode preferences.

However, modality can also be viewed as a dimension of DAP type. This view on modality assumes that there are several facets of it that can be better captured in combination with the chosen DAP type that restricts the activity and travel agenda of the person. In this context, modality of a worker on a day with a mandatory pattern (when he/she goes to work) can be manifested differently from his/her modality when he/she chooses a non-mandatory pattern (when he/she does not go to work). Thus, modality can also be embedded in the daily choice of DAP type, in addition to its possible treatment as a mid-term choice of person attributes. For example, in the mid-term, a person may choose to own a transit pass because he/she typically commutes by rail. However, on days when the person is not working, they may not have the same preference for transit.

The Jerusalem ABM already included a detailed mobility attribute model that models modality implicitly as a mid-term choice of the combination of mobility attributes for each household and person (*11*). The focus of the current paper is on the additional explicit treatment of modality as manifested in a daily choice of DAP type. The paper specifically deals with the modality of people with non-mandatory patterns, due to the greater mode flexibility associated with non-mandatory activities. The daily modality is an important driver of mode choice for any city, but it plays an especially important role for Jerusalem where a non-motorized travel share is as high as 30% of daily travel, with many persons whose entire DAP is local and non-motorized. This is especially frequent among the ultra-Orthodox Jewish and Arab population sectors for which the non-motorized mode share constitutes more than 50% of trips. Additionally, it is likely that if one household member prefers non-motorized travel modes, then other members from the same household are likely to prefer non-motorized travel modes as well. In other words, the mode choice preferences of one household member are correlated with those of other household members. The estimation results statistically confirm such correlation of modality among household members.

This paper demonstrates how modality can be included in coordinated DAP modeling. To the authors' knowledge, this is the first attempt to model daily modality coordinated at the household level.

Third, there is a substantial body of research on at-home time utilization, especially to analyze telecommuting and associated home workspaces (*12–14*). However, most ABMs in practice do not deal with the details of in-home activities for inactive people, as they do not participate in any daily travel sub-model. Still, the analysis of inactive people may help in understanding the reason for not traveling and may provide important insights on market segments likely to be sensitive to policies specific to reducing travel (such as promoting telecommuting) or changing behaviors over time (such as online shopping). It should also be mentioned that details of the inactivity of certain household members can be important to predict activity-travel patterns of the other (travel-active) household members. For example, it was well established in the previous research on DAP types that at-home DAP types of different household members are strongly positively correlated (*4*). This can be logically explained by such factors as child care and coordinated days off for vacations and family events. However, it seems that work from home was not properly treated by combining it with other non-work reasons for at-home DAP type. In this regard it can be beneficial to define work from home as a separate DAP type. The model described in this paper is a small step toward bridging the gap between the extensive research on at-home time-use and modeling of inactive (i.e., non-traveling) people in operational ABMs in practice.

Data on Observed DAP Types in Jerusalem

The population of the Jerusalem metropolitan region (1.1 million) consists of three major sectors that are characterized by significant differences in socio-economic characteristics, lifestyles, mobility attributes, and travel behavior:

- *Secular Jewish* population (45%): The socio-economic mix, household composition, and travel behavior of secular Jewish people are quite similar to those of most urban metropolitan areas in the Western world and the U.S. in particular.
- *Ultra-Orthodox Jewish* population (27%): They are characterized by striking differences relative to the secular Jewish population, including significantly greater household size with multiple children, significantly lower income levels and car ownership, as well as a lower degree of employment where many male household heads attend Jewish religious institutions (Yeshivas) on a full-time basis and are thus classified as students. More specifically, for various reasons, male ultra-Orthodox Jewish adults tend to attend Yeshivas until roughly the age of 35, after which their employment levels rise. Conversely, many ultra-Orthodox females are employed at younger ages and once the male household head becomes employed they resume their traditional role as homemakers.
- *Arabs* (28%), primarily citizens of Eastern Jerusalem: They also represent a very distinct population group. In terms of (large) household size and (relatively low) income, they are somewhat close to the ultra-Orthodox Jewish people. However, they have a unique household structure that frequently combines several generations and married siblings living together. In terms of employment, Arabs more frequently have female household heads playing a housewife role while the degree of male employment status is close to that of the secular Jewish people.

The analysis discussed in this paper is conducted using a comprehensive GPS-assisted Jerusalem Household Travel Survey completed in 2010. The survey consisted of 8,230 households (approximately 3% of Jerusalem's population) distributed across the three sectors, with a sampling rate of 3%, 4%, and 2% for the secular, ultra-Orthodox, and Arab sectors, respectively. The survey included all activities and trips recorded for one usual workday for all household members with a rich set of collected person and household attributes. Thus, the survey provided a reliable basis for identification of the DAP types for all household members.

Model Structure

In order to resolve the issue of explosion in number of alternatives in trinary choice in the framework of coordinated DAP (CDAP) modeling for bigger households, the model was formulated as a series of sequential conditional sub-models structured as multinomial logit models, as shown in Figure 2.

The first level represents a binary choice model – mandatory vs. non-mandatory DAP type – coordinated at the household level. This means that all household members simultaneously choose between mandatory and non-mandatory patterns. This binary choice structure is capable of handling big

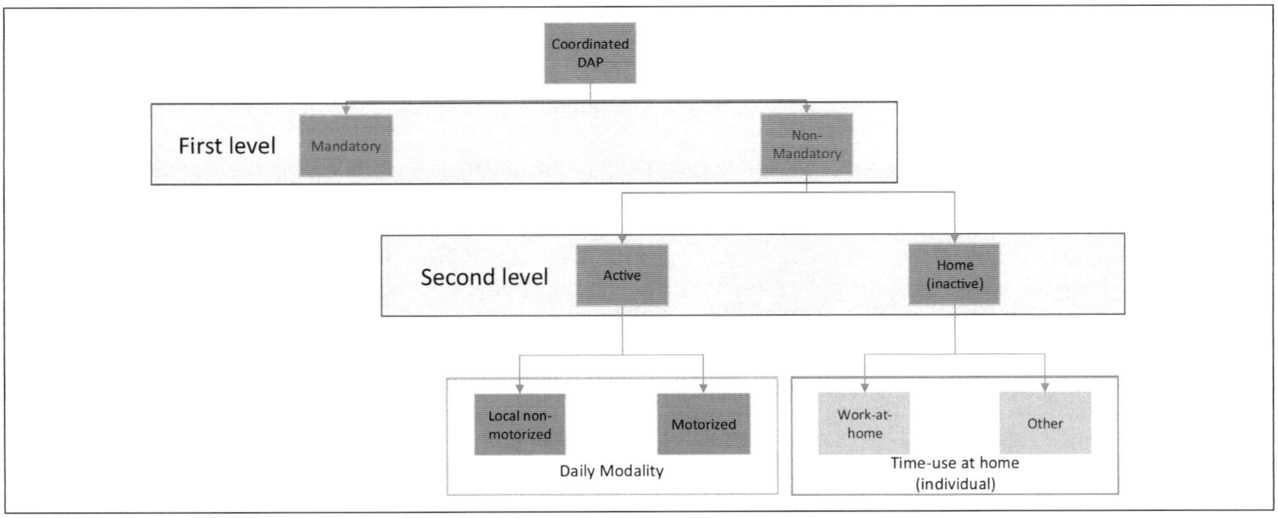

Figure 2. Model structure.

households. The number of alternatives can be calculated as $\sum_{n=1}^{N} 2^n$, *where n = household size, N = maximum household size*. The estimation was made using a maximum household size of nine. Thus, the number of alternatives at each level of DAP modeling is 1,024 and can be easily managed by any estimation software package. If a household had more than nine members, then the members were assigned priority based on person type and age, and the top nine prioritized members were selected for the model estimation. The first level gives a critical breakdown of household members between those who participate in work/school/university activities and those who do not. It is logical, and was confirmed in the previous studies on DAP type choice, that if one worker in the household does not go to work then the other worker from the same household has a higher probability not to go to work (to utilize non-work time productively and possibly for joint non-work activities). Several such correlations in DAP patterns among household members were observed in previous research across all DAP types. In order to capture these intra-household interactions, pair-wise terms segmented by person characteristics were included in the utility functions for this choice model.

The second level choice was then applied to the household members who have a non-mandatory pattern chosen for the day. For such household members, there is again a binary choice for being active (i.e., either undertake at least one out-of-home non-mandatory activity) or choose the at-home pattern. In order to introduce the interdependency between household members with a mandatory pattern and second level choice for household members with a non-mandatory pattern, some characteristics of the household members with a mandatory pattern were included as explanatory variables in this model. This way household members with a non-mandatory pattern become informed of the characteristics of the household members with a mandatory pattern. This factor may influence the person decision of being travel active or not. For example, if a child is going to school then an adult non-worker of the household may have a higher likelihood to be active as he/she may escort the child to or from school as well as undertake some household errands out of home while the child is at school.

The decomposition of the standard trinary choice DAP model into two levels of binary choice models was primarily necessary to handle large households. However, this decomposition of the model into binary choices also paved the way for behaviorally enriched CDAP. Unlike most urban metropolitan areas in the Western world, Jerusalem has an unusually high share of non-motorized travel, especially in the ultra-Orthodox and Arab sectors. These populations have special lifestyle preferences and a strong tendency to choose activity locations locally within their community or neighborhood. This results in a highly localized non-motorized pattern of behavior that is rarely observed in the large cities in the Western world. This behavior is even more prominent in the ultra-Orthodox Jewish population sector than among Arabs due to the specific role that the local religious community plays for them. This non-motorized behavior affects their activity locations and mode choice preferences directly. In order to model such behavior, daily modality for active people was introduced in CDAP. Daily modality was described by two DAP categories: (1) local non-motorized modality, and (2) motorized modality. A person with a non-mandatory pattern has a local non-motorized modality if all trips in the day are by non-motorized modes (walk/bike). This model was not applied for individuals with mandatory patterns because of a relative inflexibility in mode choice due to the work location. Their mode choice preferences are mostly dictated by mid-term modality (transit pass, driving license, etc.). Also, most of the work locations are beyond the reasonable non-motorized distance threshold. Non-mandatory activities have more degrees of freedom in terms of location and mode, and thus are interesting market segments to analyze with respect to daily modality. Similar to the models on the first two levels, this model was estimated jointly for nine household members.

The last component of CDAP for Jerusalem relates to analysis and reasons for being inactive. This component corresponds to the time-use at home for those who do not travel on the given day. Instead of analyzing the time-use as a continuous variable, the model was reduced to a binary discrete choice framework where the reasons for not traveling were specified as work-at-home vs. other. Even such a simplified approach helps in analyzing work-at-home segments and may help in analyzing the impact of specific transportation policies, such as promoting telecommuting or compress work weeks, on daily travel. Unlike other submodels in CDAP, this model was structured as a person-level model (i.e., applied to each person sequentially rather than simultaneously).

Model Estimation Results

Model Segmentation

Three population sectors in Jerusalem exhibit very different travel patterns. In order to analyze the effects pertinent to each population, the models were fully segmented by three sectors. Moreover, the model was essentially fully segmented by eight person types since each explanatory variable was specified for each person type separately: 1 = full-time worker (FW), 2 = part-time worker (PW), 3 = university student (US), 4 = non-worker (NW), 5 = retiree (RT), 6 = high school student (HS), 7 = elementary school student (ES), and 8 = pre-school child (PS). Due to the space constraints, only explanatory variables specific to full-time and part-time workers are shown for active vs. inactive non-mandatory models. Non-workers and retirees are selected as representative person types for the daily modality model, as a non-mandatory pattern is most common among these person types. Also, for the same reason, only the results for the second-level choice – active vs. inactive non-mandatory DAPs – and daily modality components of CDAP are included in this paper since they represent a unique analysis, while the upper-level choice (mandatory vs. non-mandatory DAP) in terms of behavioral findings is largely similar to the previously applied trinary DAP configurations.

Estimation Results for Choice between Active and Inactive Non-Mandatory Pattern

Estimation results segmented by population sector for active vs. inactive non-mandatory DAPs are summarized in Table 1. The coefficients for active utility components are shown while the inactive alternative served as the reference with a zero utility. It should be kept in mind that the behavioral interpretations of coefficients are subject to the fact that this model is conditional on the first-level choice outcome and it was applied only to the household members with a non-mandatory pattern (i.e., excluded persons with work, university, or school activity out of home). It is especially important for workers and students whose majority is characterized by a mandatory pattern and a non-mandatory pattern for them is somewhat a non-typical pattern of behavior on a regular weekday. As can be seen in Table 1, this model includes a rich set of explanatory variables discussed below.

Person Variables. Several person variables were tested in this model and statistically significant variables include age, gender, education attainment, and occupation. Younger full-time workers (18–35 years old) from Arab households were observed to be less active but the trend was opposite for younger part-time workers in ultra-Orthodox households. Female full-time workers were found to be more active in Arab households. This could be because, culturally, women from Arab households share more responsibility of running errands, such as shopping (*15*). Full-time workers with higher education in secular Jewish households were found to be more active, and a similar trend was observed for part-time workers in Arab households. Interestingly, both secular Jewish and Arab full-time workers with clerical jobs tend to be less active if they do not have a mandatory pattern.

Household Variables. Households' characteristics found to be statistically significant in this model are presence of pre-school child, car sufficiency, and household income. Similar to the findings in previous research (*1*, *7*, *16*), individuals from car-sufficient households were found to be more active due to a relative ease of traveling by car vs. transit. Interestingly, the effect of income was found to be different for different population sectors. Relative to people in higher income households, a person in a lower income Arab household is less likely to have an active travel pattern, but a person in a lower income secular Jewish household is more likely to have an active travel pattern. This is a manifestation of some deep cultural and lifestyle differences between population sectors.

Accessibilities. In order to capture the effect of activity supply side and travel conditions on individual DAP, it is important to include transportation accessibilities to different activities in the model. Accessibilities to retail employment (as the most probable generator of non-work activities) by different modes (auto, transit, and non-motorized) from the home location were tested, and accessibility by transit mode was found to be statistically significant for workers. In particular, it was observed that workers who live in places with a better accessibility to retail by transit mode logically have a higher

Table 1. Selected Estimation Results for Active vs. Inactive Model (Full-Time and Part-Time Workers)

	Secular Jewish		Ultra-Orthodox Jewish		Arab	
Explanatory variables	Coeff	*t*-stat	Coeff	*t*-stat	Coeff	*t*-stat
FW, active	0.03	0.12	−3.88	−2.23	0.70	2.98
PW, active	1.11	3.09	1.31	5.68	−0.84	−0.94
Age						
FW, 18–35 years					−0.49	−3.36
PW, 18–35 years			0.67	2.20		
Gender						
FW, female, active					0.28	2.05
Higher education						
FW, active	0.55	2.09				
PW, active					0.65	2.77
Occupation type						
FW, clerical worker, active	−0.96	−2.58			−0.54	−2.45
Female and presence of pre-school child						
FW, active					0.49	1.95
PW, active					0.41	0.95
Car sufficiency w.r.t. workers						
FW, zero cars, active					−0.28	−1.42
FW, fewer cars than workers, active					−0.37	−2.36
FW, more cars than workers, active	1.24	1.70				
Household income (NIS)						
FW, 0–3,500, active					−0.80	−1.72
FW, 11,001–19,000, active	−1.29	−2.25				
FW, >19,001, active	−1.55	−1.22				
Retail accessibility by modes						
FW, transit accessibility, active			0.52	2.49		
PW, transit accessibility, active					0.16	1.51
Presence of child in mandatory pattern						
FW, active	0.35	1.55	0.82	2.54	0.33	2.25
PW, active					0.47	1.69
Two-person interactions						
FWXFW, active	0.81	2.29				
FWxPW, active			1.44	1.79		
FWxUS, active	0.79	2.79				
FWxNW, active	0.42	3.47				
FWxHS, active			0.76	1.86		
FWxPS, active			0.82	3.49		
FWXFW, inactive	1.22	3.74			0.51	2.10
FWxPW, inactive	1.63	2.36			0.85	2.81
FWxUS, inactive			1.37	2.67		
FWxNW, inactive	0.51	4.51			0.49	2.44
FWxRT, inactive			1.77	3.24	0.84	5.05
FWxHS, inactive	0.57	3.26			0.62	3.01
FWxPS, inactive	0.64	3.32	1.25	4.43	0.94	4.87
PWxUS, active					1.26	2.56
PWxPS, active			1.30	4.16		
PWxPW, inactive					0.98	2.20
PWxNW, inactive					0.62	1.92
PWxES, inactive					2.07	2.63
PWxPS, inactive	1.16	2.32	1.61	4.65	1.77	3.68
Log-likelihood with constants only	−8,209.62	−3,241.36	−2,083.76			
Final log-likelihood	−6,044.73	−2,968.66	−1,799.92			

propensity to have an active non-mandatory DAP [similar impacts but in a context of auto accessibility were observed in U.S. metropolitan regions (*3*, *4*, *17*)].

Intra-Household Interactions with Persons with Mandatory DAP. It is important to include the characteristics of household members with mandatory patterns to capture the interdependency among household members and DAP types. Without such variables, two levels of DAP modeling would be merely sequential and perform as if they were independent sub-models except for the conditional use of the second-level model for only a subset of persons. Several characteristics of household members with a mandatory pattern were tested. A statistically significant conditional variable was found to be the presence of a child with a mandatory pattern (i.e., going to school) on the given day. It was found that workers are more active if there is at least one child with a mandatory pattern. This effect can be explained by escorting the going-to-school child to school (the escorting of children to school by parents is very frequent in Jerusalem). In a case in which all household children go to school this might also capture a certain freedom for the non-working adults to be engaged in some other non-work out-of-home activities.

Intra-Household Interactions with Persons with Non-Mandatory DAP. Pair-wise person-type-specific constants were added in this model to capture these intra-household interactions. For example, it was observed that for a household with at least two full-time workers, if one worker is active then the other worker is more likely to be active as well (frequently assuming a major shopping or family event that they attend jointly). Statistical significance of these pair-wise constants in Table 1 shows the presence of substantial correlation across household members in DAP choices and the necessity to model this choice dimension jointly across the relevant household members. Without such terms, this model would have performed as an independent DAP type choice for each person.

Estimation Results for Choice of Modality Pattern

Model estimation results for coordinated daily modality component for non-workers and retirees are shown in Table 2. Coefficients for the local non-motorized modality utility are included in Table 2 while motorized modality served as the reference category with a zero utility. The estimation results can be summarized as follows.

Person Variables. The constants in this model are completely segmented by person types. On comparison of coefficients for non-workers and retirees combined with the coefficient on age across population sectors, it can be observed that non-workers and retirees from ultra-Orthodox Jewish households have the highest propensity to have local non-motorized modality. This was expected because of a traditional neighborhood-community-oriented activity and travel behavior of the ultra-Orthodox Jewish population in general (*15*). In addition to the constants, this model includes several variables to incorporate effects of mid-term mobility attributes such as driving license, if usual driver, and so forth. It is important to incorporate these attributes in order to make daily modality properly sensitive to mid-term modality (*11*). A logical effect can be observed from the coefficients on the individual mobility attributes that with better individual mobility, individuals are less likely to have non-motorized daily modality.

Household Variables. Household income and car sufficiency were found to be statistically significant in affecting modality. Individuals from low-income households were observed to be more non-motorized oriented and a similar trend was observed for individuals from fewer-cars households. Researchers have found similar effects of these household characteristics on mid-term mobility attributes and modality (*11*, *17–19*).

Accessibility. It is important to include measures reflecting the travel conditions to non-work activities as they are expected to affect the mode choice preferences. Similar to the previous sub-model, accessibilities to retail by different travel modes were added in this model. The estimation results confirmed that individuals residing in areas with good transit accessibility exhibit a lesser tendency toward non-motorized modality. On the other hand, individuals living in areas with good non-motorized accessibility are more likely to be oriented toward non-motorized modality. Thus, a non-motorized pattern of daily travel behavior can be effectively induced by mixed land-use policies.

Intra-Household Interactions. Pair-wise person type constants were added to this model, similar to the previous model, to capture a correlation in daily modality across household members. These statistically significant constants confirmed that there is a substantial correlation in modality across household members, that is, if one household member is oriented toward non-motorized modes then the other household members are likely to prefer non-motorized modes as well. It should be noted that this intra-household effect manifested itself after controlling for the sector-specific cultural differences and zonal accessibilities.

Conclusion

This paper illustrated a modified approach to DAP modeling coordinated at the household level. Although the model was primarily introduced to handle large households that are typical for the city of Jerusalem, the developed methods proved useful in adding more behavioral aspects to the model. The paper introduced daily modality and added the emphasis on modeling it for all household members simultaneously. It is of special practical value for Jerusalem since such population sectors as ultra-Orthodox Jewish and Arabs are characterized

Table 2. Estimation Results for Daily Modality Model (Non-workers and Retirees)

	Secular Jewish		Ultra-Orthodox Jewish		Arab	
Explanatory variables	Coeff	*t*-stat	Coeff	*t*-stat	Coeff	*t*-stat
NW	−3.75	−3.27	−2.39	−1.93	1.12	0.95
RT	−12.21	−1.44	−35.77	−1.70	−0.92	−0.34
Usual driver						
NW	−1.35	−3.40	−1.65	−2.83	−2.29	−2.99
RT	−0.73	−1.88	−1.65	−2.83	−2.29	−2.99
Driving license						
NW	−0.83	−3.35			−0.51	−2.09
RT	−0.43	−1.78			−0.85	−1.26
Age						
NW, age	0.14	2.46	0.12	2.04	−0.10	−2.18
NW, age*age	0.00	−2.63	0.00	−2.66	0.00	2.07
RT, age	0.26	1.15	0.93	1.63		
RT, age*age	0.00	−1.04	−0.01	−1.61		
Gender						
NW, female					0.37	1.54
NW, female					0.37	1.54
Household income (NIS)						
NW, 0–3,500					0.68	2.15
NW, 3,501–7,000					0.43	1.52
NW, 0–7,000	0.34	2.25	0.51	1.90		
NW, 11,001+			−0.77	−1.39		
RT, 0–3,500			1.05	1.74	0.68	2.15
RT, 3,501–7,000					0.43	1.52
RT, 0–7,000	0.34	2.25				
RT, 19,001+			−0.77	−1.39		
Car sufficiency w.r.t. workers						
NW, zero cars			0.53	2.32	0.43	1.93
NW, fewer cars than workers					0.58	2.26
NW, more cars than workers	−0.35	−1.05				
RT, zero cars			0.53	2.32	0.82	1.46
RT, fewer cars than workers					0.58	2.26
RT, more cars than workers	−0.85	−3.20				
Retail accessibility						
NW, transit	−0.10	−2.64	−0.20	−2.24	−0.31	−2.78
NW, non-motorized	0.18	2.42	0.31	3.83	0.30	3.30
RT, transit	−0.10	−2.64	−0.38	−1.78	−0.54	−1.41
RT, non-motorized	0.14	2.17	0.35	1.65	0.61	2.47
Two-person interactions						
FWxNW, non-motorized	1.46	2.77			0.75	2.38
FWxRT, non-motorized			6.00	3.53		
PWxNW, motorized					1.03	1.78
USxNW, non-motorized					1.64	2.48
NWxNW, non-motorized	1.02	2.56			1.31	5.62
NWxRT, non-motorized	0.91	2.09	2.56	2.17	1.24	1.95
NWxNW, motorized					0.60	2.98
NWxRT, motorized			2.17	2.01	0.80	1.64
RTxRT, non-motorized	1.58	4.93			2.28	2.93
Log-likelihood with constants only	−1,328.3	−1,041.78	−600.43			
Final log-likelihood	−1,195.94	−1,023.04	−608.75			

by a large share of persons who have an entire day of travel implemented in a non-motorized fashion. The identification of such individuals in the ABM chain at the DAP stage helped make the subsequent conditional choices of trip destinations and modes more realistic. It would have been much more difficult to achieve the same level of replication of non-motorized DAP share through a sequence of destination and mode choices applied for each tour and trip. Also, the estimation results confirmed that daily modality patterns of household members are correlated.

In addition to daily modality, this paper presented a simple but useful approach to understand the at-home time-use. A binary choice model was formulated for the main reason of being inactive (working-at-home or other). While this model doesn't have a direct impact on travel model, as inactive people do not go through the subsequent chain of daily travel sub-models, this model helps in understanding such population segments as telecommuters and their impact on travel behavior of other household members.

Future Directions

The model described in this paper is a sequential and conditional model and is currently implemented without any feedback from the lower levels to upper levels. This model can be easily extended to better address the joint nature of the choices. A Gibbs sampler can be used in which slightly modified models could be applied in an iterative manner, similar to the way in which the mobility attributes model is structured in the Jerusalem ABM (*11*). With the Gibbs sampling approach, the CDAP type could be further enriched with more attributes.

MDCEV represents a promising alternative direction to model the most important travel and activity characteristics for each person up-front in the ABM model chain. Recent MDCEV developments incorporated multiple budget constraints and nested structures that are needed for handling individual DAPs with intra-household interactions (*20*).

Another emerging direction for modeling DAP is a machine learning approach. Instead of using econometric models with an observed set of daily patterns, a machine learning approach can be envisioned which will generate patterns based on the observed data, and pattern recognition methods may help in a better classification of DAPs and forecasting evolving daily patterns. The authors plan to compare the two approaches and evaluate their performance.

Finally, it is important to keep in mind that this type of model can be framed in many ways and with different levels of detail. Daily DAP type is largely an arbitrary definition that compresses the detailed description of individual DAP (sequence of activities and trips with all pertinent characteristics) into a manageable number of choice alternatives that capture the most important characteristics. Thus, the choice of the DAP model is subject to the ultimate model system design. If DAP captures more details up front, fewer sub-models will be needed afterwards. If DAP is defined in very aggregate terms, more sub-models will be needed down the model stream. In terms of mimicking the real-world individual decision-making, DAP should be formulated as the most strategic decision that the individual makes for the next day. In this regard such DAP classification as going to work or school or staying at home for the whole day are still appealing as they give rise to the trinary DAP choice widely accepted in practice (*4*). Modeling intra-household interactions and separating of work-from-home from other inactive patterns also look as essential. Handling large households and bringing in modality are additional examples that proved to be necessary for the Jerusalem ABM. However, there are many more potential directions for additional classification of DAP types. For example, it is logical to include duration of the mandatory activity and associated commuting as a dimension for DAP type. For example, it is appealing to consider a pattern with long work activity and long commuting that most frequently would not have any other out-of-home activities as a separate pattern type. Each DAP type that streamlines the subsequent chain of sub-models can be useful for ABM design. This direction is intertwined with the general evolution of ABM designs.

Author Contributions

The authors confirm contributing to the paper as follows: study conception and design: Gaurav Vyas, Peter Vovsha; data collection: Peter Vovsha, Danny Givon, Yehoshua Birotker; analysis and interpretation of results: Gaurav Vyas, Peter Vovsha; draft manuscript preparation: Gaurav Vyas, Peter Vovsha; All authors reviewed the results and approved the final version of the manuscript.

References

1. Bowman, J. L. 1995. Activity Based Travel Demand Model System with Daily Activity Schedules, Master of Science Thesis in Transportation, Massachusetts Institute of Technology.
2. Bowman, J. L., and M. Ben-Akiva. Activity-Based Disaggregate Travel Demand Model System with Activity Schedules. *Transportation Research Part A: Policy and Practice*, Vol. 35, No. 1, 2001, pp. 1–28.
3. Vovsha, P., E. Petersen, and R. Donnelly. Impact of Intrahousehold Interactions on Individual Daily Activity-Travel Patterns. *Transportation Research Record: Journal of the Transportation Research Board,* 2004. 1898: 87–97.
4. Bradley, M., and P. Vovsha. A Model for Joint Choice of Daily Activity Pattern Types of Household Members. *Transportation*, Vol. 32, No. 5, 2005, pp. 545–571.
5. Lemp, J. Understanding Joint Daily Activity Pattern Choices across Household Members Using a Latent Class Model Framework. Presented at 93rd Annual Meeting of the Transportation Research Board, Washington, D.C., 2014.
6. Vuk, G., J. Bowman, A. J. Daly, and S. Hess. Impact of Family In-Home Quality Time on Person Travel Demand. *Transportation*, Vol. 43, No. 4, 2016, pp. 705–724
7. Bhat, C. R., K. G. Goulias, R. M. Pendyala, R. Paleti, R. Sidharthan, L. Schmitt, and H. H. Hu. A Household-Level Activity Pattern Generation Model with an Application for Southern California. *Transportation*, Vol. 40, No. 5, 2013, pp. 1063–1086.

8. Paulssen, M., D. Temme, A. Vij, and J. L. Walker. Values, Attitudes and Travel Behavior: A Hierarchical Latent Variable Mixed Logit Model of Travel Mode Choice. *Transportation*, Vol. 41, No. 4, 2014, pp. 873–888.
9. Dugundji, E., and J. Walker. Discrete Choice with Social and Spatial Network Interdependencies: An Empirical Example Using Mixed Generalized Extreme Value Models with Field and Panel Effects. *Transportation Research Record: Journal of the Transportation Research Board*, 2005. 1921: 70–78.
10. Outwater, M., B. Sana, N. Ferdous, W. Woodford, C. Bhat, R. Sidharthan, R. Pendyala, and S. Hess. *Characteristics of Premium Transit Services That Affect Mode Choice: Key Findings and Results*. TCRP H-37. Resources Systems Group, University of Texas at Austin, Arizona State University, and University of Leeds, 2013.
11. Vovsha, P., G. Vyas, D. Givon, and Y. Birotker. Individual Mobility Attributes and Their Impact on Modality Style: Comparisons across Three Population Sectors in Jerusalem, Israel. *Transportation Research Record: Journal of the Transportation Research Board,* 2013. 2382: 132–142.
12. Bernardo, C., R. Paleti, M. Hoklas, and C. R. Bhat. An Empirical Investigation into the Time-Use and Activity Patterns of Dual-Earner Couples With and Without Young Children. *Transportation Research Part A: Policy and Practice*, Vol. 76, 2015, pp. 71–91.
13. Kaufman-Scarborough, C. Time Use and the Impact of Technology. *Time & Society*, Vol. 15, No. 1, 2006, pp. 57–80.
14. Vyas, G., P. Vovsha, D. Givon, and Y. Birotker. Usual Work Arrangements: Statistical Analysis and Model Implementation for Jerusalem. Presented at 93rd Annual Meeting of the Transportation Research Board, Washington, D.C., 2014.
15. Vyas, G., C. Bernardo, P. Vovsha, D. Givon, Y. Birotker, A. Mossek, and E. Bluer. Differences in Travel behavior Across Population Sectors in Jerusalem, Israel. *Transportation Research Record: Journal of the Transportation Research Board*, 2015. 2495: 65–73.
16. Bhat, C. R., and V. Pulugurta. A Comparison of Two Alternative Behavioral Choice Mechanisms for Household Auto Ownership Decisions. *Transportation Research B: Methodological*, Vol. 32, No. 1, 1998, pp. 61–75.
17. Cervero, R., and J. Murakami. Effects of Built Environments on Vehicle Miles Traveled: Evidence from 370 US Urbanized Areas. *Environment and Planning A*, Vol. 42, No. 2, 2010, pp. 400–418.
18. Potoglou, D., and Y. O. Susilo. Comparison of Vehicle-Ownership Models. *Transportation Research Record: Journal of the Transportation Research Board*, 2008. 2076: 97–105.
19. Scott, D. M., and K. W. Axhausen. Household Mobility Tool Ownership: Modeling Interactions between Cars and Season Tickets. *Transportation*, Vol. 33, No. 4, 2006, pp. 311–328.
20. Castro, M., C. R. Bhat, R. M. Pendyala, and S. R. Jara-Diaz. Accommodating Multiple Constraints in the Multiple Discrete-Continuous Extreme Value (MDCEV) Choice Model. *Transportation Research Part B: Methodological*, Vol. 46, No. 6, 2012, pp. 729–743.

The Standing Committee on Traveler Behavior and Values (ADB10) peer-reviewed this paper (18-01480).

Article

TRR
JOURNAL OF THE TRANSPORTATION RESEARCH BOARD

Transportation Research Record
2018, Vol. 2672(47) 68–80

DOI: 10.1177/0361198118772723
journals.sagepub.com/home/trr
SAGE

Travel Behavior Classification: An Approach with Social Network and Deep Learning

Yu Cui[1], Qing He[2], and Alireza Khani[3]

Abstract

Uncovering human travel behavior is crucial for not only travel demand analysis but also ride-sharing opportunities. To group similar travelers, this paper develops a deep-learning-based approach to classify travelers' behaviors given their trip characteristics, including time of day and day of week for trips, travel modes, previous trip purposes, personal demographics, and nearby place categories of trip ends. This study first examines the dataset of California Household Travel Survey (CHTS) between the years 2012 and 2013. After preprocessing and exploring the raw data, an activity matrix is constructed for each participant. The Jaccard similarity coefficient is employed to calculate matrix similarities between each pair of individuals. Moreover, given matrix similarity measures, a community social network is constructed for all participants. A community detection algorithm is further implemented to cluster travelers with similar travel behavior into the same groups. There are five clusters detected: non-working people with more shopping activities, non-working people with more recreation activities, normal commute working people, shorter working duration people, later working time people, and individuals needing to attend school. An image of activity map is built from each participant's activity matrix. Finally, a deep learning approach with convolutional neural network is employed to classify travelers into corresponding groups according to their activity maps. The accuracy of classification reaches up to 97%. The proposed approach offers a new perspective for travel behavior analysis and traveler classification.

As car ownership increases rapidly, together with increasing environmental concerns, there has been more interest in services that enable people to share their automobiles. Ride-sharing or carpooling services are the ultimate way to make better use of the empty seats in personal passenger cars in order to reduce fuel consumption, transportation cost, and emissions. Cici et al. found that if individuals were willing to carpool with others who live and work within 1 km, the traffic in the city of Madrid would decrease by 59% (*1*). There are many existing smartphone apps providing ride-sharing services. However, this kind of one-time sharing mode offers only limited benefits. Long-time ride-sharing services can provide more advantages. There are many kinds of travelers, including ones who depart at the same time of day, ones who do not have regular departure times, ones who start work later than others, and so forth. The assumption is that travelers with similar behaviors may enjoy their shared rides better given more common behaviors. Long-term ride sharing matched according to travel behaviors will provide steady benefits for transportation, environment, and society.

Household travel survey data, or travel itinerary data, is a major input to travel behavior modeling. It can be used in many areas, particularly in travel behavior research and activity-based travel demand forecasting. The traditional methods used for collecting these individual travel data are telephone-based or computer-assisted interviews and activity logs recorded from study participants. The typical drawbacks of these methods include high recruitment cost, low response and sampling rates, undersampling or oversampling on certain types of trips, inaccuracies in times, surrogate reporting and confusion of appropriate trip purpose (*2*). Nowadays time–location

[1]Department of Civil, Structural and Environmental Engineering, University at Buffalo, The State University of New York, Buffalo, NY
[2]Department of Industrial and Systems Engineering and Department of Civil, Structural and Environmental Engineering, University at Buffalo, The State University of New York, Buffalo, NY
[3]Department of Civil, Environmental, and Geo-Engineering, University of Minnesota, Minneapolis, MN

Corresponding Author:
Address correspondence to Qing He: qinghe@buffalo.edu

data becomes accessible with the development of new techniques. Travel activities can be traced by various sensors such as GPS, GSM, Wi-Fi, RFID, and Bluetooth that are commonly available in smartphones or cars. Such data are usually collected when an event is triggered such as making a phone call, passing a toll booth, or turning on Bluetooth devices. Conducting a household travel survey with GPS devices is a complementary way of collecting reliable and accurate data. GPS provides high-resolution time–space data. The main advantages of the high-resolution GPS data include near-continuous location tracking, high temporal resolution, and minimum report burden for participants, which may significantly improve the understanding of travel activities in both spatial and temporal dimensions.

The essential goal of this study is to classify travelers based on the characteristics of their historical travel data. Therefore, travelers within the same category could be potentially paired and recommended with ride-sharing services. To pursue this goal, this paper constructs a social network of travelers based on the Jaccard similarity coefficient, and employs a community detection algorithm to cluster travelers into groups. In this paper, the social network concept is borrowed to describe people who have similar behavioral patterns rather than construct a real social network. After detecting travelers' groups, labels are manually assigned to each group according to trip and activity information. Further, an image is built of the activity map for each traveler and a deep-learning-based approach is employed to perform image classification. Therefore, the travelers are classified accordingly into different groups depending on their activity maps.

The rest of this paper is structured as follows: the literature review summarizes previous studies in travel behavior analysis, community detection and deep learning. Next the datasets and initial result of the data analysis are introduced. Then the methodology is presented. The results are demonstrated with some numerical examples. Finally, the conclusions and future research topics are presented.

Literature Review

Travel Behavior Data Analysis and Classification

Household trip data is crucial for travel demand forecasting and transportation system planning. The survey-based methods used for trip data collection went through the stages of paper and pencil interviews (PAPI), computer-assisted telephone interviews (CATI), and computer-assisted-self- interviews (CASI) (*3*). Although the computer-assisted interviews tried to help respondents to understand questions and recall trips they had during a day, these methods are restricted by the accuracy of recall, reliability, and compliance (*4*). Recently, GPS and GIS technologies have been used to supplement the traditional survey data. GPS and GIS land use data can be used for trip identification, travel characteristics identification, trip end clustering, and trip purpose prediction (*3*, *5*, *6*). However, the accuracy is influenced by the dilution of precision of the GPS logs and inaccuracy in the GIS database (*3*). Kim et al. developed an activity travel data collection method facilitated by a smartphone application and an interactive web interface. The data collected by this method in Singapore was further implemented with an ensemble-learning-based classification method to recognize travel patterns (*7*). Schumpeter et al. devised a multi-stage hierarchical matching procedure to calculate a cluster center of stop ends by combining trip ends and identifying trips with obvious purposes with the sociodemographics of the respondents (*8*). Some researchers also used decision-tree-based classifiers to derive trip purposes and implemented the methods in C4.5, C5.0, or an adaptive boosting environment (*9*, *10*). Some of the previous studies above also require the socioeconomic characteristics of respondents (such as age, gender, and household income) for travel behavior analysis.

Researchers have also been making great efforts to classify travelers by using daily travel data and sociodemographic data. The criteria to select similarity measures depend on the analysts' importance ranking of various affecting attributes and the situations to be dealt with (*11*). Consequently, the resulting similarity measures could be subjective and case sensitive and thus derive quite inconsistent results. Hanson et al. divided individuals into five homogeneous travel behavior groups by using complex multi-day travel data and explained variability in individuals' daily travels (*12*). Shoval et al. implemented a sequence alignment method based on GPS data and clustered the data into three temporal–spatial time geographies (*13*). Kitamura and van der Horn showed that daily participation could be very stable in different types of activities (based on the categories of working, leisure, shopping and other activities) (*14*). Axhausen et al. collected six weeks' continuous travel diaries from about 300,000 inhabitants in Germany in Fall 1999 (*15*). Hazard models were used to analyze this high-quality data. A low degree of spatial variability of daily activities was also found from the analysis. Jiang et al. employed the K-Means algorithm via principal component analysis (PCA) to cluster daily patterns of human activities in Chicago (*16*). They separated more than 3,000 individuals who participated in a 1-day or 2-day survey conducted by "Travel Tracker Survey" from January 2007 to February 2008 into 8 groups on weekdays and 7 groups on weekends, respectively. The same methodology applied to kernel density estimation, allowed them to analyze and explore diverse urban spatial–temporal structures. This research indicated how individuals in different activity pattern clusters make use of different sub-regions for different activity types (*17*). Travel behavior classification not only differentiates individuals with different travel patterns, but also uncovers human mobility patterns. Gonzalez et al. found that travel trajectories show lévy flight or random walk pattern to a large extent. Individuals show a high probability of returning to a few highly frequented locations (*18*). This means humans are following highly predictable mobility patterns (*19*). Ponieman et al. also retrieved social phenomena information (e.g. commute and major sport events) from call detail records (CDR) data by just counting how many phone calls were made by users in two different time windows (from 21:00 to 05:00, and from noon to 16:00 during weekdays) from inside and outside Buenos Aires city. They found the average radius of commute (ROC) was approximately 7.8 km (*19*). Nevertheless, that study was limited by the accuracy of CDR data. With the combination of several machine learning algorithms, Ma et al. identified travel patterns for transit riders from smart transit card data in order to attract more users, retain loyal users, and finally improve overall transit services performance (*20*). Williams et al. developed a new method derived from the neural coding concept of synchrony and measured regularity of visiting a specific location for individuals (*21*). However, this research only explored three places. Investigating how location types will influence visit patterns of individuals will provide more information to understand people travel patterns.

Community Detection

Nowadays, lots of complex systems can be represented by networks. For instance, each user can be a node in the social network. Then the friendship can be represented by edges. Researchers aim to understand the network by finding community structures. A community is a collection of nodes that are homogeneous within the group and heterogeneous with other groups in the network, and this kind of network is known as a community structure. Newman and Girvan (*22*) employed centrality indices to find boundaries of communities. They tested this method on two networks which are collaboration networks and food web networks. Both cases retrieved significant and informative community segments. The same authors also developed a community detection algorithm and proposed a new community structure strength measurement (*23*). Their algorithm showed highly effective performance for both computer-generated and real-world networks while detecting communities. Radicchi et al. (*24*) developed a fully self-contained new local algorithm for community detection and tested it on both artificial and real-world network graphs. This new method demonstrated the potentials of implementing community detection algorithms in large-scale technological and biological applications. In a community structure, groups do not need to be necessarily mutually exclusive; they can overlap. Palla et al. (*25*) proposed an algorithm which can uncover the overlapping community structure of complex networks in nature and society. Community detection techniques can also be used for habitat preservation, animal genetics and wildlife corridors (*26*). Moreover, it can be implemented in the transportation field. For example, Lin et al. employed a community detection algorithm to study vehicle accident causative factors (*27*).

Compared with traditional clustering algorithms, the community detection algorithm provides several advantages. First, it is easy to implement, and steps are intuitive. Second, final networks can be decomposed into communities for different levels. Third, this algorithm runs fast even for large and high dimensional datasets.

Deep Learning

With technological innovation, artificial intelligence (AI) emerges and integrates into everyday life rapidly. From education to finance, from marketing to health care, and from communication to transportation, with the advent of AI, individuals can save plenty of time, reduce mistakes, relieve pressures, and stay safe (*28*).

Deep learning or deep neural network, as a branch of machine learning and AI, is an artificial neural network (ANN) that contains more than one hidden layer. This kind of algorithm has shown superior performance especially in automatic speech recognition, image recognition, natural language processing, and recommendation systems. Readers can refer to Schmidhuber et al. (*29*) for more details of deep learning. Deep neural networks typically can be categorized into two types, recurrent neural networks (RNN) and convolutional neural networks (CNN). Long short-term memory RNNs is a widely used algorithm in the field of speech recognition (*30*, *31*). However, in the field of image recognition, the CNN is the most prevailing algorithm. Krizhevshy et al. developed a CNN consisting of five convolutional layers, some of these followed by max-pooling layers and three fully-connected layers. This network achieved an error rate of 15.3% while implementing the ImageNet Large Scale Visual Recognition Challenge (ILSVRC)-2012 dataset (*32*). Simonyan and Zisserman improved CNNs with utilizing very small convolution filters and reached an error rate of 6.8% on the ILSVRC-2014 dataset. He et al. won the ILSVRC-2015 competition with 3.57% error using deep residual learning. Further, the winning team Trips-Soushen of ILSVRC-2016 produced a 2.99% error rate (*33*, *34*). This technique also has been applied in transportation, especially in the visual sensor data (images, videos) processing domain. Only a few studies implemented deep learning in travel behavior. Dong et al. employed deep learning to model driving behavior based on GPS data. They combined CNN and RNN to extract features to represent driver behaviors. A driver classification task was also conducted and they achieved significant outstanding performance compared with traditional machine learning algorithms (*35*).

Data Description and Preliminary Analysis

In this paper, the raw data is acquired from the California Household Travel Survey (CHTS) conducted by the California Department of Transportation (Caltrans) from February 2012 to January 2013. The CHTS is designed to collect household travel information across all 58 counties of California and three adjacent counties in Nevada by using computer-assisted telephone interviewing (CATI), website, and GPS devices. The entire household survey uses three types of GPS devices, wearable GPS device, in-vehicle GPS device, and in-vehicle GPS device plus an on-board diagnostic (OBD) unit. A total of 108,778 individuals belonging to 42,431 households participated in this survey, and 10,474 respondents from 5,460 households carried GPS devices.

Households who conducted the survey with GPS devices generated both survey and GPS data. The survey data includes activities they completed on the assigned travel date only. Each participant has one assigned travel date; this indicates that every participant only has one day of survey data for both non-GPS households and GPS households. However, besides survey data, GPS households also collect 7 days of GPS data as well.

There are 39 different trip purposes included in this household travel survey. These trip purposes are categorized into eight groups, which are home, work, school, transportation or transitions (transit), shopping or errands (shop), personal business (person), recreation or entertainment (rec), and other as shown in Table 1. Table 2 verifies the common knowledge that there are more work and school activities during weekdays, while people have more recreation, shopping, and personal business activities on weekends. It is found that the average time individuals spend at home is 13.98 hours per day during weekdays and 16.77 hours on weekends, respectively. The average time participants spend in the workplace is 7.11 hours during the weekday and 5.51 hours on the weekend. Also, the average time that participants spend at school is 5.18 hour during weekdays and 2.92 hours on weekends.

1. This study only utilizes data of participants who have both valid survey and GPS data. Several rules are created as follows in order to screen the feasible data. Both the first and last activity location should be 'home'.
2. Participants need to have more than one activity on the assigned travel date. If individuals stay at home for a whole assigned travel day, travel information cannot be retrieved. This indicates that participants should also make more than one trip on the assigned travel date.
3. From the household survey data, it is noticed that there might be more than one driver utilizing the same vehicles within a household. It is difficult to distinguish travelers who share the same car since GPS trips are recorded at the vehicle level, not the person level. To avoid sampling errors arising from multiple drivers when they were sharing the same vehicle, the records of the vehicles with multiple drivers are removed.

After these rules are applied, 8,849 unique individuals remain with 50,103 trips in this research. The radius of gyration (ROG) in transportation is a measure to describe the activity territory for each participant. According to Kang (*36*), the ROG for each participant's trajectory up to time t can be calculated by using the formula

$$r_g^{\alpha}(t) = \sqrt{\frac{1}{n_c^{\alpha}(t)} \sum_{i=1}^{n_c^{\alpha}(t)} (x_i - x_c)^2 + (y_i - y_c)^2} \quad (1)$$

Where coordinates x_i, y_i denote the ith $(i = 1, 2, \ldots, n_c^{\alpha}(t))$ position recorded for user α, and x_c and y_c represent the center of the mass of trajectories.

In Figure 1, the blue line is the ROG calculated for weekdays, and the red dashed line is the ROG for weekends. It can be seen that these two ROG distributions are almost the same. Therefore, individuals follow similar travel habits whether weekday or weekend. They seldom travel farther on weekends. Figure 1 also shows that there is a significant effect of distance decay. Moreover, based on the probability distribution of ROG, more than 95% of the participants have ROG values of less than 10 miles, and the mode is around 5 miles. This result is similar to the ROC calculated in Ponieman et al. (*19*) which is 7.8 km (4.85 miles).

Figure 2*a* and *b* depict the distribution of the first trip departure time and the last trip return time, respectively. As can be seen, there is a significant pattern of departure time in the morning, and return time in the evening during weekdays. For weekends, individuals depart in the morning and return home in the evening later than weekdays. Moreover, the distribution of departure times on

Table 1. Taxonomy of Activity Purposes

Activity purpose category	Activity purpose
Home	Personal activities (sleeping, personal care, leisure, chores)
	Preparing meals or eating
	Hosting visitors or entertaining guests
	Exercise (with or without equipment) or playing sports
	Study or schoolwork
	Work for pay at home using telecommunications equipment
	Using computer, telephone, cell, smartphone, or other communications device for personal activities
	All other activities at my home
Work	Work or job duties
	Training
	Meals at work
	Work-sponsored social activities (holiday or birthday celebrations, etc.)
	All other work-related activities at my work
	Work-related (meeting, sales call, delivery)
School	In school, classroom, or laboratory
	Meals at school or college
	After school, or non-class-related sports or physical activity
	All other after school or non-class related activities (library, band rehearsal, clubs, etc.)
Transportation or Transitions (Transit)	Change type of transportation or transfer (walk to bus, walk to or from parked car)
	Pickup or drop off passenger(s)
	Loop trip (for interviewer only-not listed on diary)
Shopping or Errands (Shop)	Routine shopping (groceries, clothing, convenience store, household maintenance)
	Shopping for major purchases or specialty items (appliance, electronics, new vehicle, major household repairs)
	Household errands (bank, dry cleaning, etc.)
Personal Business (Personal)	Volunteer work or activities
	Drive through other (ATM, bank)
	Service private vehicles (gas, oil, lube, repairs)
	Personal business (visit government office, attorney, accountant)
	Healthcare (doctor, dentist, eye care, chiropractic)
Recreation or Entertainment (Rec)	Non-work-related activities (social clubs, etc.)
	Exercise or sports
	Drive through meals (snacks, coffee, etc.)
	Eat meal at restaurant or diner
	Outdoor exercise (playing sports, jogging, bicycling, walking, waking the dog, etc.)
	Indoor exercise (gym, yoga, etc.)
	Entertainment (movies, watch sports, etc.)
	Social, or visit friends or relatives
Other	Other (specify) Note: listed on diary
	Don't know or refused

Table 2. Activity Purpose Statistics

	Weekday		Weekend	
Activity purpose	Number of activity	Proportion	Number of activity	Proportion
Home	199554	54.33%	85920	57.16%
Work	35424	9.64%	4051	2.70%
School	12410	3.38%	381	0.25%
Transit	42142	11.47%	11258	7.49%
Shop	24774	6.75%	14998	9.98%
Person	15299	4.17%	8440	5.62%
Rec	34766	9.47%	24036	15.99%
Other	2911	0.79%	1222	0.81%
Total	367280	100.00%	150306	100.00%

the weekend is flatter which means the departure times on the weekend involve more uncertainty than those on weekdays. The pattern of return time is similar. The mean first departure times on weekdays and weekends are 08:56 and 10:54, respectively. However, the last trip return times are 18:28 and 17:55 for weekdays and weekends, respectively. As shown in Figure 2*b*, there is a small hump near noon. This hump is more flat and larger on weekends than weekdays. Individuals have more flexible times on weekends, and some of them may have eating, recreation, or shopping activities and return home at noon then stay home until the next day. On the contrary, individuals' departure and return times are restricted by work or study on weekdays.

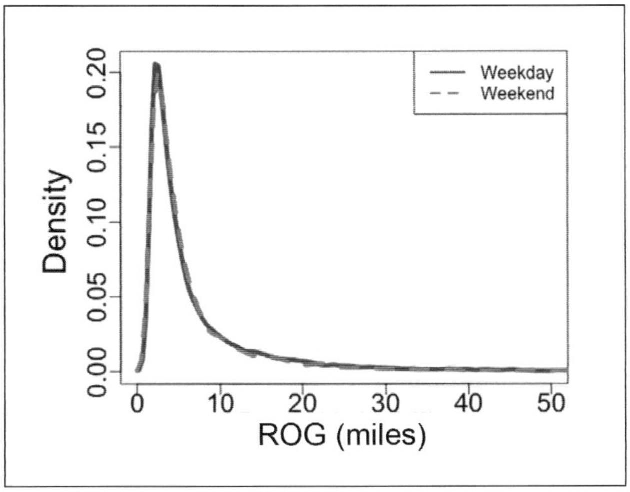

Figure 1. ROG plot.

Methodology

Matrix Similarities

In this paper, individuals' daily activities are represented using a matrix. In order to construct this matrix, all activity purposes are categorized into eight groups and 24 hours are divided into 288 five-minute bins. Therefore, the dimension of the matrix is 8 × 288. Each activity starts from the trip starting time for this activity and ends when the activity ends. For instance, an individual departs from home at 17:30 for the supermarket and arrives at the supermarket at 17:40. The individual finishes shopping at 18:20 and leaves the supermarket. Therefore, this shopping activity is from 17:30 to 18:20.

Individuals only can conduct one of eight activities shown in Table 3 at one time. So,

$$x_{a,t} = \begin{cases} 1 & \text{when activity a occurrs at time interval } t \\ 0 & \text{otherwise} \end{cases} \tag{2}$$

where a represents activity and $a \in \{\text{Home, Work, School, Trans, Shop, Person, Rec, Other}\}$, . t represents time bin number, and $t \in [1, 288]$. In this fashion, a binary matrix is generated to represent daily activities. The two plots in Figure 3 are examples of color-coded daily activity matrices for two individuals. Figure 3*a* is a daily activity map of a student. This participant departs from home to school at approximately 07:30 and stays there until 14:30. After going back home at 18:30, the student again attends a recreation activity, and returns home at 20:30. Figure 3*b* describes a full time employee's weekday activities. The employee departs from home at 06:00,

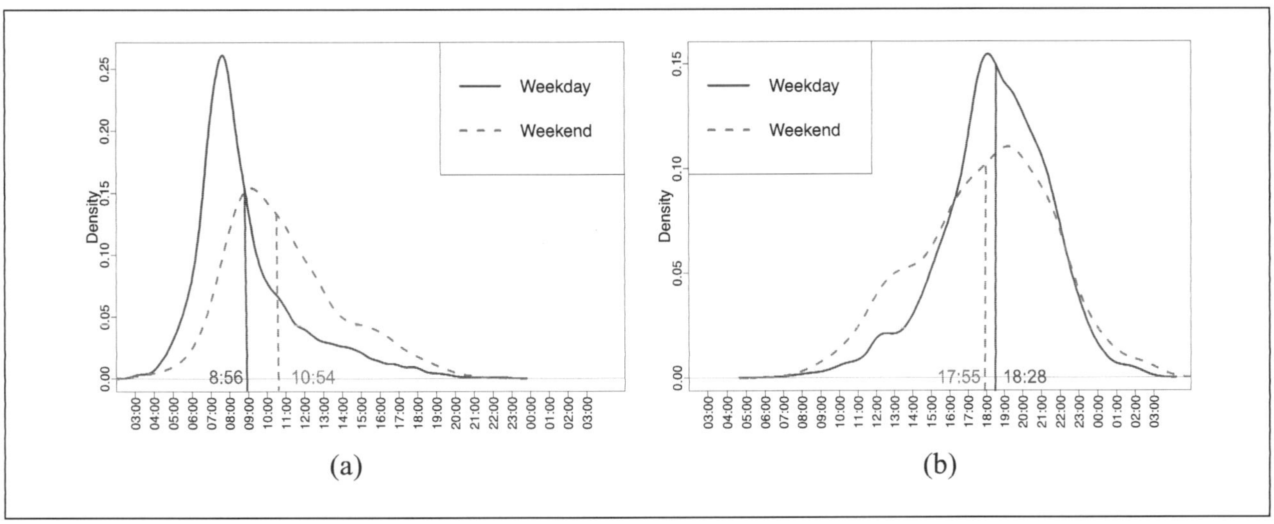

Figure 2. Distributions of (*a*) first trip departure time and (*b*) last trip return time.

Table 3. Statistics of Clusters

	Cluster1		Cluster2		Cluster3		Cluster4		Cluster5		Cluster6	
Activity purpose	#	%	#	%	#	%	#	%	#	%	#	%
Home	2201	55.74	545	56.83	3426	48.53	548	54.69	377	56.61	311	56.75
Work	37	0.94	5	0.52	1936	27.43	285	28.44	183	27.48	1	0.18
School	1	0.03	1	0.10	1	0.01	0	0.00	0	0.00	155	28.28
Transit	351	8.89	51	5.32	696	9.86	61	6.09	35	5.26	36	6.57
Shop	650	16.46	71	7.40	347	4.92	46	4.59	24	3.60	12	2.19
Person	234	5.93	24	2.50	143	2.03	15	1.50	12	1.80	8	1.46
Rec	465	11.78	260	27.11	504	7.14	44	4.39	34	5.11	24	4.38
Other	10	0.25	2	0.21	6	0.08	3	0.30	1	0.15	1	0.18

Note: # = number of activities; % = percentage; Transit = Transportation/Transitions; Person = Personal Business.

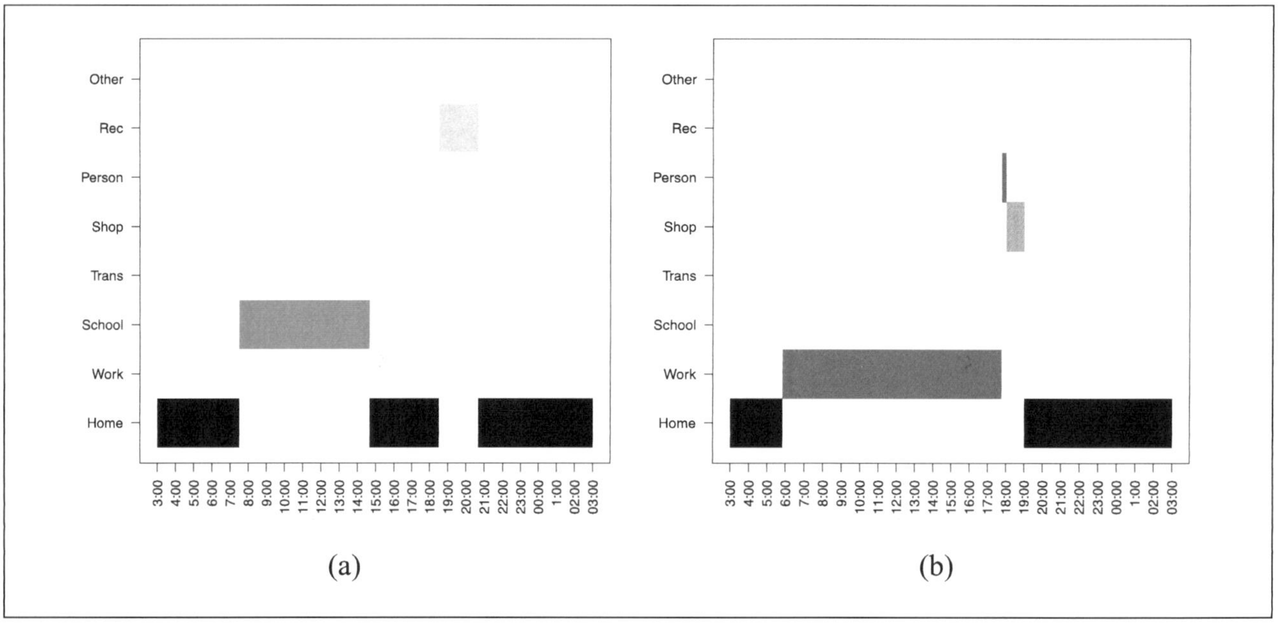

Figure 3. Activity maps.

conducts a short personal business activity and a shopping trip after work at 18:00.

Given an individual's activity matrix, the similarity between every pair of participants is now calculated for constructing the community structure. In this paper, the Jaccard similarity coefficient is implemented to measure similarities for binary matrices of individuals' daily trips.

$$J(A,B) = \frac{|A \cap B|}{|A \cup B|} = \frac{|A \cap B|}{|A| + |B| - |A \cap B|} = \frac{M_{11}}{M_{01} + M_{10} + M_{11}} \quad (3)$$

where A and B represents n binary entries, respectively. M_{01} represents the total number of entries where the entries of A is 0 and the entries of B is 1; M_{10} represents the total number of entries where the entries of A is 1 and entries of B is 0; M_{11} represents the total number of entries where the value of entries of A and B are both 1. Jaccard similarity coefficient ranges from 0 to 1. If $J(A,B) = 0$, this means A and B are totally different. And if $J(A,B) = 1$, this means A and B are exactly the same. Moreover, if A and B are both empty, $J(A,B)$ is defined as 1.

Another similarity measure is the simple matching coefficient (SMC), shown in Equation 4. Unlike the Jaccard similarity coefficient, SMC includes M_{00} which represents the total number of entries where the entries of A and B are both 0. SMC is appropriate when 0 and 1 represent equivalent information, such as gender (*37*).

However, in this paper, 0 and 1 do not carry symmetrical information, and a majority of entries is 0 in the dataset. If SMC is used to measure similarity, M_{00} will dominate the similarity and force it close to 1. Therefore, the Jaccard similarity coefficient is more suitable to describe similarity for asymmetrical attributives than SMC.

$$\text{SMC} = \frac{M_{00} + M_{11}}{M_{00} + M_{01} + M_{10} + M_{11}} \tag{4}$$

Community Detection

In this subsection, first the data is converted into a social network. Each participant in the dataset can be considered as a node. Jaccard similarity coefficient is used to measure the similarity between two users. If Jaccard similarity coefficient is greater than a threshold θ, it indicates two users have similar travel behaviors. In consequence, an edge is created between these two users

$$\begin{cases} \text{Edge}_{ij} = 1 & W_{ij} \geq \theta \\ \text{Edge}_{ij} = 0 & W_{ij} < \theta \end{cases} \tag{5}$$

and the weight of the edge of user i and j can be represented by W_{ij}, this value equals to Jaccard similarity coefficient between these two users. After building the social network, the fast unfolding community detection method is employed to cluster nodes (*38*). This is a kind of modularity maximization algorithm which is one of the most widely used methods for community detection.

In community detection, the modularity of a partition measures the density of edges within communities while comparing with edges between communities and scales between −1 and 1. In the network where edges have weights, the formula of modularity is shown as

$$Q = \frac{1}{2m} \sum_{i,j} \left[W_{ij} - \frac{k_i k_j}{2m} \delta(c_i, c_j) \right] \tag{6}$$

where
W_{ij} is the weight of the edge between node i and j,
$k_i = \sum_j W_{ij}$ represents the summation of the weights for edges attached to node i,
c_i is the community index assigned to this node in this iteration,
$\delta(c_i, c_j)$ equals 1 if $c_i = c_j$ and 0 otherwise, and
$m = \frac{1}{2} \sum_j W_{ij}$.

This algorithm consists of two stages. First, each node is assigned to different communities so that each node is a community, and the initial number of communities is as many as nodes. Then, for each node i, the gains of modularity are measured when removing i from its community and adding in neighboring communities respectively. Then for each isolated node i, the gain of modularity is measured when removing i from the origin community and adding in neighboring communities respectively. And the node i will be allocated into the community with the maximal gain, but only if this gain is positive. If the maximum gain is not positive, node i will stay in the original community. This stage will proceed recursively until a local maximum of the modularity is achieved. The second stage of this algorithm is to build a new network with communities found during the first part. After the second stage is complete, the first stage of this algorithm should be reactivated to reconstruct weighted network. Then these two stages will run alternatively until there are no more changes and the global modularity maximum ($\max(q)$) is found.

Deep Learning

Clustering procedure assigns the same labels for individuals in the same groups, and these labels are considered as the true label in the classification task. In order to classify travelers according to their activity maps, the CNN is implemented to complete this task. CNN, which is a kind of feed-forward artificial neural network, is recognized as a powerful and prevalent tool in image reorganization. Compared with the traditional neural network, CNN not only contains more layers, but also learns from filters, represented by a vector of weights with which the input is convolved. Filters are implemented to slide across all the areas of images, and the size of the commonly used filter is 2 × 2 which is also utilized in this paper. A CNN is composed of input and output layers, as well as multiple hidden layers between input and output layers. There are three kinds of hidden layers, including convolutional layers, pooling layers, and fully connected layers. The function of convolution layers is applying a convolution task to the input and passing the result to the next layer. Pooling layers combine the output from one layer and pass it into a single neuron in the other layer. In the CNN, every convolutional layer is followed by a pooling layer. Moreover, the max pooling is implemented in the pooling layer which will pass the maximal value from the previous layer to the next layer. Fully connected layers aim to connect all neurons in the previous layer to all neurons in the next layer. Figure 4 illustrates a CNN with two convolutional layers and two fully collected layers. This CNN is constructed in the next section.

In the traditional ANN networks, a sigmoid function is utilized to process data. However, in this paper, Rectified Linear Units (ReLU) are implemented as

$$f(x) = \max(0, x) \tag{7}$$

where x is the input to a neuron.

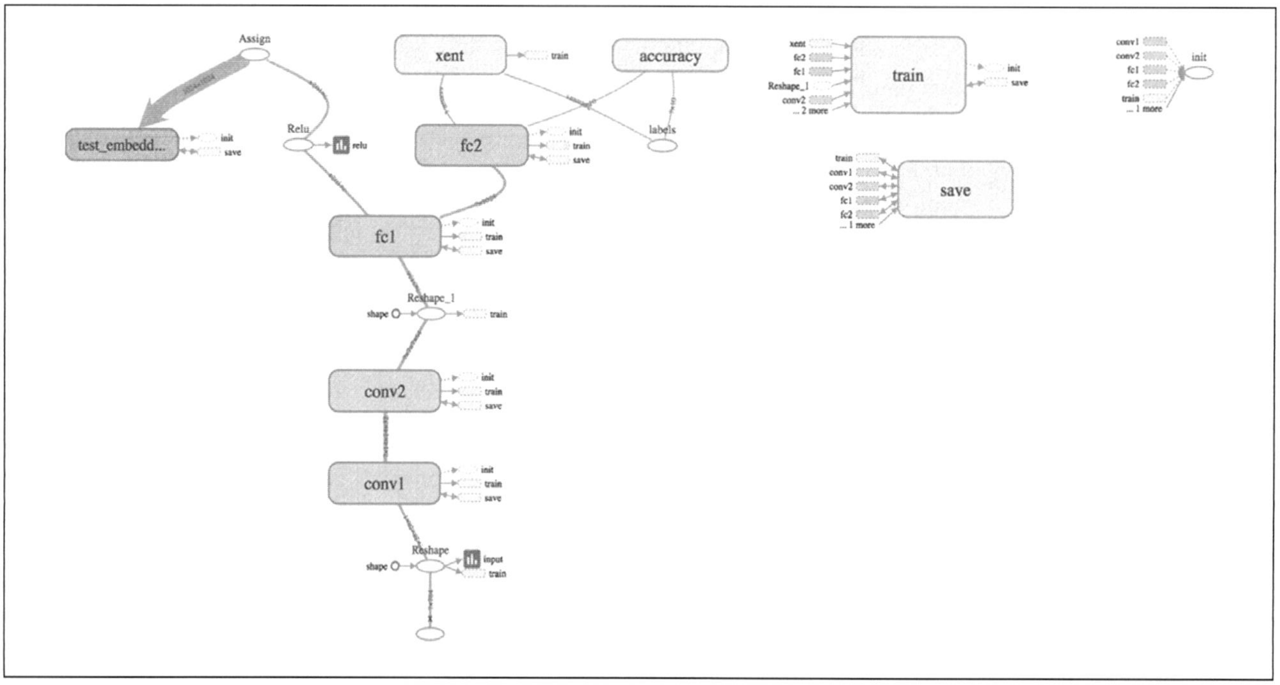

Figure 4. The layout of a CNN.

The sigmoid function contains three problems: the first one is that saturated neurons kill the gradients that might always be zero. Second, the sigmoid output is not zero-centered. The last one is that the computation of exponential is very expensive. However, ReLU will not saturate in positive region and it is very computationally efficient (approximately 6 times faster than sigmoid function) (*32*).

In this paper, the proposed CNN includes two fully connected layers, and this may result in overfitting since fully connected layer occupies most of the parameters. Dropout process is added between two fully connected layers in order to prevent overfitting. Moreover, this process also can speed up the training process. Finally, cross entropy is utilized as the loss function as shown in Equation 8.

$$\mathrm{H(p, q)} = -\sum_{x} p(x)\log q(x) \qquad (8)$$

where $p(x)$ and $q(x)$ are two probability distributions over discrete variable x, and $q(x)$ is the estimate distribution for true distribution $p(x)$. $H(p, q)$ is the cross entropy for the distribution p and q.

Numerical Examples

In the first step, a $1 \times 2{,}034$ ($= 288 \times 8$) binary vector is constructed to represent the activity chain on the assigned travel date for each participant. Then the Jaccard similarity coefficient is calculated for every pair of individuals according to their activity matrices. The Jaccard similarity coefficient becomes the link weight between a pair of individuals. The Jaccard similarity coefficient ranges from 0 to 1. In this paper, an edge is built between two individuals in the community structure when the matrix similarity is greater than 0.9 ($\theta = 0.9$). An undirected graph is constructed with 3,887 nodes and 98,026 edges, shown in Figure 5. An experiment is conducted to find an appropriate threshold θ of matrix similarity on 0.85, 0.9, and 0.95. When θ is 0.85, there are too many edges and the ratio of edge to node is too high. In this case, there is a very big cluster and several really small clusters. Therefore, the clustering result is not good for 0.85. When θ is set as 0.95, the number of the node is too low and the ratio of edge to node is too low as well. Each cluster contains not too many nodes. Moreover, most nodes are not clustered in any group. Therefore, the clustering result is not good for 0.95 either. Under 0.90, the number of node, edge and edge to node is reasonable. The number of nodes within clusters is reasonable and there are significant differences among each cluster. Moreover, the number of nodes which do not belong to any cluster is not too high. Therefore, the threshold θ is set as 0.9. In this graph, community detection algorithm is applied with Gephi, a graph visualization tool. Finally, 7 clusters in green (Cluster 1, 905 nodes), blue (Cluster 2, 229 nodes), pink (Cluster 3, 1583 nodes), orange (Cluster 4, 255 nodes), red (Cluster 5, 172 nodes), and yellow (Cluster 6, 147 nodes) are detected.

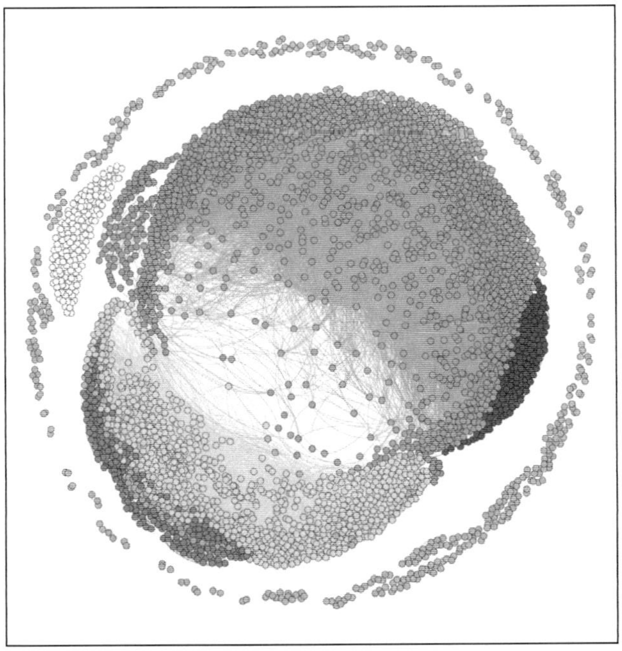

Figure 5. Clustering results.

The nodes isolated with others or not clustered in any groups are also displayed in the periphery in gray color, and they are considered as Cluster 7. Each node represents an individual, and nodes in same colors belong to the same clusters.

After identifying clusters, characteristics are summarized and concluded for each cluster. From Table 3, it can be seen that Cluster 1 and Cluster 2 represent individuals who do not go to work. On the contrary, individuals in Cluster 3, Cluster 4 and Cluster 5 need to work a lot. And participants in Cluster 6 take lots of trips to school. Furthermore, Cluster 1 and Cluster 2 can be differentiated from each other according to shopping and recreation activities in Table 3. From Table 3, it can be seen that individuals in Cluster 1 conduct more shopping activities, whereas individuals in Cluster 2 do many more recreation activities. Then Cluster 3, Cluster 4, and Cluster 5 can be distinguished with the support of Figure 6. Figure 6*a* shows the starting time (including travel time) of the first working activity. As can be seen, individuals in Cluster 3 and Cluster 4 go to work mostly at 07:00. However, the departure time of individuals in Cluster 5 is much later and shows more uncertainty than

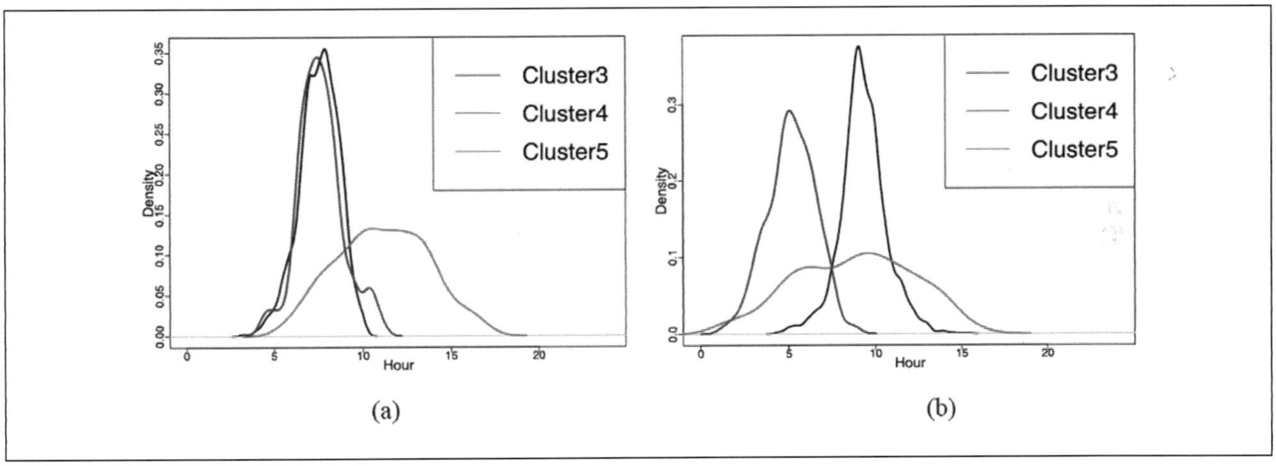

Figure 6. Distributions of (a) the starting time of the first working activity, and (b) working duration.

Table 4. Accuracy for Traditional Machine Learning Algorithm and CNN

	SVM	KNN	RF	CNN
Cluster 1	97.5%	100.0%	97.5%	100.0%
Cluster 2	57.5%	87.5%	60.0%	95.0%
Cluster 3	95.0%	90.0%	92.5%	97.5%
Cluster 4	90.0%	97.5%	90.0%	100.0%
Cluster 5	97.5%	97.5%	100.0%	97.5%
Cluster 6	92.5%	100.0%	95.0%	95.0%
Cluster 7	17.5%	35.0%	30.0%	85.0%
Average accuracy	78.2%	86.8%	80.7%	95.7%

individuals in the other two clusters. Regarding Cluster 5, the departure time for the first working activity is distributed from 10:00 to 14:00 more evenly. To separate Cluster 3 and 4, refer to Figure 6*b* that presents the working duration distribution. It can be seen that Cluster 3 shows a high probability of working approximately 9 hours (including travel time to work) per day. However, for Cluster 4, the working duration, which is around 5 hours, is much shorter than Cluster 3. The working duration distribution for Cluster 5 is also flat and ranges from 4 to 12 hours.

In summary, individuals can be clustered into 7 groups:

- Cluster 1: Non-working individuals with more shopping activities;
- Cluster 2: Non-working individuals with more recreation activities;
- Cluster 3: Individuals with normal working start time and a full-time job;
- Cluster 4: Individuals with part-time job;
- Cluster 5: Individuals with late start working time;
- Cluster 6: Individuals who need to attend school; and
- Cluster 7: Individuals that are not in any of the first 6 clusters.

With identified travel behavior clusters, a travel behavior classification task is performed with CNN to classify new travelers according to their activity maps. TensorFlow (*39*), one of most powerful deep learning tools, is employed to construct CNN in this paper. According to the aforementioned clustering results, clusters have different sizes. To create balance training and testing datasets, 100 records are randomly selected from each cluster for training, and 40 records from each cluster for test. There are 7 classes in classification task, in addition to the 6 classes which are identified by community detection algorithm. Class 'other' is added which represents users that are not in any of the 6 clusters. Moreover, several traditional machine learning algorithms are also implemented as benchmarks.

After training a CNN, accuracy as high as approximately 95% is finally achieved on the test dataset, and detailed accuracy information for all algorithms is shown in Table 4. Therefore, CNN is an appropriate and efficient algorithm for classifying participants into groups with similar travel behavior according to their activity maps.

Conclusion

In this paper, CHTS data is used to analyze and classify travel behavior. After processing raw data, some interesting observations are made. First, travelers make more commute trips to work and school during weekdays but more recreation, shopping, and personal business activities on weekends. Second, most participants' travel territory is around 5 miles, and 95% of participants travel within 10 miles for each trip. Third, most individuals have a low degree of spatial variability.

Based on 5-minute interval and eight activity types, an activity matrix is further constructed for each participant. Jaccard similarity coefficient is employed to calculate the similarity between every pair of participants. The output similarities are utilized to construct a social network for participants. After this, a community detection algorithm is adopted to cluster individuals into groups with the same travel behavior. The algorithm produces seven clusters: 1) non-working people with more shopping activities, 2) non-working people with more recreational activities, 3) individuals with normal working start time and a full-time job, 4) Individuals with part-time job, 5) individuals with late start working time, 6) individuals who need to attend school, and 7) individuals that are not in any of the first 6 clusters. Then individuals are classified by using a CNN and approximately 95% is achieved in accuracy.

In future, a variety of data sources, including passive trip trajectory data, transit smart card data, and social media data can be collected to provide more information about individual travel behavior. Therefore, the analysis of similar driver behavior can go a step further.

Acknowledgments

This study was partially supported by National Science Foundation award CMMI-1637604 and Region 2 University Transportation Research Center faculty-initiated research project.

Author Contributions

The authors confirm contribution to the paper as follows: study conception and design: Qing He, and Alireza Khani; data collection: Yu Cui; analysis and interpretation of results: Yu Cui, Qing He, and Alireza Khani; draft manuscript preparation: Yu Cui, Qing He, and Alireza Khani. All authors reviewed the results and approved the final version of the manuscript.

References

1. Cici, B., A. Markopoulou, E. Frias-Martinez, and N. Laoutaris. Assessing the Potential of Ride-Sharing Using Mobile and Social Data: A Tale of Four Cities. *Proc., 2014 ACM International Joint Conference on Pervasive and Ubiquitous Computing*, Seattle, Washington, ACM, New York, 2014. pp. 201–211.
2. Gong, L., T. Morikawa, T. Yamamoto, and H. Sato. Deriving Personal Trip Data from GPS Data: A Literature Review on the Existing Methodologies. *Procedia-Social and Behavioral Sciences*, Vol. 138, 2014, pp. 557–565.

3. Wolf, J., R. Guensler, and W. Bachman. Elimination of the Travel Diary: Experiment to Derive Trip Purpose from Global Positioning System Travel Data. *Transportation Research Record: Journal of the Transportation Research Board*, 2001. 1768: 125–134.
4. Wu, J., C. Jiang, D. Houston, D. Baker, and R. Delfino. Automated Time Activity Classification Based on Global Positioning System (GPS) Tracking Data. *Environmental Health*, Vol. 10, No. 1, 2011, p. 101.
5. Bohte, W., and K. Maat. Deriving and Validating Trip Destinations and Modes for Multiday GPS-Based Travel Surveys: Application in the Netherlands. Presented at 87th Annual Meeting of the Transportation Research Board, Washington, D.C., 2008.
6. Chen, C., H. Gong, C. Lawson, and E. Bialostozky. Evaluating the Feasibility of a Passive Travel Survey Collection in a Complex Urban Environment: Lessons Learned from the New York City Case Study. *Transportation Research Part A: Policy and Practice*, Vol. 44, No. 10, 2010, pp. 830–840.
7. Kim, Y., F. C. Pereira, F. Zhao, A. Ghorpade, P. C. Zegras, and M. Ben-Akiva. Activity Recognition for a Smartphone and Web Based Travel Survey. *arXiv preprint arXiv:1502.03634*, 2015.
8. Schönfelder, S., K. W. Axhausen, N. Antille, and M. Bierlaire. Exploring the Potentials of Automatically Collected GPS Data for Travel Behaviour Analysis-A Swedish Data Source. In *GI-Technologien für Verkehr und Logistik* (J. Möltgen, and A. Wytzisk, eds.), IfGIprints, Münster, Germany, 2002, pp. 155–179.
9. Deng, Z., and M. Ji. Deriving Rules for Trip Purpose Identification from GPS Travel Survey Data and Land Use Data: A Machine Learning Approach. *Proc., 7th International Conference on Traffic and Transportation Studies 2010*, ASCE, Kunming, China, 2010. pp. 768–777.
10. Griffin, T., and Y. Huang. A Decision Tree Classification Model to Automate Trip Purpose Derivation. *Proc., ISCA 18th International Conference on Computer Applications in Industry and Engineering*, Honolulu, Hawaii, ISCA, Honolulu, 2005. pp. 44–49.
11. Jones, P. Developments in Dynamic and Activity-Based Approaches to Travel Analysis. Oxford Studies in Transport, Avebury, Aldershot, 1990.
12. Hanson, S., and J. Huff. Classification Issues in the Analysis of Complex Travel Behavior. *Transportation*, Vol. 13, No. 3, 1986, pp. 271–293.
13. Shoval, N., and M. Isaacson. Sequence Alignment as a Method for Human Activity Analysis in Space and Time. *Annals of the Association of American Geographers*, Vol. 97, No. 2, 2007, pp. 282–297.
14. Kitamura, R., and T. Hoorn. Regularity and Irreversibility of Weekly Travel Behavior. *Transportation*, Vol. 14, No. 3, 1987, pp. 227–251.
15. Axhausen, K. W., A. Zimmermann, S. Schönfelder, G. Rindsfüser, and T. Haupt. Observing the Rhythms of Daily Life: A Six-Week Travel Diary. *Transportation*, Vol. 29, No. 2, 2002, pp. 95–124.
16. Jiang, S., J. Ferreira, and M. C. González. Clustering Daily Patterns of Human Activities in the City. *Data Mining and Knowledge Discovery*, Vol. 25, 2012, pp. 478–510.
17. Jiang, S., J. Ferreira, Jr., and M. C. Gonzalez. Discovering Urban Spatial-Temporal Structure from Human Activity Patterns. *Proc., ACM SIGKDD International Workshop on Urban Computing*, Beijing, China, ACM, New York, 2012, pp. 95–102.
18. Gonzalez, M. C., C. A. Hidalgo, and A.-L. Barabasi. Understanding Individual Human Mobility Patterns. *Nature*, Vol. 453, No. 7196, 2008, pp. 779–782.
19. Ponieman, N. B., A. Salles, and C. Sarraute. Human Mobility and Predictability Enriched by Social Phenomena Information. *Proc., 2013 IEEE/ACM International Conference on Advances in Social Networks Analysis and Mining*, Niagara, Ontario, ACM, New York, 2013, pp. 1331–1336.
20. Ma, X., Y.-J. Wu, Y. Wang, F. Chen, and J. Liu. Mining Smart Card Data for Transit Riders' Travel Patterns. *Transportation Research Part C: Emerging Technologies*, Vol. 36, 2013, pp. 1–12.
21. Williams, M. J., R. M. Whitaker, and S. M. Allen. Measuring Individual Regularity in Human Visiting Patterns. *Proc., Privacy, Security, Risk and Trust (PASSAT), 2012 International Conference on and 2012 International Confernece on Social Computing (SocialCom)*, Amsterdam, Netherlands, IEEE, New York, 2012. pp. 117–122.
22. Girvan, M., and M. E. Newman. Community Structure in Social and Biological Networks. *Proceedings of the National Academy of Sciences of the United States of America*, Vol. 99, No. 12, 2002, pp. 7821–7826.
23. Newman, M. E., and M. Girvan. Finding and Evaluating Community Structure in Networks. *Physical Review E*, Vol. 69, No. 2, 2004, p. 026113.
24. Radicchi, F., C. Castellano, F. Cecconi, V. Loreto, and D. Parisi. Defining and Identifying Communities in Networks. *Proceedings of the National Academy of Sciences of the United States of America*, Vol. 101, No. 9, 2004, pp. 2658–2663.
25. Palla, G., I. Derényi, I. Farkas, and T. Vicsek. Uncovering the Overlapping Community Structure of Complex Networks in Nature and Society. *Nature*, Vol. 435, No. 7043, 2005, pp. 814–818.
26. Soundarajan, S., and C. Gomes. Using Community Detection Algorithms for Sustainability Applications. *Proc., 3rd International Conference on Computational Sustainability*, Biocenter, University of Copenhagen, Copenhagen, Denmark, 2012.
27. Lin, L., Q. Wang, and A. Sadek. Data Mining and Complex Network Algorithms for Traffic Accident Analysis. *Transportation Research Record: Journal of the Transportation Research Board*, 2014. 2460: 128–136.
28. Wikipedia. *Applications of Artificial Intelligence*. https://en.wikipedia.org/wiki/Applications_of_artificial_intelligence-Transportation.
29. Schmidhuber, J. Deep Learning in Neural Networks: An Overview. *Neural Networks*, Vol. 61, 2015, pp. 85–117.
30. Gers, F. A., N. N. Schraudolph, and J. Schmidhuber. Learning Precise Timing with LSTM Recurrent Networks. *Journal of Machine Learning Research*, Vol. 3, 2002, pp. 115–143.

31. Fernández, S., A. Graves, and J. Schmidhuber. An Application of Recurrent Neural Networks to Discriminative Keyword Spotting. In *Artificial Neural Networks–ICANN 2007*, 2007, pp. 220-229.
32. Krizhevsky, A., I. Sutskever, and G. E. Hinton. Imagenet Classification with Deep Convolutional Neural Networks. In *Advances in Neural Information Processing Systems*, Lake Tahoe, Nevada, December 3–6, 2012. pp. 1097–1105.
33. He, K., X. Zhang, S. Ren, and J. Sun. Deep Residual Learning for Image Recognition. *Proc., IEEE Conference on Computer Vision and Pattern Recognition*, Las Vegas, Nev., IEEE, New York, 2016. pp. 770–778.
34. IMAGENET. *Large Scale Visual Recognition Challenge*, 2016. http://image-net.org/challenges/LSVRC/2016/results. Accessed May 8, 2018.
35. Dong, W., J. Li, R. Yao, C. Li, T. Yuan, and L. Wang. Characterizing Driving Styles with Deep Learning. *arXiv preprint arXiv:1607.03611*, 2016.
36. Kang, C., X. Ma, D. Tong, and Y. Liu. Intra-Urban Human Mobility Patterns: An Urban Morphology Perspective. *Physica A: Statistical Mechanics and its Applications*, Vol. 391, No. 4, 2012, pp. 1702–1717.
37. Wikipedia. *Jaccard index*. https://en.wikipedia.org/wiki/Jaccard_index.
38. Blondel, V. D., J.-L. Guillaume, R. Lambiotte, and E. Lefebvre. Fast Unfolding of Communities in Large Networks. *Journal of Statistical Mechanics: Theory and Experiment*, Vol. 2008, No. 10, 2008, p. P10008.
39. TensorFlow. *TensorFlow*. https://www.tensorflow.org/. Accessed May 8, 2018.

The Standing Committee on Traveler Behavior and Values (ADB10) peer-reviewed this paper (18-01761).

Article

TRR
JOURNAL OF THE TRANSPORTATION RESEARCH BOARD

Transportation Research Record
2018, Vol. 2672(47) 81–91

DOI: 10.1177/0361198118784135
journals.sagepub.com/home/trr
SAGE

Modeling the Structural Relationships of Activity-Travel Participation of Working Women

Rezwana Rafiq[1] and Michael G. McNally[1]

Abstract

The structural relationships between work and non-work activity participation and associated travel engagements are developed for single working mothers and married working mothers. Multiple-group structural equation modeling (SEM) is used to analyze the relationships between work and non-work time usage of the two groups for both in-home and out-of-home contexts. Non-work activities are childcare, maintenance, and discretionary activities. In addition to revealing interrelationships between work and non-work activity participation for each group, this multiple-group SEM model allows for the comparison of how these relationships differ across the two study groups. Data were obtained from the American Time Use Survey for 2012. The resulting model shows that these two groups demonstrate statistically significant differences in their activity interactions in terms of how participation in an activity affects the duration of other activities and travel. The model also captures the different degree of impacts of household and person-level socio-demographic characteristics (exogenous effects) of these two groups on activity participation (activity duration) and travel engagement (travel time). The findings of this research advance the understanding of the activity participation and time-allocation decisions made by working mothers, which can lead toward policy formulation for them in terms of social, land-use planning, and activity-based approach perspectives.

During the past three decades, the United States has experienced several considerable demographic trends such as an increasing number of women participating in the paid labor force and changes in household structure (*1*). Demographic trends are directed toward the continuation of such changes (*1*, *2*). The rate of labor force participation of working women with children less than 18 years old has increased from 47.4% in 1975 to 70.3% in 2013 (*3*). Changes in household demographic structure have resulted in a large and growing number of single-parent families (*1*, *2*). The number of total family households including married and single families with children less than 18 years has increased about 17% from 1974 to 2014, whereas the number of one-parent families has increased around 146% (more than twice) over the same period (*4*).

Recent research has examined the unique time-use pattern and time pressures faced by dual-earner couples with children (*5*). In dual-earner families with children, both men and women face challenges, albeit differently from each other (*6*, *7*), to allocate their daily schedule to meet the demand of work, household, and childcare responsibilities. Most of the studies suggest that women take more child and adult responsibilities for their households, and make more trips than men (*1*, *8*). Single working mothers, being the lone earner, usually have full responsibility for their families, including financial support, childcare, and other household activity needs, which makes their daily activity-travel pattern even more complicated and constrained than mothers with partners (*1*, *9*).

Women are usually the primary caregiver for their children. The childcare activities and household responsibilities of women have not yet evolved to be commensurate with their increased involvement in the labor force. From an activity-based travel behavioral perspective, a better understanding of activity-travel participation of both single and married working mothers will provide a more accurate examination of their daily decision-making processes in balancing work schedules with non-work activities. Several studies in transportation planning (*1*, *10*, *11*) focused on this growing segment of the labor force, but they did not cover extensive time-allocation analysis of their in-home and out-of-home work and non-work activity participation and travel engagement.

As the work activity is usually pursued at a fixed location and for a fixed start time and duration, it functions as a mandatory activity for working people on which the flexibility of

[1]Institute of Transportation Studies (ITS), University of California, Irvine, CA

Corresponding Author:
Address correspondence to Rezwana Rafiq: rrafiq@uci.edu

other non-work activities depends. In this research, the authors are interested in examining the structural interdependency between work and non-work activity-travel participation of working women. In particular, the study explores the time-allocation and activity-travel participation decision of working women in the context of two different types of households: single-parent households and two-parent households. This will help identify which working mother group has more constraints in terms of balancing time between work and non-work activity purposes. Multiple-group structural equation modeling (SEM) is used to investigate the relationships among socio-demographic characteristics, different activity participation, and travel engagement of these two working mother groups.

The paper is organized as follows. After brief review of relevant literature, the data and variables used in this model are discussed, followed by a model description. The next section provides a discussion of the empirical results. Finally, the paper concludes with a summary of key findings and policy implications.

Literature Review

Several research studies in the social science area have focused on the time-use of working women. Douthitt et al. identified socio-economic factors (by using ordinary least squares regression) that can influence the time-use decisions by single and married working mothers (*12*). Using the 2003 American Time Use Survey (ATUS) data and the 2000 United Kingdom Time Use Study data, Kalinkoski et al. found that primary and passive child care among single mothers does not differ greatly from that among married or cohabiting mothers (*13*). Kendig and Bianchi applied bivariate and multivariate regression on ATUS 2003 and 2004 data to observe the relationship between household structure and time engagement in childcare by mothers (*14*). They found that single mothers spend less time with their children than married mothers as a result of their time constraints. Harvey and Mukhopadhyay showed that there is high incidence of time deficit among employed single parents with their children (*9*). Several other papers have provided general descriptive analysis of the time-use of individuals in single and dual-earner parent households (*15–17*).

These research studies analyzed the time-use patterns of working mothers in terms of their time poverty from a social science perspective; they did not focus on the activity-travel behavior of these groups. This paper, in contrast, examines the interrelationship between work and non-work activity-travel time allocation for working mother groups. SEM is a potential tool to analyze this kind of interrelationships in a comprehensive framework. SEM can capture the direct relationships between activity demand and the need for travel to reach activity locations, the interrelationships between in-home and out-of-home activity participation, and the feedback from travel time to activity time. SEM can also handle the missing values and can correlate error terms (*18*).

Transportation research has often used SEM to analyze activity participation and travel behavior of individuals. Kitamura et al. and Golob et al. applied SEM in modeling joint demand for activity duration and travel (*19*, *20*). Golob developed a choice model for household heads (male and female) and explained the causal relationships between household activities and travel to out-of-home activity sites (*21*). Gould and Golob explored the impact of online shopping on travel behavior and found that the time saved by online shopping might induce latent demand for out-of-home maintenance activities for working women (*22*).

Lu and Pas used SEM to capture the relationships among socio-demographics, activity participation, and travel behavior (*23*). Similar work conducted by Kuppam and Pendyala examined the relationships between activity duration and trip generation for workers and found that a substitution effect exists between in-home and out-of-home activity participation (*24*). Simma and Axhausen investigated the interaction between household heads in terms of their out-of-home activity-travel behavior in nuclear families, and observed that working women more often own a car, make more total trips but fewer maintenance trips, and participate less in other activities than housewives (*25*).

More recently, Ren and Kwan explored the impact of the Internet on human activity-travel participation by using a multi-group SEM across two gender groups (*26*). Whitehead-Frei and Kockelman investigated how time spent with children influences other activity-travel time allocations for single household head and non-single household head mothers using a SEM model on ATUS 2007 data (*27*). Their results suggested that the amount of time spent at home with children has a statistically significant and negative impact on work, leisure, and personal activities at home. As the target groups in our research are more restricted and different from the above-mentioned research, the findings from this research are not directly comparable.

A seminal work exploring the complex relationships between activity participation and associated travel demand using SEM was conducted by Golob and McNally (*6*). The authors have used several of Golob and McNally's elements in the construction of this study's SEM, such as arranging different activity purposes in hierarchical order, connecting out-of-home activity purposes with their associated travel, and associating different personal and household-level socio-demographic characteristics with activity-travel participation. In addition, in-home and out-of-home substitution effects between different activity purposes were considered. As this study's target groups are mothers, who are usually the primary caregivers of their children, childcare activities were separated from other maintenance activities to identify the structural relationships between work and each of the three non-work activity purposes.

Data and Sample

Source of Data and Sample Size

ATUS 2012 data were used. ATUS collects time-use information for various activity categories (i.e., personal care, household activities, work, socializing, traveling, etc.) performed by individuals for one complete day (1440 minutes). ATUS data also contain socio-demographic information about the respondents. In this work, single working mothers were define as a female who is at least 18 years old, who is single (not living with a spouse or partner), who is employed (in-home or out-of-home, or both), and who has her own children (<18 years old) living in her home. Married working mothers are defined by the attributes of being married, employed, with age of at least 18 years, having at least one child (<18 years old) and live in a dual-earner household (employed partner). According to the above-mentioned criteria, among 12,443 respondents in ATUS 2012 dataset, 555 respondents are single working mothers and 1,003 respondents are married working mothers.

Descriptive Statistics of Data

Table 1 shows the socio-demographic and economic characteristics of both single and married working mothers. Most of the single working mothers (44.86%) have only one child, and most of the married working mothers (48.55%) have two children. Generally, children of age range 11–17 years dominate in these two groups of households. Again, the majority of each target groups is employed full time, whereas a higher number of married mothers do part-time jobs than that of single mothers. Compared with single working mother households, the households of married working mothers are more financially solvent (observed from their household income).

Table 2 shows the mean and standard deviation of time spent in different activities by the two target groups. Independent sample *t*-test at 5% significance level was performed in order to evaluate whether the mean difference (unweighted means) of activity durations differ significantly across the two groups. The mean difference is found significant (p-value < 0.05), across all activity durations except for consumer purchase and travel.

Model Specification

Multiple-group SEM was used to investigate the structural time-use interrelations between work and non-work activities. The model also tests the invariance in causal structure across the two different working mother groups.

The Exogenous and Endogenous Variables

The exogenous and endogenous variables specified for the model are shown in Table 3. The exogenous variables include household and personal socio-demographic characteristics of the target groups. The endogenous variables represent in-home, out-of-home and associated travel-related time-use for four different activity purposes: (i) *work*: working, work-related activities, (ii) *childcare*: tasks related to children's care, education, health, (iii) *maintenance*: housework, preparing meals, shopping, and so forth, and (iv) *discretionary*: religious activities, visiting relatives and friends, attending parties, watching TV, sports, exercise, and so on.

The Structural Equation Path Model

Let us denote measured exogenous variables as **X** and measured endogenous variables as **Y**. The equation for the endogenous variables is given by (*28*):

$$\mathbf{Y} = \Gamma\mathbf{X} + \mathbf{B}\mathbf{Y} + \zeta \tag{1}$$

where **Y** is an ($m \times 1$) column vector of endogenous variable and **X** is an ($n \times 1$) column vector of exogenous variables.

The structural parameters are the elements of the matrices are (*6*):

Γ ($m \times n$) matrix of direct causal (regression) effects from the (n) exogenous variables to the (m) endogenous variables;

B ($m \times m$) matrix of causal links between the m endogenous variables;

and:

$\Psi = E(\zeta\zeta')$ = variable-coyariance matrix of the (m) error terms

Equation 1 can be expressed in matrix form as (*28*):

$$\begin{bmatrix} Y_1 \\ Y_2 \\ \cdots \\ Y_m \end{bmatrix} = \begin{bmatrix} \gamma_{11} & \cdots & \gamma_{1n} \\ \cdots & \cdots & \cdots \\ \gamma_{m1} & \cdots & \gamma_{mn} \end{bmatrix} \begin{bmatrix} X_1 \\ X_2 \\ \cdots \\ X_n \end{bmatrix} + \begin{bmatrix} 0 & \cdots & \beta_{1m} \\ \cdots & \cdots & \cdots \\ \beta_{m1} & \cdots & 0 \end{bmatrix} \begin{bmatrix} Y_{11} \\ Y_2 \\ \cdots \\ Y_m \end{bmatrix} + \begin{bmatrix} \zeta_1 \\ \zeta_2 \\ \cdots \\ \zeta_m \end{bmatrix} \tag{2}$$

For identification of system (*1*), **B** must be chosen such that (**I-B**) remains non-singular, where **I** is an identity matrix of dimension m. It can be shown that the total effects of the endogenous variables on each other are given by (*6*):

$$T_{yy} = (I - B)^{-1} - I \tag{3}$$

The total effects of the exogenous variables on the endogenous variables are given by (*6*):

$$T_{xy} = (I - B)^{-1}\,\Gamma \tag{4}$$

The Initial Conceptual Model

The initial conceptual structure of the proposed model is shown in *solid lines* and the additional links to improve the

Table 1. Socio-Demographic and Economic Characteristics of the Target Groups

Variable	Single working mothers (%)	Married working mothers (%)
Total respondents	555	1,003
Number of children		
Number of children = 1	44.86	27.72
Number of children = 2	36.22	48.55
Number of children >2	18.92	23.73
Number of children by age		
Age of children 0–5	36.22	40.38
Age of children 6–10	42.34	45.96
Age of children 11–17	53.33	47.56
Residence status		
Metropolitan area	80.36	86.04
Non-metropolitan area	18.74	13.36
House ownership type		
Owned	45.77	85.84
Rented for cash	52.97	13.46
Occupied without cash rent	1.26	0.70
Living arrangements		
With children and others (parents, siblings, other relatives and non-relatives)	17.48	6.78
Only with children	82.52	93.22
Holding multiple jobs		
Yes	9.55	7.38
No	90.45	92.62
Full-time/part-time status		
Full-time	73.69	64.91
Part-time	26.31	35.09
Class of workers		
Private	75.32	62.51
Government	16.58	21.34
Family income distribution (per year)		
<$20K	29.55	3.19
$20K–$40K	35.14	8.18
$40K–$60K	16.58	14.56
$60K–$100K	13.15	31.90
>$100K	5.59	42.17

Table 2. Mean Activity Durations of Single and Married Working Mothers

Activities (both in home and out of home except for travel)	Single working mothers (n = 555)		Married working mothers (n = 1003)		Difference in mean (α = 5%)	
	Mean duration (min)	SD	Mean duration (min)	SD	t-value	p-value
Work	424.57	172.08	380.27	197.66	3.4048	0.0007
HH childcare	110.33	92.48	144.44	118.19	−5.2922	0.0000
HH maintenance	135.65	124.13	156.69	131.50	−2.9315	0.0034
Consumer purchases	57.97	58.96	59.82	56.77	−0.4158	0.6777
Socializing and recreation	233.40	162.92	212.06	141.37	2.5028	0.0125
Travel	64.25	62.68	69.68	67.89	−1.5043	0.1328

model results are shown in *dotted lines* in Figure 1. The rectangular boxes represent the endogenous variables and the arrows represent postulated non-zero direct effects. Only the direct effects between endogenous variables are shown in the figure, and *not* the direct effects among exogenous and endogenous variables.

Table 3. List of Variables

Variables	Abbreviation
Exogenous	
No. of children 0–5 years of age	nchild0_5
No. of children 6–10 years of age	nchild6_10
No. of children 11–17 years of age	nchild11_17
Presence of relatives in the HH [yes = 1, no = 0]	caregiver
Resides in metropolitan region [yes = 1, no = 2]	region
Type of worker [private = 1, government = 2]	wrktype
Employment type [full time = 1, part time = 2]	emptype
Work mode [private car = 1, others = 2]	wmode
Household ownership [own = 1, others = 0]	hhown
Household income less than $20K [dummy]	inc_low
Household income $20K to $60K [dummy] (ref. group)	inc_mid
Household income $60K or more [dummy]	inc_high
Endogenous	
In-home activity duration (min)	
Work	durwork_inhome
Childcare	durcc_inhome
Maintenance	durmaintain_inhome
Discretionary	durdiscre_inhome
Out-of-home activity duration (min)	
Work	durwork_outhome
Childcare	durcc_outhome
Maintenance	durmaintain_outhome
Discretionary	durdiscre_outhome
Travel times (min) to access out-of-home	
Work	durtravel_work
Childcare	durtravel_cc
Maintenance	durtravel_maintain
Discretionary	durtravel_discre

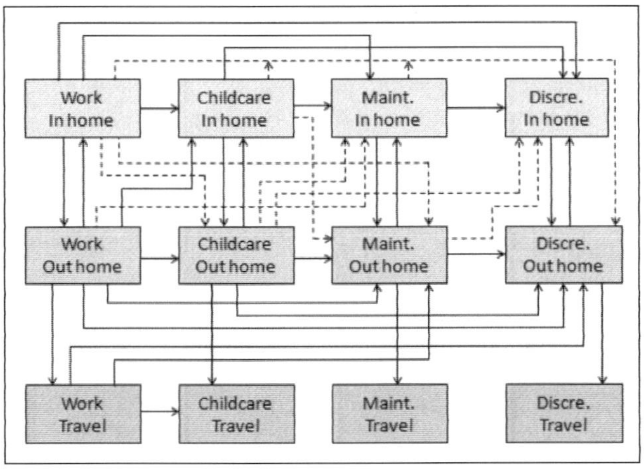

Figure 1. Final structural equation model (solid links belong to the initial model).

The first row of variables corresponds to the activity duration of four different types of activities performed in home. Similarly, the second row represents the activities performed out of home. The last row corresponds to the travel time associated with each of the four out-of-home activity purposes. The specific hypothesis behind the direct effects between the endogenous variables is discussed below.

In-Home Activity Interactions. This study follows a hierarchical order among activities based on their degree of flexibility. For example, work is a mandatory activity and the least flexible (because it is usually pursued at a fixed location with a fixed time), whereas discretionary activities are more flexible and are constrained by the availability of time imposed by the mandatory activities (*6*, *29*). So, in the activity hierarchy, work is positioned at the top, followed by childcare, then by maintenance, and finally by discretionary tasks. This leads to six hierarchical activity links: (1) work to childcare, (2) work to maintenance, (3) work to discretionary, (4) childcare to maintenance, (5) childcare to discretionary, and (6) maintenance to discretionary. All the coefficients of these connections are expected to be *negative*. This is because spending more time in one activity will potentially reduce the time for another activity, as the total duration of a day is fixed (24 hours).

Out-of-Home Activity Interactions. The same hierarchical order is also maintained in out-of-home activities. Therefore, another set of six links between the out-of-home activities are defined. In addition, one direct connection is provided from out-of-home work to in-home childcare. This is based on the assumption that a mother spending more time in out-of-home work might end up spending less time at in-home childcare (e.g., helping their education, etc.). All these coefficients are expected to be *negative*.

Interactions between In-Home and Out-of-Home Activities. It is expected that there is a substitution effect between in-home and out-of-home activities of the same type (*6*, *24*). That means, each in-home activity duration may affect the out-of-home activity duration of the same activity type and vice versa. So, non-recursive connections are defined between in-home and out-of-home activity purposes in the model.

Activity-Travel Interactions. In the model, a direct connection is assigned from each of the out-of-home activity to their associated travel. These direct connections represent travel as a derived demand (*6*). That means, the demand for travel for a particular activity is created to meet the demand of that particular out-of-home activity. Each of the coefficients is expected to be *positive*. In addition, two feedback effects from work travel to out-of-home maintenance and discretionary activities are expected. It is anticipated that the more travel time involved in work, the less time a person can spend for out-of-home maintenance and discretionary activities. Moreover, it is expected that work travel time may directly influence the childcare travel time.

Effects of Exogenous Variables. For each of the endogenous variables in the model, a subset of exogenous variables is selected that can potentially affect the endogenous variable. For example, the work travel time may be affected by the type of employment, type of worker, and the nature the of work mode.

Error-Term Covariance. It is assumed that certain endogenous variable error terms may be correlated. For example, the unexplained part of work travel time can be correlated with the unexplained part of childcare travel time. Mothers often pick up or drop off their kids on their way to or from work, thus, these two activity travels are usually performed in a joint manner. Similarly, the error terms of the travel time for maintenance activities may be correlated with that of discretionary activities.

Estimation of the Model

Based on the conceptual structure of endogenous variables (Figure 1 with only solid links) and the best possible combination of exogenous variables, two *multi-group structural models* were estimated by using the maximum likelihood missing value (MLMV) estimation method. One model was specified by constraining all the corresponding parameters to be equal for single and married mothers and another model without having such constraints. χ^2 values of 3470 (df 478) and 1039 (df 348) were obtained for the constrained and unconstrained model, respectively, both having a p-value < 0.05. The high χ^2 value with low p-value indicates that the model fits are not satisfactory for the both models.

To improve the model, a sequence of hypothesized direct effects was introduced between endogenous variables, and it was found that 10 additional direct connections (shown in *dotted lines* in Figure 1) improved the model significantly. The models result in χ^2 (398) = 2805.45 (p-value = 0.000) for the constrained model and χ^2 (395) = 389.74 (p-value = 0.565) for the unconstrained model. The constrained model has unsatisfactory model fit, which implies that parameters are not equal across the two study groups (*28*). On the other hand, the unconstrained model has lower χ^2 value with high p-value. Moreover, the same model results in zero root-mean-square error value (and also exactly 1.0 Comparative Fit Index (CFI) and Tucker Lewis Index (TLI)) indicating a good model fit (*28*). Therefore, the unconstrained one was used as the final model.

Model Results and Discussion

The model has three layers: in-home, out-of-home activities, and associated travel, and each layer has the same four sets of activity types. Each of the (unstandardized) coefficients of direct effects in each layer is discussed separately.

Effects between Endogenous Variables

The statistically significant direct effects between endogenous variables are shown in Figure 2 for single and married working mothers.

In-home Activity Interactions. In terms of the in-home activity interactions for both the target groups (Figure 2), it was observed that for single mothers, in-home work activities negatively affect maintenance in-home activities (coefficient of −0.200). This implies that the more time a mother is involved in work activity at home, the less time she can spend, as expected, in household maintenance tasks at home. In contrast, working at home does not significantly affect the duration of the other three in-home activities for married working mothers. In both of the groups, however, the time spent in in-home maintenance tasks *positively* affects the time spent in discretionary activities at home. The coefficient value between these two activities are observed to be 0.368 for single working mother and 0.173 for married working mothers. That means, the effect is higher for single mothers than the married mothers. The positive correlation between these two activity purposes can be attributed to joint participation in these two activities when they are at home. For example, preparing meals (maintenance task) can accompany watching TV or talking on the phone (discretionary). In addition, childcare

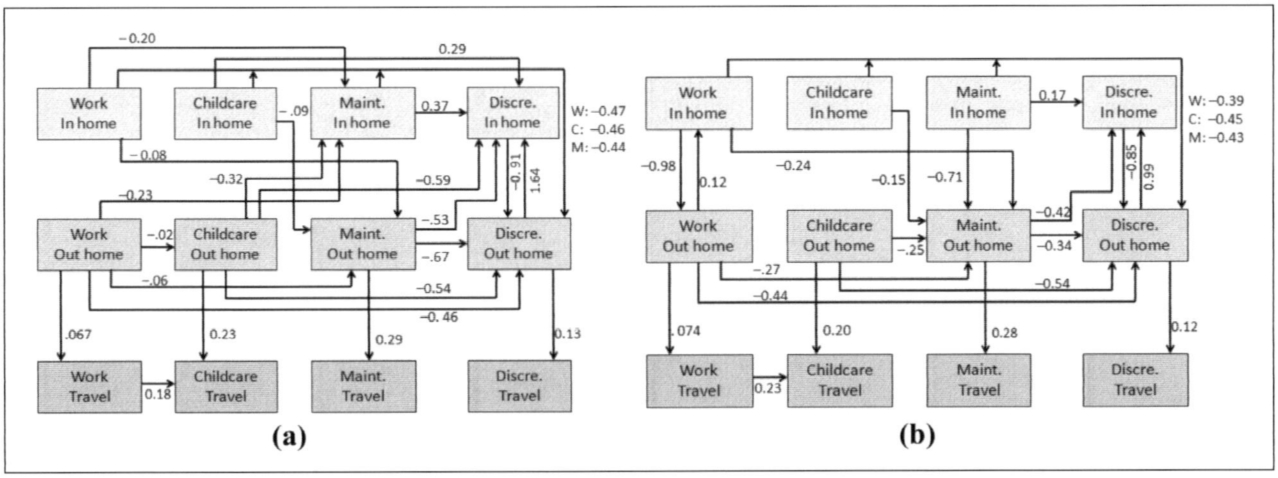

Figure 2. Significant direct effects (at ≤ 5% sig. level) between endogenous variables: (*a*) Single working mothers and (*b*) Married working mothers.

duration at home *positively* affects discretionary activities at home only for single working mothers, which is in contrary to the initial hypothesis. One possible reason may be childcare at home is performed while doing another discretionary task, but this joint activity type is not identified in ATUS data.

Out-of-Home Activity Interactions. In terms of out-of-home activity interactions, the duration of out-of-home activities of higher order (e.g., work) indeed affects the duration of lower order out-of-home activities (e.g., discretionary), and all coefficients are negative as expected. More precisely, all non-work out-of-home activity participations are controlled by the duration of out-of-home work activity for both of the groups. The only exception is that out-of-home work duration does not affect out-of-home childcare duration for married mothers. One possible explanation can be married mothers can have their male partners taking their children out of home for outdoor activities. Again, a greater involvement in out-of-home childcare activities leads to married mothers spending less time in out-of-home maintenance activities, but this effect was not seen for single mothers.

Interactions between In-Home and Out-of-Home Activities. The authors postulated interactions between in-home and out-of-home activity participation for each corresponding activity purposes by placing feedback effect or non-recursive relationship between them in the path model. The model results reveal that for single working mothers the feedback effect only exists between in-home and out-of-home discretionary activity participation, whereas for married mothers two such feedback effects exist between work in home and out of home as well as discretionary in-home and out-of-home activities. Also note that for single mothers, childcare out-of-home activities significantly affect maintenance in-home activities but not out-of-home maintenance activities, whereas for married mothers out-of-home childcare effects are reversed (maintenance out of home but not in home). Furthermore, in-home participation in all three higher-order activities significantly affects the out-of-home participation of discretionary activities for both married and single mothers, and the size of the effects are also very close for both groups.

Activity-Travel Interactions. According to the hypothesis, out-of-home activity participation for all activity purposes has direct *positive* effect on their associated travel. In the model results, positive coefficients were indeed obtained for all four activity-travel interactions and the size of effects are all very similar. These activity-travel interactions represent travel as a derived demand to access the associated activity, and the corresponding coefficients for each of the effects can be explained in terms of hours of travel time required to access one hour of that activity (*6*). For example, the coefficient values for the direct effect from out-of-home work to the associated travel are 0.067 and 0.074 for single working mothers and married working mothers, respectively. This translates to eight hours of out-of-home work activity requiring about 32 minutes of travel time for single mothers (the travel duration for married mothers is around 35 minutes). Moreover, the significant positive direct effect from work travel time to childcare travel time for both the mother groups implies that the associated trips to access these two activities are made jointly in most of the cases. For example, mothers usually pick up or drop off their kids on their way to home from work or way to work from home.

Error-Term Covariance. The postulated error-term covariance between work travel time and childcare travel time is statistically significant for both the target groups. No significant relationships were found between the error terms of maintenance travel time and discretionary travel time for any of the groups.

Table 4. Direct Effects of the Exogenous Variables for Single Working Mothers – MLMV Estimation (*t*-statistic in parentheses)

	Endogenous variable											
Exogenous variables	Work in home	Childcare in home	Maint. in home	Discre. in home	Work out home	Childcare out home	Maint. out home	Discre. out home	Work travel	Childcare travel	Maint. travel	Discre. travel
nchild0_5		34.221		−37.451**								
		(4.06)		(−1.49)								
nchild6_10		10.719*										
		(2.02)										
nchild11_17						4.913*						
						(2.01)						
caregiver			−42.975*			−12.163*						
			(−2.65)			(−2.41)						
inc_low							−12.629**					
							(−1.87)					
inc_high							18.449*					
							(2.36)					
region						7.985*						
						(2.12)						
wrktype					−79.905*							
					(−2.51)							
emptype					−90.602*							
					(−2.48)							
wmode									9.633**		25.197**	
									(1.74)		(1.66)	
hhown												

Note: All reported coefficients are significant at 0.001 level expect the (*) marked ones. * = 5% level of significance; ** = 1% level of significance.

Effects of Exogenous Variables

Next, the *direct* effects of exogenous variables on each of the endogenous variables for each of the two groups (Tables 4 and 5) are analyzed. Recall there are 11 exogenous variables that represent the socio-demographics characteristics of the individuals at the household and personal levels. First, the effects of exogenous variables in different in-home activity purposes will be described. For single working mothers, the number of children aged between 0 to 5 significantly positively affects (coefficient 34.22) the amount of time spent in childcare. For the presence of the same children group, the discretionary activity duration significantly reduces (coefficient −37.45). In contrast, the number of children in this particular age group does not affect in any of the in-home activities of married working mothers. They, however, are affected by more younger children. For example, the presence of children aged 11–17 reduces childcare time for mothers, which may be because children within this age group are grown enough and may not require a higher time demand for their care from mothers. This reduction in childcare time in turn allows mothers to spend time in discretionary in-home activities (effect size 18.15). No such relation was seen for single working mothers. The presence of related individuals in single-mother households significantly reduced their household maintenance time (coefficient −42.98).

Next the impact of exogenous variables on different endogenous variables is assessed. Employment type (full time vs. part time) and the type of worker (private vs. government) affect out-of-home activity durations in both groups. If the mothers are part-time employees or work in government, they spend less time in work than full-time or private sector employees. It is also observed that single mothers happen to spend more time in out-of-home childcare activity if they have children in the age group 11–17 years in their households (effect size 4.91). On the contrary, the presence of children in the same group causes married mothers to spend more time in out-of-home maintenance activities (coefficient 11.75). Again, the duration of childcare is reduced for single mothers if they have one of their relatives at home. Interestingly, single mothers living in non-metropolitan areas tend to spend more time in out-of-home childcare activities. For both mother groups, higher income leads to more time spent in outdoor maintenance activities (more shopping, grocery, etc.).

Only a very few exogenous variables affect the time spent in travel for different activity purposes. No single exogenous variable affecting the travel time of married working mothers was found. For single mothers, however, the nature of their work travel mode significantly affects their work and maintenance travel time. If they do not use a private vehicle for going to work, they end up spending more time in work and maintenance-related travel (coefficients are 9.63 and 25.19, respectively).

Conclusion and Policy Implications

This study investigated the structural relationships between the time allocation in work and non-work activities and associated travel engagement for two groups of working mothers

Table 5. Direct Effects of the Exogenous Variables for Married Working Mothers – MLMV Estimation (*t*-statistic in parentheses)

	Endogenous variable											
Exogenous variables	Work in home	Childcare in home	Maint. in home	Discre. in home	Work out home	Childcare out home	Maint. out home	Discre. out home	Work travel	Childcare travel	Maint. travel	Discre. travel
nchild0_5												
nchild6_10								8.958**				
								(1.83)				
nchild11_17		−27.135*		18.145**			11.747**					
		(−2.08)		(2.14)			(1.89)					
caregiver												
inc_low												
inc_high							13.632**					
							(1.87)					
region												
wrktype					−111.767							
					(−6.78)							
emptype					−40.475*							
					(−2.11)							
wmode												
hhown												

Note: All reported coefficients are significant at 0.001 level expect the (*) marked ones. * = 5% level of significance; ** = 1% level of significance.

(single and married) and revealed interesting relations between their work and non-work time usage in both in-home and out-of-home contexts. Several policy implications can be drawn from this research.

Social Science Perspective

An important finding presented in this paper is that both single and married working mothers who spend more time in out-of-home work can spend less time in out-of-home maintenance and discretionary activities. Bernardo et al. indicate similar kinds of time-poverty issues for dual-earner parents (*5*). In addition, spending more time in work outside home constrains single working mothers to spend less time with their children in their out-of-home activities, such as education, games, competitions, and other outdoor events. These time-constraint issues can lead toward formulating work-friendly social policies for working mothers to help balancing their work, family, and personal well-being issues.

Urban Planning and Accessibility Perspective

This research also provides important insights in terms of urban planning and transportation accessibility for the two study groups. For example, it is found that for both groups, 1 hour of out-of-home discretionary activities requires around 7 minutes of travel time. This indicates they mostly participate in discretionary activities that are only 7 minutes away (that is, very close to their locations). These activities mostly include visiting nearby friends and relatives. From the descriptive analysis it was also observed that they do not undertake long-distance travel to enjoy outdoor art, entertainment, events, and other such outdoor recreational facilities. To address this time-poverty issue, important land-use policy decisions should be taken to facilitate different outdoor relaxing and entertaining facilities within accessible distances.

Activity-Based Behavioral Analysis Perspective

From an activity-based approach perspective, this research provides valuable information on daily time-allocation and activity-travel participation decisions made to participate in work and non-work activities by the study groups. The division of activity participation based on in-home and out-of-home helps to understand the pattern of substitution effect between these two activity locations. Again, the developed model reveals how the activity participation is related to travel by explaining the amount of travel time required to access that particular activity. Furthermore, this research identifies the important individual and household-level socio-demographic factors that drive the time-use decisions in activity participation as well as travel engagement by our target groups. This activity-based behavioral information will provide important insights in time-allocation models of activity-based travel-demand modeling. After analyzing the multi-group SEM, it was found that our two particular working mother groups demonstrated statistically significant differences in their activity interactions in terms of how participation in one activity or the other affects the duration of another activity (and travel). This finding will help to consider these two population segment differently instead of pooling them while modeling the activity patterns of individuals at different households. Although a few prior works (*12–14*) have studied the activity time allocation of these two groups of women, they did not examine the interdependencies among different activities as well as travel through a structural relationship.

Since ATUS data do not provide location, accessibility, land-use, and household-level data, these factors are not incorporated into this research. However, the next research phase will incorporate the household interaction and spatial components by considering travel diary data.

Author Contributions

The authors confirm contribution to the paper as follows: study conception and design: Rezwana Rafiq, Michael G. McNally; analysis and interpretation of results: Rezwana Rafiq, Michael G. McNally; draft manuscript preparation: Rezwana Rafiq. All authors reviewed the results and approved the final version of the manuscript.

References

1. Rosenbloom, S. Understanding Women's and Men's Travel Patterns: The Research Challenge. In *Conference Proceedings 35: Research on Women's Issues in Transportation: Report of a Conference, Volume 1: Conference Overview and Plenary Papers*. Transportation Research Board of the National Academies, Washington, D.C., 2006, pp. 7–28.
2. Cherlin, A. J. Demographic Trends in the United States: A Review of Research in the 2000s. *Journal of Marriage and Family*, Vol. 72, No. 3, 2010, pp. 403–419.
3. Bureau of Labor Statistics Reports. *Women in the Labor Force: A Databook*. Report 1052. U.S. Bureau of Labor Statistics. 2014.
4. U.S. Census Bureau. *Current Population Survey, March and Annual Social and Economic Supplements, 2014 and Earlier*. http://www.census.gov/hhes/families/files/fm1.xls. Accessed November 8, 2015.
5. Bernardo, C., R. Paleti, M. Hoklas, and C. R. Bhat. An Empirical Investigation into the Time-Use and Activity Patterns of Dual-Earner Couples with and Without Young Children, *Transportation Research Part A: Policy and Practice*, Vol. 76, 2015, pp. 71–91.
6. Golob, T. F., and M. G. McNally. A Model of Activity Participation and Travel Interactions Between Household Heads. *Transportation Research Part B: Methodological*, Vol. 31, No. 3, 1997, pp. 177–194.
7. Nobis, C., and B. Lenz. Gender Differences in Travel Patterns: Role of Employment Status and Household Structure. In *Conference Proceedings 35: Research on Women's Issues in Transportation, Report of a Conference, Volume 2: Technical Papers*, Transportation Research Board of the National Academies, Washington, D.C., 2005, pp. 114–123.
8. Offer, S., and B. Schneider. Revisiting the Gender Gap in Time-Use Patterns: Multi-tasking and Well-Being among Mothers and Fathers in Dual-Earner Families. *American Sociological Review*, Vol. 76, No. 6, 2011, pp. 809–833.
9. Harvey, A. S., and A. K. Mukhopadhyay. When Twenty-Four Hours is Not Enough: Time Poverty of Working Parents. *Social Indicators Research*, Vol. 82, No. 1, 2007, pp. 57–77.
10. Farber, S., A. Páez, R. G. Mercado, M. Roorda, and C. Morency. A Time-Use Investigation of Shopping Participation in Three Canadian cities: Is There Evidence of Social Exclusion? *Transportation*, Vol. 38, No. 1, 2011, pp. 17–44.
11. Lee, Y., M. Hickman, and S. Washington. Household Type and Structure, Time-Use Pattern, and Trip-Chaining Behavior. *Transportation Research Part A: Policy and Practice*, Vol. 41, No. 10, 2007, pp. 1004–1020.
12. Douthitt, R. A., C. D. Zick, and J. McCullough. The Role of Economic and Demographic Factors in Explaining Time-Use of Single and Married Mothers. *Lifestyles*, Vol. 11, No. 1, 1990, pp. 23–51.
13. Kalenkoski, C. M., D. Ribar, and L. S. Stratton. *Parental Child Care in Single Parent, Cohabiting, and Married Couple Families: Time Diary Evidence from the United States and the United Kingdom*. Working Paper No. 440, The Levy Economics Institute of Bard College, Annandale-on-Hudson, New York, 2006.
14. Kendig, S. M., and S. M. Bianchi. Single, Cohabitating, and Married Mothers' Time with Children. *Journal of Marriage and Family*, Vol. 70, No. 5, 2008, pp. 1228–1240.
15. Jacobs, J. A., and K. Gerson. Overworked Individuals or Overworked Families? Explaining Trends in Work, Leisure, and Family Time. *Work and Occupations*, Vol. 28, No. 1, 2001, pp. 40–63.
16. Sayer, L. C., S. M. Bianchi, and J. P. Robinson. Are Parents Investing Less in Children? Trends in Mothers' and Fathers' Time with Children. *American Journal of Sociology*, Vol. 110, No. 1, 2004, pp. 1–43.
17. Bianchi, S. M. Maternal Employment and Time with Children: Dramatic Change or Surprising Continuity? *Demography*, Vol. 37 No. 4, 2000, pp. 401–414.
18. Golob, T. F. Structural Equation Modeling for Travel Behavior Research. *Transportation Research Part B: Methodological*, Vol. 37, No. 1, 2003, pp. 1–25.
19. Kitamura, R., J. P. Robinson, T. F. Golob, M. A. Bradley, J. Leonard, and T. van der Hoorn. A Comparative Analysis of Time Use Data in the Netherlands and California. *Proc., 20th PTRC Summer Annual Meeting: Transportation Planning Methods*, PTRC Education and Research Services Ltd., London, 1992, pp. 127–138.
20. Golob, T. F., R. Kitamura, and C. Lula. Modeling the Effects of Commuting Time on Activity Duration and Non-Work Travel. Presented at 73rd Annual Meeting of the Transportation Research Board, Washington, D.C., 1994.
21. Golob, T. F. A Model of Household Demand for Activity Participation and Mobility. In *Theoretical Foundations of Travel Choice Modeling* (T. Gärling, T. Laitilla, and K. Westin, eds.), Pergamon, Oxford, 1998, pp. 365–398.
22. Gould, J., and T. F. Golob. Shopping without Travel or Travel without Shopping? An Investigation of Electronic Home Shopping. *Transport Reviews*, Vol. 17, No. 4, 1997, pp. 355–376.
23Lu, X., and E. I. Pas. Socio-Demographics, Activity Participation and Travel Behavior. *Transportation Research Part A: Policy and Practice*, Vol. 33, No. 1, 1999, pp. 1–18.
24. Kuppam, A. R., and R. M. Pendyala. A Structural Equations Analysis of Commuters' Activity and Travel Patterns. *Transportation*, Vol. 28, No. 1, 2001, pp. 33–54.

25. Simma, A., and K. Axhausen. Within-Household Allocation of Travel: Case of Upper Austria. *Transportation Research Record: Journal of the Transportation Research Board*, 2001. 1752: 69–75.
26. Ren, F., and M. P. Kwan. The Impact of the Internet on Human Activity–Travel Patterns: Analysis of Gender Differences Using Multi-Group Structural Equation Models. *Journal of Transport Geography*, Vol. 17, No. 6, 2009, pp. 440–450.
27. Whitehead-Frei, C., and K. Kockelman. Americans' Time Use: A Focus on Women and Child Rearing—Structural Equations Modeling. *Transportation Research Record: Journal of the Transportation Research Board*, 2010. 2163: 32–44.
28. Kline, R. B. *Principles and Practice of Structural Equation Modeling*. Guilford Publications, New York, 2016.
29. Golob, T. F., and R. M. Pendyala. *Structural Models and the Effects of the Commute Trip on Travel and Activity Participation*. North American Meeting of the Regional Science Association, New Orleans, 1991, pp. 7–10.

The Standing Committee on Traveler Behavior and Values (ADB10) peer-reviewed this paper (18-01794).

Article

TRR
JOURNAL OF THE TRANSPORTATION RESEARCH BOARD

Transportation Research Record
2018, Vol. 2672(47) 92–102

DOI: 10.1177/0361198118791865
journals.sagepub.com/home/trr
SAGE

Effects of Traffic Congestion on Vehicle Miles Traveled

Reza Sardari[1], Shima Hamidi[1], and Raha Pouladi[1]

Abstract

The effects of traffic congestion on travel behavior are complex and multidimensional because they are related to various factors such as density, land use patterns, network connectivity, and individual preferences. Traffic congestion is a phenomenon that not only affects transportation systems but also influences commuters' quality of life and population mobility. The present research aims to analyze the effects of traffic congestion on individuals' travel behaviors, addressing both direct and indirect effects of congestion on vehicle miles traveled (VMT) per driver by implementing structural equation modeling (SEM) techniques. In addition to the causal analysis between traffic congestion and VMT, this study examined the complex relationship between an individual's socioeconomic characteristics, the built environment, congestion, and VMT. Measuring local congestion at a national level is also a key contribution of this research. This study used the same methodology as the Texas A&M Transportation Institute to compute a road congestion index and quantify local congestion for 93,769 drivers within 337 metropolitan areas. Our findings suggest that congestion is the main driver of VMT reduction. The findings also confirm that residents in compact development regions have lower daily VMTs because of the proximity of origins and destinations in denser areas with higher job–population balances. Therefore, rather than expanding highway networks, public transit investment might address traffic congestion more efficiently—not only by providing residents with more equitable and sustainable means of transportation, but also by encouraging people to reside in more compact and location-efficient areas.

Travel behavior is a broad concept that includes various transportation planning outcomes such as person miles traveled (PMT), vehicle miles traveled (VMT), mode split, route choice, and trip frequencies. These elements of travel behavior can be influenced by factors such as the built environment, urban form, and travelers' socioeconomic characteristics. Numerous studies have examined travel behavior and its relationship with the built environment, urban form, and urban spatial structure (*1–6*). Researchers who analyzed the built environment have summarized its relationship to travel behavior through a series of "D" factors: density, diversity, design, destinations, and distance to transit (*3, 7*). A review of these studies indicates that built environmental factors such as land use diversity, density, and urban design influence travel behavior outcomes such as travel mode choice, vehicle trips (VT), and VMT (*8*). In addition to the built environment, other studies have focused on travelers' socioeconomic characteristics and concluded that gender, race, income, and age affect travel behavior (*9–14*).

Another key but mostly neglected factor that may affect travel behavior—especially vehicle miles traveled—is traffic congestion. Traffic congestion provides negative feedback on transportation system by increasing travel time and travel cost. Due to the negative impacts of traffic congestion, it has been argued that commuters change their travel behaviors to maximize benefits and save time by avoiding traffic congestion (*15*). Although measuring the effects of traffic congestion on travel behavior is a critical factor for travel demand modeling and highway improvements studies, there is a lack of strong evidence with regard to the influence of traffic congestion on travel behavior. According to Ewing et al, although congestion can affect travel behavior, this variable has been largely absent from previous studies of travel behavior (*5*).

The purpose of the present study is to address this gap in the literature and investigate how traffic congestion could affect travel behavior outcomes with reference to VMT per driver. This study aims to contribute to the literature by taking a closer look at data from 400,000

[1]College of Architecture, Planning and Public Affairs, University of Texas at Arlington, Arlington, TX

Corresponding Author:
Address correspondence to Reza Sardari: reza.sardari@outlook.com

individual trips within 16 states in the United States. In doing so, this study makes two main contributions to the travel behavior literature: first, it estimates traffic congestion around households' home locations, which has not been studied previously; second, it investigates the effects of congestion on VMT per driver using a household-level analysis.

The only widely used measure of congestion is the Urban Mobility Report developed by the Texas A&M Transportation Institute (TTI). However, this measure is applied at the urbanized area level; a valid and reliable measure of congestion at a smaller scale, such as neighborhoods, has not been developed.

Using National Household Travel Survey (NHTS) disaggregated data and estimating traffic congestion around households' home and work locations, the present study measures traffic congestion and other control variables—including built environment variables and the socioeconomic characteristics of travelers—and analyzes these variables with structural equation modeling (SEM) techniques to determine whether traffic congestion positively or negatively affects VMT.

Literature Review

Travel Behavior and the Built Environment

"Travel behavior" encompasses several factors, including mode choice, average trip length, number of trips per day, and ridership (*15*). Due to the broad context of travel behavior, previous studies have each analyzed only a subset of travel behavior factors. For example, Ewing and Cervero reviewed the following travel outcomes: trip frequencies (rates of trip making), trip lengths (either in distance or time), mode choices or modal splits, and cumulative PMT, VMT, and VHT (*7*). Additionally, travel behavior elements can be investigated separately for work trips and non-work trips and can be analyzed at different levels such as person- or household-level (*16*).

Travel behavior outcomes have a different relationship with socioeconomic and built environment factors. According to Handy, "characteristics of the built environment are significant predictors of VMT, which is the outcome of the combination of trip lengths, trip frequencies, and mode split" (*17*). However, Ewing and Cervero discussed socioeconomic characteristics and the built environment separately in their comprehensive synthesis (*7*); their results can be summarized as follows:

- The socioeconomic characteristics of travelers are the primary factor that affects trip frequencies.
- Built environment characteristics of the neighborhood are the primary factors that influence trip length and VMT.
- Both socioeconomic characteristics of travelers and the characteristics of the built environment affect mode choice (though they argue that socioeconomic characteristics are the more important factor).

Previous research has indicated that the built environment and urban form can influence travel behavior (*3*, *7*, *16–26*). A review of these studies revealed that certain built environment factors such as land use diversity, density, and urban design affect travel mode choice, VT, and VMT (*8*). The relationship between the built environment and travel behavior has been summarized as a series of "D" factors, including density, diversity, design, destination accessibility, and distance to transit (*3*, *7*). Meanwhile, VMT has been shown to depend on trip frequencies, trip lengths, and individuals' preferences about how to get around (*7*). A study by Diao and Ferreira Jr found that built environment factors such as accessibility to work and non-work destinations, transit accessibility, and connectivity have a stronger relationship with VMT than demographic factors (*18*).

The missing variable in previous studies is traffic congestion and its impact on travelers' commuting patterns. Congestion is considered a major urban transportation problem as it has grown not only in the largest cities, but in cities of every size. Congestion and delays are often defined as "clog," "excessive fullness," and the "impedance" for the movement of people and goods, which can be measured in terms of travel distance, travel time, and speed. Traffic congestion causes negative outcomes such as increased travel times, travel costs, and air pollution, all of which can affect commuters' travel behavior.

Although traffic congestion is an important factor affecting travel behavior, only a few studies have quantified the impact of congestion on travel behavior (*27*, *28*). To examine the impact of traffic congestion on household behavior, Davis focused on commuting time as the outcome of joint residential and employment decisions, concluding that congestion plays an important role in affecting travel behavior by influencing households' decisions in selecting a residential location (*27*).

Sweet and Chen investigated the effects of traffic congestion on taking public transit (*28*). The results of this study indicate that higher traffic congestion and unreliable auto travel conditions induce mode-switching to public transit when reliable transit service is available (*28*).

According to Kuzmyak, residents in higher-density neighborhoods with traffic congestion on adjacent streets make substantially shorter trips (*29*). This research—conducted in Phoenix, AZ—indicates that work trips are about 7 mi long in higher-density neighborhoods, compared with 11 mi in suburban neighborhoods. Kuzmyak also found that shopping trips averaged <3 mi in length compared with over 4 mi in suburban areas.

A recent study investigated the relationship between compact development, VMT, and traffic congestion using SEM techniques (*30*). Although this research explored the relationship between VMT and total delay (as a measure of traffic congestion), it does not present individuals' travel behaviors because it relied on TTI's Urban Mobility Report, with aggregated data from urbanized areas in the United States.

To determine the effects of congestion on multiple components of travel behavior, the present study investigates how traffic congestion affects individuals' PMT, VMT, and trip frequencies.

Measuring Traffic Congestion

According to Lomax et al, the scale of traffic congestion can be classified by different geographic levels: intersection or interchange, facility segment, route or corridor, sub-region or subarea, region, and state or nation (*31*). Based on geographic level, there are several approaches to developing congestion measures, as outlined below.

$$Road\ Congestion\ Index = \frac{Prin.Arterial\ VMT(perln.mi) \times Prin.Arterial\ VMT + Freeways\ VMT(perln.mi) \times Freeway\ VMT}{5,000 \times Prin.Arterial\ VMT + 14,000 \times Highways\ VMT} \quad (4)$$

Congested Travel is a measure that captures the extent of congestion by quantifying the extent to which the system is influenced by congestion. Equation 1 illustrates the computation of Congested Travel by vehicle miles as the product of the congested segment length and the vehicle volume summed across all congested segments (*31*).

$$Congested\ Travel(vehicle\ mile) = \sum\{Congested\ Segment\ Length(miles) \times Vehicle\ Volume(vehicles)\} \quad (1)$$

Percent of Congested Travel is an extension of the Congested Travel measure that also measures the extent of congestion (*31*). This measure can be computed when speed and occupancy data are available for each roadway segment. It is computed as the ratio of the congested segment person-hours of travel to the total person-hours of travel, as shown in Equation 2.

$$Percent\ of\ Congested\ Travel = \left[\frac{\sum_{i=1}^{m}\left(\left(\begin{matrix}Actual\ Travel\\ Time_i\ (minutes)\end{matrix} - \begin{matrix}FFS\ or\ PSL\ Travel\\ Time_i(minutes)\end{matrix}\right)\left(\begin{matrix}Vehicle\ Volume_i\\ (vehicles)\end{matrix} \times \begin{matrix}Vehicle\ Occupancy_i\\ (persons/vehicle)\end{matrix}\right)\right)_{Each\ congested\ segment}}{\sum_{i=1}^{n}\left(\begin{matrix}Actual\ Travel\ Rate_i\\ (minutes\ per\ mile)\end{matrix} \times \begin{matrix}Length_i\\ (miles)\end{matrix} \times \begin{matrix}Vehicle\ Volume_i\\ (vehicles)\end{matrix} \times \begin{matrix}Vehicle\ Occupancy_i\\ (persons/vehicle)\end{matrix}\right)_{All\ segments}}\right] \times 100 \quad (2)$$

Congested Roadway is another measure of traffic congestion (*31*). As shown in Equation 3, it is the sum of the mileage of roadways that operate under free-flow speed (FFS) or posted speed limit (PSL) conditions.

$$Congested\ Roadway(miles) = \sum Congested\ Segment\ Length(miles) \quad (3)$$

Additionally, a volume/capacity (V/C) ratio can be used to calculate the total mileage of congested roads. This ratio is often used as a measure of the sufficiency of existing or proposed capacity.

Congestion can be quantified as a continuous variable using the road congestion index (RCI) methodology developed by the TTI. The final index is a continuous variable that represents a higher level of traffic congestion if the index value is greater than or equal to 1.0. This continuous congestion index has the advantage of being comparable to discrete factors because it can be easily measured, and it has a straightforward interpretation.

In addition to geographic level measures, "individual" measures are those measures that consider the individual traveler by exploring delay per traveler, travel time, and travel time index (*31*).

In this research, we focused on geographic level, or "area mobility," measures. Area mobility measures can be computed using traffic count data, average daily traffic data, and road capacity. Measures such as the RCI (Equation 4) can be computed with national public data such as Highway Performance Monitoring System (HPMS) or national highway system data. However, measures such as Percent of Congested Travel (Equation 2) require detailed data, including actual travel time and FFS or PSL conditions. Although measures with actual travel times are more accurate than average daily traffic, intensive detailed data, GPS data, or probe vehicle techniques are required to develop time-related mobility measures.

Methodology

The relationship between the built environment, congestion, and travel behavior is complex. To examine these interrelationships, the following steps were completed:

first, built environment data, traffic congestion data, and individuals' travel survey data were gathered and cleaned from various sources such as Longitudinal Employer-Household Dynamics (LEHD) program, NHTS, and HPMS. The main component of this step was developing a mesoscopic-level traffic congestion measure that examines the level of congestion within neighborhoods. To understand transportation systems and quantify mesoscopic-level traffic congestion, different approaches were evaluated to examine congestion metrics with the available traffic data sets such as Freight Analysis Framework (FAF) network version 4, the HPMS data sets, and National Performance Management Research Data Set (NPMRDS).

The second step of this research involved developing a structural equation model. SEM is a statistical methodology for testing complex hypotheses and evaluating causal relationships involving multiple, interacting variables (*32*, *33*). Due to the causal relationship between the built environment and traffic congestion, SEM was used to examine the relationships between these factors simultaneously.

Data

In this study, data sets were gathered from several secondary sources, including the NHTS, the U.S. Census, the LEHD program, the HPMS, the Transit-Oriented Development (TOD) database, and the National Historical Geographic Information System. To overcome the issue of aggregation bias, the present study incorporated disaggregated travel data from 16 states.

Household Travel Survey Data. The NHTS is the main source of transportation studies conducted under the sponsorship of the FHWA. These data were gathered from individual and household surveys, which include vehicle and daily trip information as well as socioeconomic information such as age, race, income, and so forth. Aside from these surveys conducted across the entire country, 16 states partnered with the FHWA to provide more detailed add-on data for their respective jurisdictions. The add-on data for these 16 states were requested and gathered from the FHWA and state departments of transportation. After gathering the data, they were cleaned, merged, geocoded in ArcGIS, and reorganized in Microsoft Access and Excel.

To better understand mobility patterns, all home, work, and trip-end locations were geocoded in ArcGIS for further analysis. Based on the purpose of the present research, five criteria were implemented to determine the study's sample and remove outliers from the NHTS data set:

- *Weekday trips*: Trips during weekends do not represent typical commute travel behavior. Therefore, in this study, weekday trips were selected for the analysis.
- *Reported VMT and trip miles*: Households that did not report their trip miles were excluded from the sample.
- *Maximum limit for VMT*: According to the FHWA, households with VMTs of more than 310 mi per day should be considered outliers (*34*). Therefore, only households with VMTs of 310 mi or less were included in the sample.
- *Maximum limit for daily trips*: According to the FHWA, households with more than 20 trips per day should be considered outliers (*34*). Therefore, only households with 20 daily trips or less were included in the sample.
- *Focusing on urbanized areas*: As traffic congestion is not a common phenomenon in rural areas, households were selected from urban areas to ensure a more homogenous sample.

In sum, 93,769 persons and their corresponding data from 410,029 trips were selected for statistical modeling and further analysis. Figure 1 shows the spatial distribution of households from the 16 states included in the sample.

Congestion Data. In this study, the RCI formula (Equation 4) was selected to calculate the level of congestion around each household's home location. The main components of this index are VMT and VMT per lane mile of highways and principal arterials. To measure these components, traffic volume and network inventory data from HPMS geospatial files were extracted. The total VMT of freeways and principal arterials—as well as lane miles of roadways—was extracted within 2 mi of individuals' home locations.

The 2-mi radius network buffer was selected for two main reasons. First, traffic congestion usually occurs at major roads such as highways and principal arterials, so traffic congestion within a walkable distance from home locations cannot be a good indicator of traffic congestion. The second reason is related to data availability at the national level (covering 16 states). As the HPMS national shapefile includes highways, principal arterials, and major collectors, the 2-mi buffer was found to be the minimum area covering the HPMS network system.

After developing a congestion index around each household's home location, the index was validated by inspecting Google Earth satellite images to identify the most and least congested neighborhoods in the sample. This process was completed for all urban areas. Based on the calculations, the most congested neighborhoods were identified as downtown Manhattan in New York, San Francisco and Daly City in California, downtown Phoenix in Arizona, and southern Indianapolis in

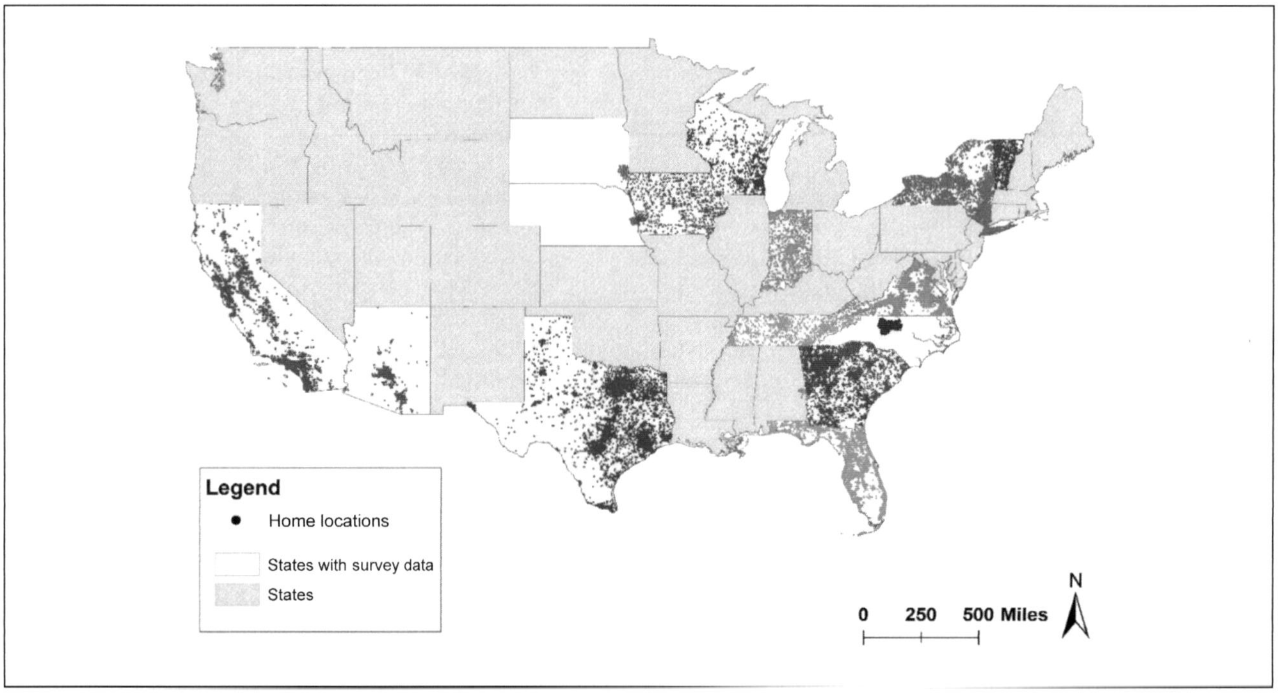

Figure 1. The spatial distribution of households in the sample.
Note: Colored dots in the figure represent households; different colors were used for each state to improve readability.

Table 1. Descriptive Statistics of Congestion Index

Variable	Number of buffers	Minimum	Maximum	Mean	SD
Congestion index	93,769	0.01	4.58	0.91	0.64

Note: SD = standard deviation.

Arizona. Table 1 presents the descriptive statistics of the traffic congestion index.

Built Environment Data. Using ArcGIS software, built environment data sets were summarized within the 2-mi network buffer around each household's home location. These data include job–population balance entropy, activity density, distance to central business district (CBD), major road density, and distance to TOD. The built environment data sets were extracted from various sources, including the block-level 2010 Census, the LEHD 2010, the HPMS 2010, and the Environmental Systems Research Institute (ESRI) Business database. It is important to note that previous studies considered a smaller radius (0.25–0.5 mi) for computing the built environmental variables. However, as mentioned above, we selected a larger radius to measure traffic congestion associated with the extended areas outside neighborhoods.

Variables

The variables in the structural equation (SE) model are defined in Table 2. These variables were classified into three general categories:

- *Outcome variable*: VMT per driver.
- *Exogenous explanatory variables*: The exogenous variables—household income, gender, race, and life cycle—were determined by examining households' characteristics and individuals' preferences. Number of highways, intersection density, road density, job–population balance, and distance to CBD were also treated as exogenous because they are the result of regional land use and transportation systems.
- *Endogenous variables*: Traffic congestion is an endogenous variable that affects daily VMT both directly and indirectly. The number of vehicles for each household is another endogenous variable that can be influenced by other socioeconomic

Table 2. Variables Included in the Analysis

Variable	Definition	Source
Outcome variable		
VMT	Daily vehicle miles traveled (VMT per driver)	(*37*)
Exogenous variable		
Age	Age of commuter	(*37*)
Workers	Number of household workers	(*37*)
Children	Household children <16 years old	(*37*)
Male	Gender: male (dummy variable)	(*37*)
hh_income	Household income	(*37*)
White	Race: white (dummy variable)	(*37*)
hh_size	Household size	(*37*)
hwy_dr	Number of times using highways for daily trips	(*37*)
cbd_dist	Distance to CBD	(*38*)
act_dens	Population and job densities (per acre)	(*39, 44*)
Job_pop	Job–population balance	(*39, 40*)
road_dens	Major road density	(*41*)
int_dens	Intersection density	(*42*)
tod_dist	Distance to TOD	(*43*)
Endogenous variables		
Congestion	Congestion index	Calculated based on TTI (*31*)
Vehicles	Number of household vehicles	(*37*)

factors such as income, age, and life cycle. Table 2 displays the study's exogenous and endogenous variables and their original sources.

Analytical Methods

Analyzing traffic congestion and its interrelationship with land use requires complex models with endogenous and endogenous variables. The present research addressed this by using SEM, instead of single-equation models, to represent complex relationships. The calculation of SE models involves solving a set of equations—one for each "response" or "endogenous" variable in the model. Variables that are solely predictors of other variables are termed "influences" or "exogenous" variables. Typically, solution procedures for SE models focus on the observed versus model-implied correlations in the data. The unstandardized correlations, or covariances, are the raw material for the analyses. Models are automatically compared with a "saturated" model (i.e., one that allows all variables to intercorrelate), and this comparison allows the analysis to discover missing pathways and, thereby, reject inconsistent models.

In SEM, correlations are characterized by curved bi-directional arrows and causal pathways are denoted by unidirectional straight arrows. Also, each endogenous (response) variable contains an error term that is visually presented in the model (*5*). The causal paths presented in the final SEM path diagram are statistically significant with non-zero values. To estimate relationships in SEM, the Amos software package and maximum likelihood procedures were used to develop model specifications (*35*).

Figure 2 presents the path diagram generated in Amos for the best-fitted model. Correlations are represented by curved bi-directional arrows. To simplify the complex causal diagram, some correlations are omitted from the figure. The full model is described in Table 3. By convention, circles represent error terms in the model, of which there is one for each endogenous (response) variable.

Results

There are different indices for evaluating an SE model's goodness of fit. According to Hox and Bechger, we can reject hypothesized models with a high chi-square value (*36*). Therefore, it is necessary that the model's chi-square value is small and translates to a large, non-significant p-value (>.05). In our model, congestion and VMT exhibited a chi-square of 18.06 ($p = 1.55$, $df = 13$). Both the low ratio of chi-square to degrees of freedom and the high p-value are indicators of the model's goodness of fit.

Traffic Congestion Index

The results of the model indicate a positive relationship between built environment factors and traffic congestion. The strongest predictors of traffic congestion are intersection density and activity density. As presented in Table 3, a 1% increase in intersection density is associated with a 0.4% increase in the traffic congestion index.

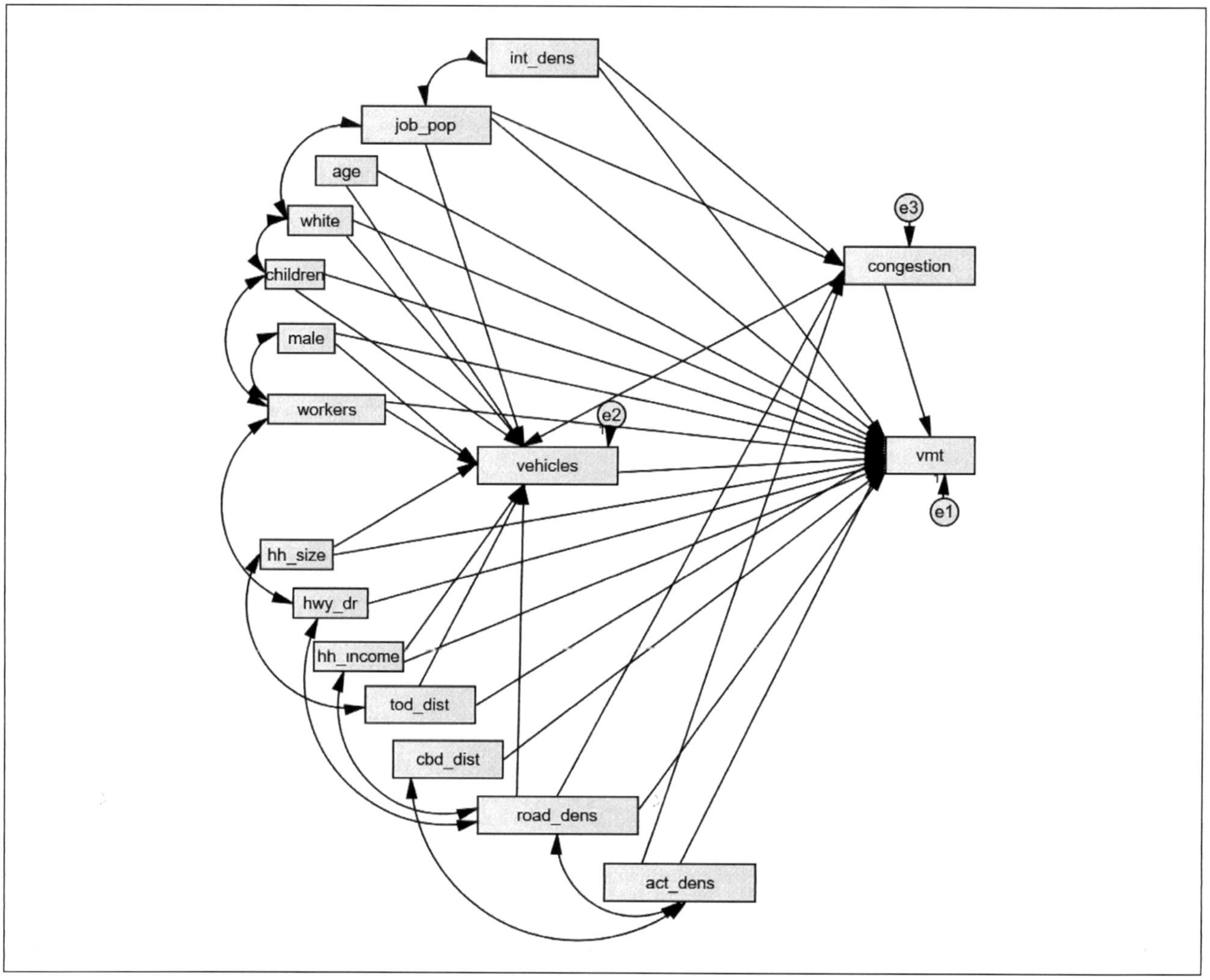

Figure 2. Causal path diagram explaining VMT per driver. e1, e2, and e3 denote error terms in the model.

Areas with a higher activity density (job and employment) and a higher job–population balance exhibit a higher level of traffic congestion. According to Table 3, a 1% increase in activity density is associated with a 0.3% increase in the traffic congestion index. The reason for this is because trip origins and destinations are more concentrated in denser areas.

In contrast to activity density, road density exhibited a negative relationship with traffic congestion. Areas with higher road density have less traffic congestion; a 1% increase in road density is associated with a 0.1% decline in the traffic congestion index.

Vehicle Counts

The results in Table 3 indicate that both the socioeconomic characteristics of drivers and built environment factors affect vehicle ownership. As shown in Table 3, the strongest predictors are household size, household with children, and household income. Considering socioeconomic characteristics of drivers, these results align with expectations. Drivers with higher household income and age have a higher number of vehicles; a 1% increase in household income is associated with a 0.3% increase in number of vehicles in each household. Also, household size and number of workers in the household are determinants of the number of vehicles in each household. Furthermore, the results indicated that white drivers and male drivers own more vehicles compared with other groups. The number of vehicles owned by households with children <16 years is lower than households without children, which can be related to other factors such as age limitations for obtaining a driver's license.

Table 3. Path Coefficient Estimates and Statistics for Direct Effects in the Model

Variables	Estimate	SE	CR	*p*-value
congestion<—int_dens	0.434	0.011	38.44	<.001
congestion<—act_dens	0.339	0.02	17.134	<.001
congestion<—job_pop	0.083	0.016	5.068	<.001
congestion<—road_dens	−0.095	0.004	−22.076	<.001
vehicles<—workers	0.311	0.004	82.575	<.001
vehicles<—age	0.156	0.006	26.894	<.001
vehicles<—white	0.117	0.008	14.105	<.001
vehicles<—children	−0.369	0.01	−36.222	<.001
vehicles<—male	0.039	0.006	6.478	<.001
vehicles<—hh_size	0.904	0.01	90.22	<.001
vehicles<—hh_income	0.338	0.005	72.778	<.001
vehicles<—congestion	−0.120	0.001	−9.925	<.001
vehicles<—job_pop	−0.025	0.006	−4.295	<.001
vehicles<—road_dens	0.120	0.001	8.363	<.001
vehicles<—tod_dist	0.039	0.002	22.856	<.001
vmt<—workers	0.226	0.005	44.918	<.001
vmt<—age	1.032	0.008	136.866	<.001
vmt<—white	−0.047	0.011	−4.343	<.001
vmt<—children	0.445	0.013	33.785	<.001
vmt<—male	0.188	0.008	24.295	<.001
vmt<—hh_size	−0.449	0.013	−33.459	<.001
vmt<—hwy_dr	0.593	0.004	148.464	<.001
vmt<—hh_income	0.197	0.006	32.163	<.001
vmt<—vehicles	0.132	0.004	31.34	<.001
vmt<—congestion	−0.107	0.002	−4.429	<.001
vmt<—int_dens	−0.230	0.005	−4.601	<.001
vmt<—act_dens	−0.12	0.006	−20.268	<.001
vmt<—job_pop	−0.620	0.008	−8.207	<.001
vmt<—cbd_dist	0.260	0.003	8.748	<.001
vmt<—road_dens	−0.130	0.002	−6.739	<.001
vmt<—tod_dist	0.140	0.003	5.185	<.001

Note: SE = standard error; CR = critical ratio.

Interestingly, households living in congested areas tend to own fewer vehicles than households living within uncongested areas. Since areas with a higher job–population balance are more congested, people living in these areas have fewer cars. In contrast to density and job–population balance, residents in areas with higher road densities have higher numbers of household vehicles in their households. Also, drivers who live far from the CBD have a higher number of household vehicles.

VMT

As expected, both socioeconomic and built environment factors influence daily VMT. The strongest predictors of VMT—in order of strength—are age, the number of times using a highway for daily trips, household size, and households with children. As shown in Table 3, drivers with higher household incomes, higher numbers of workers, and higher numbers of vehicles have higher VMTs. Age and race are other considerable factors; older drivers have higher VMTs, and white drivers have lower VMTs. There is a positive relationship between income and the proportion of a population that is white, meaning that white groups have more options with regard to home location.

Life cycle is another significant factor that affects VMT. Drivers with larger household sizes and drivers with children < 16 years old have higher VMTs. Also, male drivers have higher daily VMTs compared with females.

The effects of the built environment on travel behavior are the key findings of this research. The results present that congestion suppresses VMT. Drivers who live in areas with higher traffic congestion have lower daily VMTs. Likewise, both activity density and job–population balance result in lower VMTs. In addition, closer distance to rail stations and CBDs results in lower daily VMTs. The findings also indicate that drivers who live in areas with higher intersection densities and major road densities have lower VMTs.

The direct and indirect effects of variables on one another and the total effects of different variables on VMT per driver are summarized in Table 4. Age is the key demographic variable that affects VMT per driver, largely through its direct effect. The dominant effect of age is positive, as older drivers have higher daily VMTs.

Table 4. Direct, Indirect, and Total Effects of Variables on VMT per Driver in the Model

Variables	Direct effect	Indirect effect	Total effect
Age	1.032	0.021	1.053
hwy_dr	0.593	0	0.593
Children	0.445	−0.049	0.396
Workers	0.226	0.041	0.267
hh_income	0.197	0.045	0.242
Male	0.188	0.005	0.193
Vehicles	0.132	0	0.132
cbd_dist	0.260	0	0.260
tod_dist	0.140	0.005	0.145
Congestion	−0.170	−0.020	−0.190
rd_dist	−0.130	0.002	−0.132
int_dens	−0.230	−0.004	−0.234
White	−0.047	0.016	−0.031
job_pop	−0.062	−0.004	−0.066
act_dens	−0.12	−0.003	−0.123
hh_size	−0.449	0.12	−0.329

Using highways for daily commuting is another key factor that influences VMT per driver. This is a considerable finding that challenges policies that support highway development as a congestion reduction strategy.

Both direct and indirect effects of added number of workers and number of vehicles in the household resulted in increased daily VMT per driver. Increasing intersection density, job–population balance, and activity density mitigate VMT per driver. The final SE model presented the interrelationship between density, congestion, and VMT. This interrelationship can be summarized as follows:

- A 1% increase in activity density is associated with a 0.1% decline in VMT per driver.
- A 1% increase in activity density is associated with a 0.3% increase in traffic congestion.

The effects of congestion on VMT per driver present a noticeable result. Both direct and indirect effects of congestion have negative impacts on VMT per driver. In other words, drivers within congested areas had lower daily VMTs because of the direct effects of traffic congestion. Furthermore, traffic congestion mitigated the number of vehicles owned by households. As previously mentioned, drivers in congested areas reported a lower number of household vehicles. In other words, traffic congestion directly reduced vehicle ownership and indirectly suppressed VMT per driver.

Discussion and Conclusion

Scholars have been debating the effects of traffic congestion on greenhouse gas emissions, environmental justice, toxic air exposure, public health, and physical activities, among others, yet few studies have considered its effect on travel behavior. The goal of this study was to investigate the impacts of congestion, the built environment, and socioeconomic characteristics on VMT, both directly and indirectly. The present research provides a key contribution to the literature by exploring individuals' mobility patterns within urbanized areas using disaggregated household travel survey data.

Another advancement by this study is the development of a congestion measure that is estimated at the local level. Local congestion measures can help planners and decision makers to find the interrelationships between congestion and factors such as travel behavior and public health variables (e.g., obesity, physical activity). In this study, a traffic congestion index was calculated for each household's home location within a 2-mi buffer. With this computation, we implemented a methodology that can be used in other fields/analyses such as public health and travel demand modeling.

The findings confirm that congestion and built environment factors (density and job–population balance) mitigate VMT, both directly and indirectly. In other words, due to the proximity of origins and destinations in denser areas with higher job–population balances, residents in compact development regions have lower daily VMTs. This is a key point for promoting policies that support compact development. It has been argued that developing highways and new roads is not the best solution to overcome traffic congestion. Furthermore, developing highways encourages auto dependency and spreading activities throughout low-density suburbs. Therefore, instead of expanding highway networks, investment in public transit may better address traffic congestion—not only because it provides residents with a more equitable and sustainable means of transportation, but also because in the long run it encourages people to reside in more compact and location-efficient urbanized areas. Ultimately, the results of this study can help decision makers, transit agencies, and future travel demand models by providing modeling estimates for transportation studies.

Author Contributions

The authors confirm contribution to the paper as follows: study conception and design: Reza Sardari; data collection: Reza Sardari; analysis and interpretation of results: Reza Sardari, Shima Hamidi, Raha Pouladi; draft manuscript preparation: Reza Sardari. All authors reviewed the results and approved the final version of the manuscript.

References

1. Bento, A. M., M. L. Cropper, A. M. Mobarak, and K. Vinha. The Effects of Urban Spatial Structure on Travel

Demand in the United States. *Review of Economics and Statistics*, Vol. 87, No. 3, 2005, pp. 466–478. https://doi.org/10.1162/0034653054638292
2. Bhat, C. R., and J. Y. Guo. A Comprehensive Analysis of Built Environment Characteristics on Household Residential Choice and Auto Ownership Levels. *Transportation Research Part B: Methodological*, Vol. 41, No. 5, 2007, pp. 506–526. https://doi.org/10.1016/j.trb.2005.12.005
3. Cervero, R., and K. Kockelman. Travel Demand and the 3Ds: Density, Diversity, and Design. *Transportation Research Part D: Transport and Environment*, Vol. 2, No. 3, 1997, pp. 199–219. https://doi.org/10.1016/S1361-9209(97)00009-6
4. Cervero, R., and J. Murakami. Effects of Built Environments on Vehicle Miles Traveled: Evidence from 370 US Urbanized Areas. *Environment and Planning A*, Vol. 42, No. 2, 2010, pp. 400–418. https://doi.org/10.1068/a4236
5. Ewing, R., S. Hamidi, F. Gallivan, A. Nelson, and J. Grace. Combined Effects of Compact Development, Transportation Investments, and Road User Pricing on Vehicle Miles Traveled in Urbanized Areas. *Transportation Research Record: Journal of the Transportation Research Board*, 2013. 2397: 117–124.
6. Sarzynski, A., H. L. Wolman, G. Galster, and R. Hanson. Testing the Conventional Wisdom About Land Use and Traffic Congestion: The More We Sprawl, the Less We Move? *Urban Studies*, Vol. 43, No. 3, 2006, pp. 601–626. https://doi.org/10.1080/00420980500452441
7. Ewing, R., and R. Cervero. Travel and the Built Environment: A Synthesis. *Transportation Research Record: Journal of the Transportation Research Board*, 2001. 1780: 87–114.
8. Handy, S., R. Lee, R. Maiss, J. Walters, and T. Parker. *Improved Data and Tools for Integrated Land Use-Transportation Planning in California: Appendix A—Annotated Literature Review on Land Use-Transportation Relationships*, 2012. http://ultrans.its.ucdavis.edu. Accessed Nov. 15, 2015.
9. Blumen, O. Gender Differences in the Journey to Work. *Urban Geography*, Vol. 15, No. 3, 1994, pp. 223–245. https://doi.org/10.2747/0272-3638.15.3.223
10. Hecht, A. The Journey-to-Work Distance in Relation to the Socio-Economic Characteristics of Workers. *Canadian Geographer*, Vol. 18, No. 4, 1974, pp. 367–378. https://doi.org/10.1111/j.1541-0064.1974.tb00210.x
11. Manaugh, K., L. F. Miranda-Moreno, and A. M. El-Geneidy. The Effect of Neighbourhood Characteristics, Accessibility, Home–Work Location, and Demographics on Commuting Distances. *Transportation*, Vol. 37, No. 4, 2010, pp. 627–646. https://doi.org/10.1007/s11116-010- 9275-z
12. Mauch, M., and B. Taylor. Gender, Race, and Travel Behavior: Analysis of Household-Serving Travel and Commuting in San Francisco Bay Area. *Transportation Research Record: Journal of the Transportation Research Board*, 1997. 1607: 147–153.
13. Trowbridge, M. J., and N. C. McDonald. Urban Sprawl and Miles Driven Daily by Teenagers in the United States. *American Journal of Preventive Medicine*, Vol. 34, No. 3, 2008, pp. 202–206. https://doi.org/10.1016/j.amepre.2007.11.013
14. Zax, J. S. Race and Commutes. *Journal of Urban Economics*, Vol. 28, No. 3, 1990, pp. 336–348. https://doi.org/10.1016/0094-1190(90)90032-I
15. Ben-Akiva, M. E., and S. R. Lerman. *Discrete Choice Analysis: Theory and Application to Travel Demand*, Vol. 9. MIT Press, Cambridge, MA., 1985.
16. Boarnet, M. G., and R. Crane. *Travel by Design: The Influence of Urban Form on Travel*. Oxford University Press on Demand, New York, 2000.
17. Handy, S. Understanding the Link Between Urban Form and Nonwork Travel Behavior. *Journal of Planning Education and Research*, Vol. 15, No. 3, 1996, pp. 183–198. https://doi.org/10.1177/0739456X9601500303
18. Mi, D., and J. Ferreira, Jr. Vehicle Miles Traveled and the Built Environment: Evidence from Vehicle Safety Inspection Data. *Environment and Planning A*, 2014. 1831: 2991–3009.
19. Anderson, W., P. Kanaroglou, and E. Miller. Urban Form, Energy and the Environment: A Review of Issues, Evidence and Policy. *Urban Studies*, Vol. 33, No. 1, 1996, pp. 7–35. https://doi.org/10.1080/00420989650012095
20. Bagley, M. N., and P. L. Mokhtarian. The Impact of Residential Neighborhood Type on Travel Behavior: A Structural Equations Modeling Approach. *The Annals of Regional Science*, Vol. 36, No. 2, 2002, pp. 279–297. https://doi.org/10.1007/s001680200083
21. Cervero, R. Built Environments and Mode Choice: Toward a Normative Framework. *Transportation Research Part D: Transport and Environment*, Vol. 7, No. 4, 2002, pp. 265–284. https://doi.org/10.1016/S1361-9209(01)00024-4
22. Chatman, D. How Density and Mixed Uses at the Workplace Affect Personal Commercial Travel and Commute Mode Choice. *Transportation Research Record: Journal of the Transportation Research Board*, 2003. 1831: 193–201.
23. Giuliano, G., and K. A. Small. Is the Journey to Work Explained by Urban Structure? *Urban Studies*, Vol. 30, No. 9, 1993, pp. 1485–1500. https://doi.org/10.1080/00420989320081461
24. Golob, T. F., and D. Brownstone. *The Impact of Residential Density on Vehicle Usage and Energy Consumption*. Institute of Transportation Studies, 2005. http://www.its.uci.edu/its/publications/papers/ITS/UCI-ITS-WP-05-1.pdf. Accessed Apr. 2, 2015.
25. Holtzclaw, J., R. Clear, H. Dittmar, D. Goldstein, and P. Haas. Location Efficiency: Neighborhood and Socio-Economic Characteristics Determine Auto Ownership and Use: Studies in Chicago, Los Angeles and San Francisco. *Transportation Planning and Technology*, Vol. 25, No. 1, 2002, pp. 1–27. https://doi.org/10.1080/03081060290032033
26. Mitchell, R. B., and C. Rapkin. Urban Traffic: A Function of Land Use. 1954. https://trid.trb.org/view.aspx?id = 131510. Accessed Apr. 25. 2017.
27. Davis, A. F. *The Impact of Traffic Congestion on Household Behavior: Three Essays on the Role of Heterogeneity*. Ph.D. dissertation. North Carolina State University, Raleigh, 2004.

28. Sweet, M. N., and M. Chen. Does Regional Travel Time Unreliability Influence Mode Choice? *Transportation*, Vol. 38, No. 4, 2011, pp. 625–642.
29. Kuzmyak, J. Richard. *Land Use and Traffic Congestion*. Arizona Department of Transportation Research Center, Phoenix, AZ., 2012.
30. Ewing, R., G. Tian, T. Lyons, and K. Terzano. Does Compact Development Increase or Reduce Traffic Congestion?, 2017. https://doi.org/10.3141/2397-14.
31. Lomax, T., D. Schrank, W. Eisele., and Eisele *The Keys to Estimating Mobility in Urban Areas*. The Texas A&M University, College Station, TX., 2005.
32. e Silva, J. D. A., C. Morency, and K. G. Goulias. Using Structural Equations Modeling to Unravel the Influence of Land Use Patterns on Travel Behavior of Workers in Montreal. *Transportation Research Part A: Policy and Practice*, Vol. 46, No. 8, 2012, pp. 1252–1264.
33. Grace, J. B. *Structural Equation Modeling and Natural Systems*. Cambridge University Press, Cambridge, 2006.
34. U.S. Department of Transportation, Federal Highway Administration. *Methodology: Local Area Transportation Characteristics for Households*. Bureau of Transportation Statistics, 2009.
35. Lei, P. W., and Q. Wu. Introduction to Structural Equation Modeling: Issues and Practical Considerations. *Educational Measurement: Issues and Practice*, Vol. 26, No. 3, 2007, pp. 33–43.
36. Hox, J. J., and T. M. Bechger. An Introduction to Structural Equation Modeling. *Family Science Review*, Vol. 11, No. 4, 2007, pp. 354–373.
37. U.S. Department of Transportation. Federal Highway Administration. National Household Travel Survey. 2009. www.nhts.ornl.gov. Accessed December 12, 2016.
38. Hamidi, S., and R. Ewing. A longitudinal study of changes in urban sprawl between 2000 and 2010 in the United States. Landscape and Urban Planning, Vol. 128, 2014, pp. 72-82. https://doi.org/10.1016/j.landurbplan.2014.04.021.
39. United States Census Bureau. American FactFinder, 2010. www.factfinder2.census.gov. Accessed November 2016.
40. United States Census Bureau. Longitudinal Employer-Household Dynamics (LEHD), 2010, www.lehd.ces.census.gov. Accessed December 2016.
41. U.S Department of Transportation. Office of Highway Policy Information,. Highway Performance Monitoring System (HPMS) Public Release of Geospatial Data in Shapefile Format, 2010. https://www.fhwa.dot.gov/policyinformation/hpms/shapefiles.cfm Accessed December 2016.
42. Environmental Systems Research Institute (ESRI). ArcGIS Business Analyst Database. 2012, www.esri.com. Accessed January 2017.
43. Center for Neighborhood Technology. The National Transit Oriented Developmet Database, 2010. www.toddata.cnt.org. Accessed December 2016.
44. Manson, S., J. Schroeder, D. Van Riper, and S. Ruggles. IPUMS National Historical Geographic Information System: Version 12.0 [Database]. Minneapolis: University of Minnesota, 2017. http://doi.org/10.18128/D050.V12.0. Accessed January 2017.

The Standing Committee on Traveler Behavior and Values (ADB10) peer-reviewed this paper (18-02240).

Article

TRR
JOURNAL OF THE TRANSPORTATION RESEARCH BOARD

Transportation Research Record
2018, Vol. 2672(47) 103–113

DOI: 10.1177/0361198118777064
journals.sagepub.com/home/trr
SAGE

Estimation of a Long-Distance Travel Demand Model using Trip Surveys, Location-Based Big Data, and Trip Planning Services

Carlos Llorca[1], Joseph Molloy[2], Joanna Ji[3], and Rolf Moeckel[1]

Abstract

Long-distance trips are less frequent than short-distance urban trips, but contribute significantly to the total distance traveled, and thus to congestion and transport-related emissions. This paper develops a long-distance travel demand model for the province of Ontario, Canada. In this paper, long-distance demand includes non-recurrent overnight trips and daytrips longer than 40 km, as defined by the Travel Survey for Residents in Canada (TSRC). We developed a microscopic discrete choice model including trip generation, destination choice, and mode choice. The model was estimated using travel surveys, which did not provide data about destination attractiveness and modal level of service. Therefore, a data collection method was designed to obtain publicly available data from the location-based social network Foursquare and from the online trip planning service Rome2rio. In the first case, Foursquare data characterized land uses and predominant activities of the destination alternatives, by the number of user check-ins at different venue types (i.e., ski areas, outdoor or medical activities, etc.). In the second case, the use of Rome2rio data described the modal alternatives for each observed trip. Combining data from travel surveys, Foursquare, and Rome2rio, coefficients of the model were estimated econometrically. It was found that the Foursquare data on number of check-ins at destinations was statistically significant, especially for leisure trips, and improved the goodness of fit compared with models that only used population and employment. Additionally, Rome2rio mode-specific variables were found to be significant for mode choice selection, making the resulting model sensitive to changes in travel time, transit fares, or service frequencies.

Travel demand models predict the number, destination, and modal choice of trips. These characteristics depend on, among other factors, the frequency and length of the trips. This fact segments the demand into urban travel demand, such as short-distance daily commuting, and long-distance travel demand, such as overnight trips or tourism trips. Different distance thresholds are used to distinguish between the two demand segments, although long-distance trips generally exclude recurrent commuting trips.

Traditionally, transport modelers have paid more attention to daily short-distance urban traffic demand and its corresponding models, since these trips are much higher in number than long-distance trips. Models for long-distance trips appeared later and usually transferred parameters from urban models. Some studies have started to highlight the importance of the long-distance travel demand market (sometimes mixed with the so-called intercity travel demand) based on its contribution to the vehicle-kilometers traveled (*1*), or motivated by the interest in analyzing high-speed rail networks. The development of statewide models also required specific modules for long-distance trips, which justified the need for their improvement. Data availability is reported generally as the major concern for long-distance models (*2*), since household travel surveys are typically designed for urban travel demand.

This paper shows the development of a long-distance passenger travel demand model for the province of Ontario, Canada, as part of a province-wide transportation model. The paper analyzes the limitations of survey data regarding non-selected travel alternatives, and the acquisition of additional datasets using location-based big data and online trip planning services.

Literature Review

Compared with short-distance or daily trips, the number of research studies on long-distance travel demand is limited, probably because the number of long-distance trips per

[1]Technical University of Munich, Munich, Germany
[2]IVT, ETH Zürich, Zürich, Switzerland
[3]PTV Group, Karlsruhe, Germany

Corresponding Author:
Address correspondence to Carlos Llorca: carlos.llorca@tum.de

household is lower. However, long-distance travel demand pays a significant role in the transportation system. While around 75% of trips in U.S.A. are less than 15 km, they account for only 28.9% of vehicle-kilometers traveled (*3*). On the contrary, trips over 150 km account for 1% of trips, but represent 15.5% vehicle miles traveled. Similarly, 50% of all passenger kilometers in Europe are trips beyond 100 km (*4*).

According to Giamo and Schiffer (*5*), some statewide travel demand models do not account for long-distance travel demand. Many others only forecast future trips based on trip tables, which cannot capture the sensitivities of person, environment, or trip variables. Despite this, according to Miller (*2*), the recent statewide models have contributed to the development of long-distance travel demand sub-models.

An example in this direction is the long-distance model for Ohio (U.S.A.), which is a part of its statewide model (*6*). Although the authors formulated a complete framework for trip generation and trip destination, mode choice modules could not be estimated because of the lack of observations in the survey for travel by air or public transit. Therefore, the impact of transit frequency and schedules could not be considered. Likewise, the TRANS-TOOLS project proposed a long-distance travel demand model for Europe (*4*). According to the authors, the TRANS-TOOLS model was the largest model ever with respect to population and covered area. This resulted in a very coarse level of resolution, excessive complexity, and long runtimes, making it unsuitable for testing transport planning policies. Lu et al. (*7*) developed a long-distance travel demand model considering multinomial logit models for trip generation, destination choice, and mode choice, built as a multi-level nested structure. However, the model only applied to a small set of seven destination alternatives, focusing explicitly on intercity travel and disregarding rural areas. A national long-distance passenger travel demand model was proposed by the Federal Highway Administration (*8*), which combines long-distance travel surveys from different states. The model used multinomial logit models to predict the number of trips, trip duration, and travel party size. Destination choice and mode choices were jointly estimated as a nested logit model.

Despite these large-scale examples, most studies on long-distance travel demand can be classified as case-specific or corridor-specific, with limited extrapolation or transferability. Limtanakool et al. (*9*) considered the effect of some land-use factors, such as density and use specialization, on long-distance trips in the Netherlands. However, they could only estimate a binary choice model considering the decision between car and train. Van Nostrand et al. (*10*) proposed a model only for vacation trips in the U.S.A., without taking into account long-distance business demand. Various authors (*11*) have tested the impact of automated vehicles on long-distance travel demand, reducing the share of conventional cars and airplanes for trips under 500 km. In addition, predicting the effects of building high-speed rail networks is another classic example of case-specific models (*12*, *13*), though the California statewide model for high-speed rail (*14*) did account for all sorts of long-distance trips. Regarding the temporal distribution of long-distance trips, to the best of the authors' knowledge only one, LaMondia et al. (*15*), allocated trips temporally over the year.

As a result of these approaches to long-distance travel demand, some research has identified further data requirements and research needs. Miller (*2*) pointed out several unsolved issues regarding estimation of long-distance travel demand. First, data collection could be improved by facilitating access to private-operator transit data. Second, the use of stated preference survey is complicated, expensive, and potentially biased, making the test of new modes extremely uncertain. Third, the level of aggregation of long-distance models is generally too coarse in space, time, and modes (especially regarding access and egress), thus a finer resolution was recommended. Miller proposed additional model capabilities such as non-resident visitors or long-distance through traffic. Additionally, the datasets used in destination choice models also need further improvement. The California statewide model for high-speed rail failed when estimating trips to certain tourist attractions (*14*), as the only attractors considered were population and employments at a coarse spatial resolution. Van Nostrand (*10*), too, identified the need for improving data on destination attractiveness. Lastly, the Federal Highway Administration (*8*) recommended collecting better data on long-distance trips through household travel surveys or using smartphone-based collection methodologies. Regarding transport supply, the provision of transit network data at the country level was recommended, especially for long-distance buses. Although the amounts of available data for the other modes were quite significant, converting them into a usable transit network required very long processing times.

Based on the review of the existing literature, and motivated by the previously mentioned data availability issues, this paper has the objective of estimating, calibrating, and validating a long-distance travel demand model for the province of Ontario, Canada. Specifically, the paper will present and discuss a methodology to enrich available travel survey datasets with location-based big data and trip planning services as additional data sources of both selected and non-selected destination and mode alternatives.

Methodology

This section describes the two steps of the model development: first, the model framework, and second, the data collection.

Model Framework

The model is designed as part of a four-step travel demand model for the province of Ontario. The last step, traffic assignment, is not described here, as this step is covered

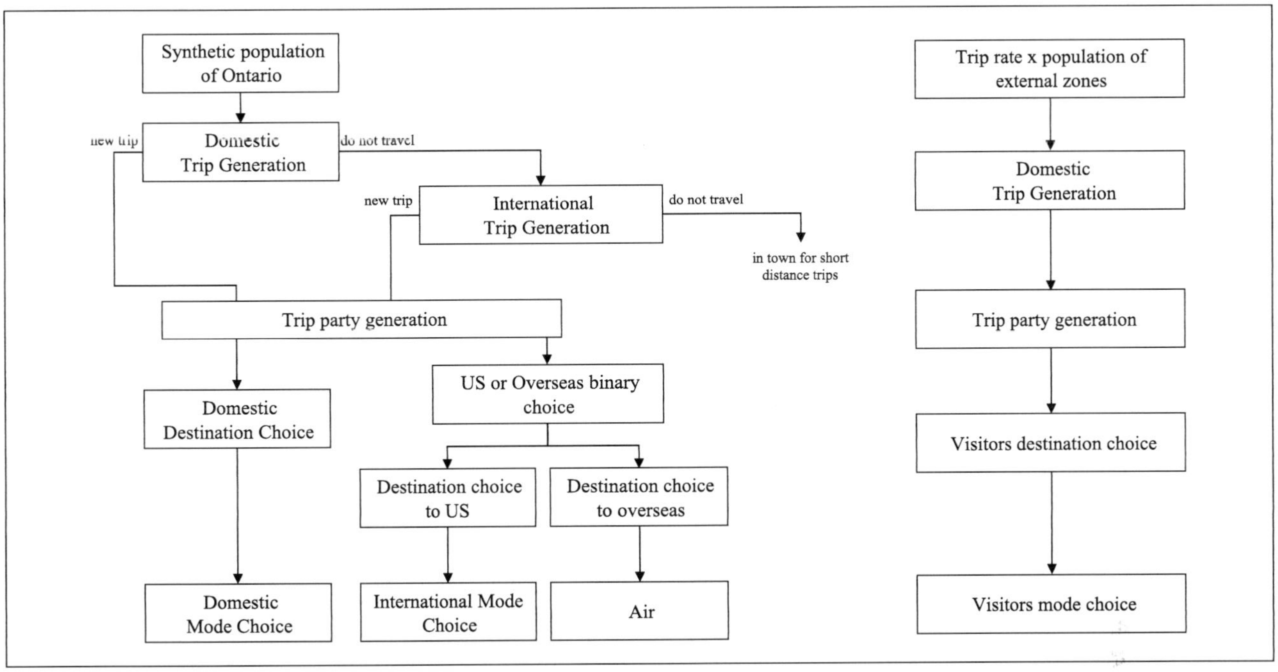

Figure 1. Model framework.

within a province-wide model where the long- and short-distance travel is assigned at the same time.

Ontario is the most populated province in Canada with more than 13 million residents, and has a land area of more than 900,000 square kilometers (roughly equal size to France and Spain combined). Although the average population density is very low, most of population is concentrated in a dense band close to the U.S. border. The proximity to heavily populated regions of Canada and the U.S. is a significant driver of travel demand between Ontario and its surroundings.

Figure 1 shows the structure of the long-distance travel demand model for Ontario residents, applied to the synthetic population that was generated for the Ontario provincial model. For visitors, either arriving from the rest of Canada or from other countries, the same structure was used except sociodemographic information about the traveler was not known. Therefore, alternative discrete choice models for visitors were estimated without considering person-related attributes. The visitors sub-model is similar to the Ontario resident long-distance model and is not described in detail in this paper due to space limitations.

The proposed approach was microscopic (individual persons, households, and trips are simulated) and the different stages were modeled using a multinomial logit formulation (*4*, *6*). Previous approaches, however, have used survey data only and not the mix of data presented here. Logit models provided the probability of selecting an alternative within a set of independent choices, and were formulated as shown in Equations 1 and 2 (*16*, *17*).

$$P_{pj} = \frac{e^{U_{pj}}}{\sum_{k=1}^{k=K} e^{U_{pk}}} \quad (1)$$

where:

- P_{pj} is the probability of selecting the alternative j by the individual p.
- U_{pj} is the utility of selecting the alternative j for the individual p, described in Equation 2.
- $k = 1, 2, \ldots K$ is the set of alternatives.

$$U_{pj} = \sum_{s=1}^{s=V} \beta_s x_s \quad (2)$$

where:

- $s = 1, 2, \ldots S$ is a set of either individual-related, alternative-related, or exogenous explanatory variables.
- β_s is the coefficient of the explanatory variable s.
- x_s is the value of the explanatory variable s.

With respect to the domestic trip generation, a multinomial logit model was proposed to predict individuals' decisions to travel or not. The alternative choice set is defined by four alternatives: "stay at home," "start a long-distance daytrip," "be away on an overnight long-distance trip," or "start or end an overnight long-distance trip." This way the model was able to select synthetic persons in Ontario to make long-distance trips or to stay at home for other trip types.

Regarding international trip generation, and as explained in the next subsection, the survey data did not provide information about persons or households. Therefore, the estimation of a logit model was not possible. Instead, the total number of international trips per person was calculated, while individuals were randomly selected. The probability of being selected was proportional to the probability of making domestic long-distance trips, assuming that the influence of person and household characteristics on the probability of traveling was comparable.

Once trips are generated, the trip party was randomly generated based on observed trip party frequencies from travel surveys.

Domestic destination choice models were estimated as logit models to predict the probability of selecting a destination zone as a function of its attraction as well as logsums obtained from the mode choice models (defined as shown in Equation 3). It accounted not only for the distance between origin and destination, but also the availability and level of service of the travel modes.

$$Logsum_{ij} = \sum_{m=1}^{m=M} e^{U_{ij,m}} \quad (3)$$

where:

- $Logsum_{ij}$ is the *logsum* between origin i and destination j.
- $U_{ij,m}$ is the utility of traveling from origin i to destination j using the mode m, obtained from an estimated mode choice model.
- $m = 1, 2, \ldots M$ is a set of alternative travel modes.

For international trips, the destination choice model is based on two decision steps. The first dealt with the decision whether to travel to the U.S.A. or overseas, and the second selected the destination within that destination. This model layout accounted for the large proportion of trips from Canada to the U.S.A. in comparison with other international trips. Both domestic and international destination choice considered single destination trips, as shown by 96% of the trips recorded in the survey.

Lastly, mode choice models were defined as multinomial logit models, where the set of alternatives were the available modes: car, air, rail, and bus. For domestic trips starting in Ontario, the mode level of service measures and personal and household attributes were included in the estimation. Level of service was the only variable group for international trips. Based on the travel surveys, no mode choice model (other than traveling by air) was proposed for trips to overseas.

Together with multinomial logit models, several alternative nested logit models were tested, although their results were not successful (no nesting structure provided a feasible nesting coefficient). Therefore, multinomial logit models included as many individual and alternative significant variables as possible in order to capture population heterogeneities and thus reduce the problems of violating independence of irrelevant alternatives (*18*).

Data Collection

The estimation of the long-distance travel demand model began with the use of two travel surveys, made specifically for long-distance trips within Canada (Travel Survey of Residents in Canada, TSRC) and trips departing or arriving to Canada (International Travel Survey, ITS). Additional data sources were used to complete the model estimation, as explained later.

Travel Surveys. Both TSRC and ITS travel surveys are designed primarily to measure domestic and international tourism, including trip characteristics, activities at the destination, and trip expenditures. The domestic travel survey, TSRC, is made by telephone and includes information about all the non-recurrent daily trips over 40 km and all the overnight trips made by the respondents during a one-month period. International trips are registered in the ITS by personal interviews at the entrances into Canada. The surveys explicitly discard domestic commuting trips, even if the distance is higher than the 40 km threshold.

These surveys were the primary source of data used for the model estimation. Data for the years 2011 to 2014 was used. Table 1 shows the main variables included from each survey. Table 2 summarizes the amount of data available for the model estimation, by both the survey records and the total number of trips from 2011 to 2014, as expanded using the survey weights.

Zone System. The long-distance travel demand model inherited the 5,000 raster cells zoning system from which the synthetic population of Ontario was obtained. Although trip generation was made at the person level, and the origin of the trip is allocated in this fine zone system, the estimation of destination choice models required a different aggregated zone system. Two main reasons motivated the aggregation: first, as the model was estimated based on survey data, it was limited to the resolution used in the dataset; second, while using logit models for destination choice, the number of alternatives could not be too big (as behavioral models, the number of choices should not exceed what a person can reasonably consider), so an aggregation or subsampling of alternatives was required (*4*). Therefore, an aggregated zone system was defined from the topographical union between Census Divisions (CD), which cover the whole Ontario region, and Census Metropolitan Area (CMA), which only cover dense urban areas. This resulted in 69 zones to be used during destination choice and mode choice (*19*).

Table 1. Variables in Survey Datasets

Object	Variable	Description, range and levels	Included in TSRC?	Included in ITS?
Person	Person age (respondent)	0–99 years	Yes	No
	Person gender (respondent)	Male Female	Yes	No
	Person education level (respondent)	Under high school High school Post-secondary University	Yes	No
	Person employment status	Employed Unemployed	Yes	No
	Household size	Number of adults Number of children	Yes	No
	Household income	Low (≤50,000 CAD) Medium low (50,000 CAD < income ≤70,000 CAD) Medium high (70,000 CAD < income ≤100,000 CAD) High (>100,000 CAD)	Yes	No
Trip	Trip purpose	Business (or work-related) Leisure (pleasure, vacation, or holiday) Visit (visit friends or relatives)	Yes	Yes
	Trip date	Day, month, year	Yes	Yes
	Trip mode	Car Air Rail Bus	Yes	Yes
	Trip origin	Province, CD, CMA	Yes	Yes
	Trip destination	Province, CD and CMA if domestic State, if to U.S. Continental region, if overseas	Yes	Yes
	Trip duration	Number of nights	Yes	Yes
	Travel party	Travel party size	Yes	Yes

Note: CAD = Canadian Dollars; CD = Census Division; CMA = Census Metropolitan Area.

Table 2. Sample Size by Purpose (Number of Records and Number of Trips—Expanded from Weights)

	Survey records for estimation			Expanded average number of trips per day		
Type of trip	Business	Leisure	Visit	Business	Leisure	Visit
Domestic from Ontario	4,771	19,686	28,378	31,665	111,907	163,471
Domestic from rest of Canada[a]	1,457	3,223	5,036	4,165	8,149	12,528
International Ontario-U.S.	9,770	31,571	5,639	28,239	93,832	16,300
International Ontario-Overseas	844	3,992	1,241	3,086	12,266	4,512
International U.S.-Ontario	4,056	13,871	5,045	11,234	30,307	14,872
International Overseas-Ontario	1,996	1,997	2,225	3,784	3,632	4,573

[a]Trips starting in other Canadian provinces were only considered if they ended in Ontario or crossed Ontario.

As a long-distance model, the zoning system also needed to include locations outside Ontario. The aggregation level of the external zones increased with their distance from Ontario: Quebec and Manitoba were represented by 38 different zones, the rest of Canada was aggregated into provinces or territories (10 zones), the U.S.A. into states (50 zones) and overseas destinations into seven continental regions.

Foursquare and Rome2rio Data. The estimation of discrete choice models required information on all the alternatives in the set, including non-selected destinations, as well as the level of service of the alternative travel modes.

With respect to measures for destination attractiveness, population and employment of the analysis zones were initially considered. These variables provided a global approach to the importance of the destinations but do not describe differences in land uses, nor consider specific touristic or business attractors. To enrich the description of destination alternatives, data was obtained from location-based social networks (LBSN). In this case, data on destination attractiveness was obtained from the Foursquare network.

The users of Foursquare check-in their visits to venues (locations such as airports, restaurants, parks, ski areas, etc.) with the goal of providing other users with information and recommendations of places to go. The data stored in this LBSN was accessed through the Foursquare public venue API, which returns a list of venues with the following relevant information:

- Name of the venue
- Venue category
- Geographical coordinates
- Number of total check-ins
- Number of unique visitors

Venues were classified into five main categories and the amount of check-ins by category within the aggregated zone was obtained. The five main categories (medical, ski area, hotel, outdoors, and sightseeing) were defined by aggregating the original list of Foursquare venue categories and sub-categories. A total of 34,041 venues and 7,981,458 check-ins were collected. Previous research provides more details on the data collection and processing methods (*19*, *20*).

Mode choice models estimation required the level of service of the different trip alternatives. The trip surveys used do not include level of service variables such as travel times, numbers of transfers, frequencies, or prices for either the selected or the non-selected modes. Therefore, level of service of the available modes was obtained from the multimodal online trip planning service Rome2rio

Using the Rome2rio API (*21*), data for all the origin and destination zones in North America was queried (a total of 167 × 167 origin–destination pairs). Origin and destination coordinates were located in the centroids of the zones. Rome2rio provided different alternatives to travel from each origin to each destination, as described by the following variables.

- Total travel time from origin to destination.
- Travel time by segment (segment is defined as a part of the trip using a single mode and vehicle; thus, the number of transfers is equal to the number of segments minus one).
- Segment mode.
- Frequency as number of services per week by segment.
- Price range and average price from origin to destination.

The data was processed according to the following criteria:

- A "main mode" was assigned based on the hierarchy: air, rail, bus, and auto. Every alternative with a flying segment was then coded as "air." If there is no air segment, every alternative with a rail segment was coded as rail, and so on.
- According to this, "main mode travel time," "access time," and "egress time" were defined.
- The frequency was defined as the minimum frequency among all the segments of the alternative, expressed in services per week.
- Driving costs were calculated by multiplying distance by fuel costs (0.072 CAD/km) instead of using Rome2rio data, as it assumed only taxi or shared vehicle costs.

Results

This section shows the results of the estimates, the technical details about model implementation, and finally the model calibration.

Model Estimation

Model estimation results for trip generation, destination choice, and mode choice are presented in this section. Table 3 shows the estimation results of the trip generation model. The columns show the coefficients of the utility functions for the three trip alternatives, with the base case being (utility equal to zero) "stay in town." As can be deduced from the negative values of the intercepts, starting a long-distance trip had a low probability.

Table 4 shows the results of the estimation of domestic destination choice models by purpose (in columns). Domestic destination alternatives were the aggregated zones (as explained above, 69 zones in Ontario and 48 in the rest of Canada). Table 4 also details the impact of distance between origin and destination as a function of the mode choice logsums (note that mode choice models were estimated before, although applied later) plus an added term of the logarithm of travel distance by car. The effect of logsums was different depending on whether the trip is a daytrip or an overnight trip. The higher coefficient of logsums for daytrips means that daytrips were more likely to select a closer destination, while the very long trips are in almost all cases overnight trips.

Additionally, the effect of the type of origin and destination zone (urban or rural) was captured using three additional dummy variables, defined as follows:

- Intra-metro: the trip starts and ends at the same urban zone.

Table 3. Domestic Trip Generation Estimation Results by Purpose

	Business			Leisure			Visit		
Variable	Away	Daytrip	Away	Daytrip	Away	Daytrip	Away	Daytrip	In/out
(intercept)	−7.929***	−6.265***	−7.707***	−5.462***	−5.326***	−5.524***	−4.691***	−4.401***	−4.448***
isYoung (age ≤25)	–	–	–	–	–	–	1.033***	0.356***	1.215***
isRetired (age >65)	–	–	–	–	–	–	–	–	–
isFemale	−0.610***	−0.945***	−0.628***	–	–	–	0.338***	−0.107*	0.158***
#adultsInHousehold	–	–	–	0.171***	−0.127***	−0.060*	−0.307***	−0.111***	−0.228***
#kidsInHousehold	–	–	–	–	–	–	−0.222***	−0.119***	−0.205***
isHighSchool	–	–	–	0.540***	0.273**	0.542***	–	–	–
isPostSecondary	–	–	–	0.520***	0.323***	0.495***	0.147*	0.212***	0.221***
isUniversity	0.783***	0.348**	0.874***	0.575***	0.322**	0.490***	0.459***	0.176**	0.456***
isEmployed	1.369***	0.923***	1.399***	–	–	–	–	–	–
hasIncomeMedLow	–	–	–	0.598***	0.392***	0.535***	0.211*	0.374***	0.231**
hasIncomeMedHigh	–	–	–	0.701***	0.561***	0.782***	0.345***	0.467***	0.464***
hasIncomeHigh	0.641***	0.452***	0.807***	1.209***	0.701***	1.146***	0.408***	0.521***	0.443***
Accessibility	−0.012**	−0.014***	−0.014**	−0.005***	−0.012**	−0.004***	−0.008***	−0.009***	−0.009***
LL		−3,736			−20,438			−20,398	
AIC		7,557			40,960			40,880	

Note: LL = Log-likelihood; AIC = Akaike information criterion; - = not significant.
****p*-value < .001; ***p*-value < .01; **p*-value < .05.
Legend for variables: "isCondition" is a dummy variable equal to 1 if the condition is fulfilled, zero otherwise; "#countVariable" is the count of the variable, "variable" is a continuous variable.

Table 4. Domestic Destination Choice Estimation Results by Purpose

Type of variable	Variable	Business	Leisure	Visit
Attraction	Log(population + employment)	0.276***	−0.193***	0.302***
	Log(Foursquare_hotel)	0.147***	0.228***	0.138***
	Log(Foursquare_sightseeing)	0.125***	0.112***	0.03***
	Log(Foursquare_outdoors)	–	0.132***	–
	Log(Fousquare_skiing)	–	0.073***	–
	isNiagara	–	1.444***	–
Origin–destination types	isIntra-metro	–	−3.053***	−1.843***
	isIntra-rural	0.98***	0.812***	0.293***
	isInter-metro	0.169***	−1.223***	–
Distance origin–destination	Logsum (if daytrip)	3.615***	9.671***	4.384***
	Logsum (if overnight trip)	0.899***	2.461***	0.909***
	Log(tripDistance)	–	−0.267***	−0.231***
LL		−17831	−73649	−110309
AIC		35,663	147,297	220,617

Note: - = not significant.
****p*-value <.001.

- Intra-rural: the trip starts and ends at the same rural zone.
- Inter-metro: the trip starts and ends at different urban zones.

According to this, intra-metro leisure and visit trips were found to be less common. On the contrary, intra-rural trips were more likely for all the purposes. Trips between different urban zones (inter-metro) were more likely for business purposes, but less likely for leisure trips.

With respect to the destination attractiveness, the use of Foursquare data greatly increased the explanatory power. The researchers found the number of Foursquare hotel check-ins was significant for all trip purposes, and sightseeing, outdoors, and skiing check-ins were significant in the case of leisure trips. Comparing the models with and without Foursquare data, the log-likelihood increased 1% for business, 2% for visit, but up to 8% for leisure, justifying the addition of the dataset, especially for leisure trips. Lastly, a dummy variable was added for the Niagara destination, as

Table 5. International Destination Choice Estimation Results by Purpose

(a) Binary choice between U.S.A. and overseas

Variable	Business	Leisure	Visit
Intercept	0.664***	0.465***	–
Accessibility to U.S.A.	0.017***	0.011***	0.023***
LL	−2018	−6477	−2328
AIC	4038	12956	4658

(b) Destination choice to U.S. states

Variable	Business	Leisure	Visit
Log(population)	1.056***	0.915***	1.026***
Logsum (if daytrip)	2.419***	3.906***	2.674***
Logsum (if overnight trip)	0.454***	0.883***	0.743***
LL	−14435	−33906	−10148
AIC	28,876	67,817	20,303

Note: - = not significant.
****p*-value <.001.

previously the model had underestimated the number of trips to that destination.

Table 5 shows the two-step destination choice models for international trips. The first step (Table 5*a*) corresponds with a binary logit model where the base alternative is to travel to an overseas destination and the second alternative is to travel to the U.S.A. As seen in the coefficients, the probability of selecting the U.S.A. increased with the accessibility to that country, resulting in an increased number of trips to the U.S.A. starting at zones close to the border, as predicted by the observations.

Trips to overseas destinations were simply distributed proportionally to the observed trip rates, since overseas zones were too big to find significant correlations between number of trips and overall sociodemographic variables. For this reason, the results are not shown in Table 5. For the trips that end in the U.S.A. a logit destination choice model is presented in Table 5*b*. The alternatives were the U.S. states. As in the case of domestic trips, the distance between origin and destination was measured by mode choice logsums, with the coefficient higher if the trip was a daytrip (usually closer destinations). Apart from this, destination attractiveness was measured by population. The correlation with more specific sociodemographic variables, such as employments by industry sector, was not possible, as the large size of the destination alternatives made their differences insignificant.

Lastly, mode choice models were estimated for both domestic and international destinations. The use of two independent models was motivated by the existing data sources (the domestic travel survey TSRC provided person- and household-related data while ITS did not) but also because observed modal shares were different for the two trip types. This could be explained by a different mode selection behavior, possibly influenced by border crossing delays or differences in public transport fares. Since both domestic trips and international trips to the U.S.A. showed a very large share for auto, auto was selected as the base alternative in the logit models.

In addition, there is a strong positive correlation between mode price and mode travel time (longer trips are also more expensive, in general). So, both variables were combined into one generalized time by using values of time, as shown in Equation 4.

$$generalized\ time = travel\ time + \frac{price}{value\ of\ time} \tag{4}$$

Later, the generalized time variable was converted to an exponential function, namely *impedance*, as shown in Equation 5.

$$impedance = e^{\alpha \cdot generalized\ time} \tag{5}$$

where α is a weighting, negative parameter, to be selected for each model. After testing different combinations, one generic coefficient for impedance was used (one coefficient for all the modes). The value of time was equal to 32 CAD/h for visit and leisure trips and 65 CAD/h for business trips, as selected for the provincial model. The values of the parameter α in Equation 5 were selected to produce the best model in terms of log-likelihood and Akaike information criterion (AIC).

As seen in Table 6, the impact of impedance was found to be always positive. As distance increases, utility decreases and shifts the modal choice from auto to air. The frequency, or number of services per week, was found to be significant in the case of domestic trips, too. Additional trip characteristics were significant as well. Bigger travel parties corresponded with trips by auto, while overnight trips increase the probability of selecting air, as this correlates with the selection of further destinations. In the case of domestic trips, the model is able to capture some person-related and household-related preferences: higher income groups had a lower probability of selecting bus and rail, young people had a higher likelihood of traveling by bus or rail, and women had less likelihood of traveling by car (auto).

Regarding mode choice models, the values of log-likelihood for the models with constants only (models with three parameters) are presented in Table 6. The comparison with the log-likelihood of the full model shows a reasonable improvement, justifying the addition of mode level of service, person attributes, and trip characteristics. The sensitivity of the model to changes in time, cost, and frequency was also captured by adding those mode-specific variables.

Implementation

The long-distance travel demand model is implemented in a JAVA program. The model is able to handle a synthetic population of 12 million with 4 million households. More than 2 million long-distance trips are processed (around one million

Table 6. Mode Choice Estimation Results by Purpose

(a) Domestic trips

	Business				Leisure				Visit			
Variable	Auto	Air	Bus	Rail	Auto	Air	Bus	Rail	Auto	Air	Bus	Rail
Intercept	0	−7.235***	−5.088***	−6.646***	0	−7.773***	−3.93***	−4.457***	0	−5.150***	−5.796***	−5.824***
impedance	2.696***	2.696***	2.696***	2.696***	6.659***	6.659***	6.659***	6.659***	2.601***	2.601***	2.601***	2.601***
#frequency	–	0.004***	0.004***	0.004***	–	0.005***	0.005***	0.005***	–	0.007***	0.007***	0.007***
isOvernight	–	2.909***	0.387***	0.387***	–	3.896***	1.356***	0.794***	–	2.766***	1.072***	0.517***
isInter_metro	–	1.724***	1.199***	2.007***	–	1.559***	1.490***	0.901***	–	1.107***	0.842***	0.842***
#tripParty	–	−0.153***	0.032***	0.032***	–	−0.055.	−0.966***	−0.494***	–	–	–	–
isIncomeLow	–	–	–	–	–	–	–	–	–	−0.878***	–	–
isIncomeHigh	–	–	−0.931***	−0.931***	–	0.272***	−0.944***	−0.993***	–	–	–	–
isYoung	–	–	0.797***	0.797***	–	−0.378***	1.466***	1.682***	–	−1.390***	1.960***	1.050***
isFemale	−0.489***	–	–	–	–	–	–	–	−0.493***	–	–	–
isUniversity	–	–	–	–	–	–	–	–	–	0.470***	0.361	1.072***
α		−0.0015 (not estimated)				−0.0004 (not estimated)				−0.0004 (not estimated)		
LL (constants only)		−3,817				−5,499				−12,880		
LL (full model)		−2,568				−4,481				−9,152		
AIC (full model)		5,172				9,018				18,346		

(b) International trips to U.S.

	Business				Leisure				Visit			
Variable	Auto	Air	Bus	Rail	Auto	Air	Bus	Rail	Auto	Air	Bus	Rail
Intercept	0	−1.310***	−4.918***	−6.381***	0	−1.207***	−2.811***	−6.576***	0	−1.616***	−5.741***	−7.191***
impedance	10.351***	10.351***	10.351***	10.351***	7.838***	10.351***	7.838***	7.838***	7.878***	7.878***	7.878***	7.878***
isOvernight	−3.242***	–	–	–	−2.305***	–	–	–	−3.891***	–	–	–
#tripParty	0.226***	–	–	–	0.195***	–	–	–	0.539***	–	–	–
A		−0.0015 (not estimated)				−0.0015 (not estimated)				−0.0015 (not estimated)		
LL (constants only)		−6,098				−15,023				−3,473		
LL (full model)		−2,214				−9,406				−1,864		
AIC (full model)		4,441				18,825				3,740		

Note: - = not significant.
****p*-value <.001.
Legend for variables: isCondition is a dummy variable equal to 1 if the condition is fulfilled, zero otherwise; #countVariable is the count of the variable.

are trips made by residents of Ontario, the rest by visitors). The model can simulate an entire average day in 5 minutes, running on a 16-core Intel® Xeon E5 PC equipped with 64 GB RAM. To reduce runtime, the code was written to allow the parallelization of certain processes.

Calibration

Two main output variables were used for calibration: the average trip distance and the modal shares, both segmented by purpose and distinguished between domestic and international travel.

With the objective of obtaining these two output variables as close as possible to the observed values, an iterative process was implemented to modify the following parameters of the models:

- Destination choice generic Logsum coefficients, using a multiplicative factor (both daytrips and overnight trips treated using the same factor).
- Mode choice mode-specific constants, using an additive factor (one factor for each mode).

After calculating both factors for each purpose and type of trip (domestic or international), the deviation of average trip distances and modal shares between model and observations were within 5%.

Discussion and Conclusion

This paper presented the design and estimation of a long-distance travel demand model for the province of Ontario, Canada. The model is operational and ready for its integration with the Ontario provincial model.

During model estimation, and confirming previous research results (*2*, *8*, *10*, *14*), several data-related issues were found. In our case, the major concern was the absence of destination and mode-specific data in the survey datasets, which limited the econometrical estimation of model parameters. By solving this issue, the major contribution of this paper is the development of two data collection methods for obtaining destination attractiveness information and mode-specific variables from LBSNs and online planning services, respectively.

In the first case, data on the number of check-ins at venues was acquired using the Foursquare social network, it was then classified and aggregated into destination alternatives. It provided valuable information about the types of attractions predominant at each location, more detailed than just population or employment—reported as insufficient previously (*14*). These data significantly improved the explanatory power of destination choice parameters, especially for leisure trips. The reason why Foursquare data are more relevant for leisure trips than for visit or business trips is probably that business and visit trips are reasonably well explained with population and employment, or that users of Foursquare are more concerned with tracking their leisure activities.

In the second case, mode-specific variables, including travel time, trip frequency, trip costs, and number of transfers, were obtained using the trip planning service Rome2rio. The Rome2rio data included almost all existing alternatives and was even able to distinguish between access and egress modes (although their impact was not statistically significant when model estimation was performed). Rome2rio data uses official operator websites and combines and post-processes their data, proving reasonable alternatives and price ranges, avoiding the need for integrating network data from different operators, which was reported as time and resource consuming by previous studies (*8*). As a result, the estimated model can capture sensitivities of modal level of service variables. An additional motivation for using trip planning services is that they replicate what real travelers do before starting a non-recurrent trip, properly reflecting what the travelers know about their alternatives.

In both cases, the developers' APIs provided by each of the websites were used to acquire large amounts of data, which allowed and significantly improved the model estimation. However, a potential limitation of the described methodology is that the data were obtained prior to model estimation. Therefore, it was used as a static data input that is not modified in model application for later years. Further work is needed to link the forecasts of destination attractiveness data with land-use and economic models. At the same time, the feedback from the traffic assignment model into the Rome2rio mode-specific travel time matrices needs to be investigated. This would integrate the effect of congestion, especially for long-term scenarios, when traffic patterns might have changed significantly.

Another limitation of LBSN data is a potential bias in user profiles. Most online data tend to be biased towards younger and more educated parts of the population. Given the lack of user-specific sociodemographic data, it is impossible to properly expand Foursquare data as is commonly done for household travel surveys. Nevertheless, LBSN data were used with the recognition that even traditional surveys are biased. LBSN data can potentially make up for its bias by the sheer volume of data.

Acknowledgments

The research was completed with the support of the Technische Universität München-Institute for Advanced Study, funded by the German Excellence Initiative and the European Union Seventh Framework Programme under grant agreement no. 291763. The authors would like to thank Sundar Damodaran, Shan Sureshan, Arthur Tai, and Rob Tardif from the Ontario Ministry of Transportation as well as Rick Donnelly, Mausam Duggal, and Bryce Sharman from WSP for their continued support and brainstorming for the development of the presented models.

Author Contributions

The authors confirm contribution to the paper as follows: study conception and design: All; data collection: Molloy, Ji; analysis and interpretation of results: All; draft manuscript preparation: Llorca. All authors reviewed the results and approved the final version of the manuscript.

References

1. Dargay, J., and S. Clark. The Determinants of Long Distance Travel: An Analysis Based on the British National Travel Survey. *Proc., World Conference on Transport Research*, Lisbon, Portugal, 2010.
2. Miller, E. The Trouble with Intercity Travel Demand Models. *Transportation Research Record: Journal of the Transportation Research Board,* 2004. 1895: 94–101.
3. Schiffer, R. G. *Long-Distance and Rural Travel Transferable Parameters for Statewide Travel Forecasting Models*. NCHRP Report 735, 2012. https://doi.org/10.17226/22661
4. Rich, J., and S. L. Mabbit. A Long-Distance Travel Demand Model for Europe. *European Journal of Transport and Infrastructure Research*, Vol. 11, No. 1, 2011, pp. 1–20.
5. Giamo, G. T., and R. G. Schiffer. Statewide Travel Demand Modeling: A Peer Exchange, Longboat Key, Florida, September 23–24, 2004. *Transportation Research Circular*, Vol. E-C075, 2005, pp. 2–212.
6. Erhardt, G. D., J. Freedman, A. Stryker, H. Fujioka, and R. Anderson. Ohio Long Distance Travel Model. *Transportation Research Record: Journal of the Transportation Research Board*, 2007. 2003: 130–138.
7. Lu, M., H. Zhu, X. Luo, and L. Lei. Intercity Travel Demand Analysis Model. *Advances in Mechanical Engineering*, Vol. 6, 2015, p. 108180. https://doi.org/10.1155/2014/108180
8. Federal Highway Administration. *Foundational Knowledge to Support a Long-Distance Passenger Travel Demand Modeling Framework Implementation Report*, 2015.
9. Limtanakool, N., M. Dijst, and T. Schwanen. The Influence of Socioeconomic Characteristics, Land Use and Travel Time Considerations on Mode Choice for Medium- and Longer-Distance Trips. *Journal of Transport Geography*, Vol. 14, No. 5, 2006, pp. 327–341. https://doi.org/10.1016/j.jtrangeo.2005.06.004
10. Van Nostrand, C., V. Sivaraman, and A. R. Pinjari. Analysis of Long-Distance Vacation Travel Demand in the United States: A Multiple Discrete–Continuous Choice Framework. *Transportation*, Vol. 40, No. 1, 2012, pp. 151–171. https://doi.org/10.1007/s11116-012-9397-6
11. LaMondia, J. J., D. J. Fagnant, H. Qu, J. Barrett, and K. Kockelman. Shifts in Long-Distance Travel Mode Due to Automated Vehicles: Statewide Mode-Shift Simulation Experiment and Travel Survey Analysis. *Transportation Research Record: Journal of the Transportation Research Board*, 2016. 2566: 1–10.
12. Gutierrez, J. Location, Economic Potential and Daily Accessibility: An Analysis of the Accessibility Impact of the High-Speed Line Madrid-Barcelona-French Border. *Journal of Transport Geography*, Vol. 9, 2001, pp. 229–242. https://doi.org/10.1016/S0966-6923(01)00017-5
13. Sung, H. G., and M. J. Choi. Determinants of Transportation Mode Choice for Long-Distance Travel in Korea. *Korea Planners Association*, Vol. 49, No. 2, 2014, pp. 245–257.
14. Outwater, M., K. Tierney, M. Bradley, E. Sall, A. Kuppam, and V. Modugula. California Statewide Model for High-Speed Rail. *Journal of Choice Modelling*, Vol. 3, No. 1, 2010, pp. 58–83. https://doi.org/10.1016/s1755-5345(13)70029-0
15. Lamondia, J. J., M. Moore, and L. Aultman-Hall. Modeling InterTrip Time Intervals Between Individuals' Overnight Long-Distance Trips. *Transportation Research Record: Journal of the Transportation Research Board*, 2015. 2525: 23–31.
16. Ben-Akiva, M. E. *Structure of Passenger Travel Demand Models*. Massachusetts Institute of Technology, 1974.
17. Train, K. E. *Discrete Choice Methods with Simulation*. Cambridge University Press, Cambridge, 2009.
18. Ben-Akiva, M. E., and S. R. Lerman. *Discrete Choice Analysis*. Massachusetts Institute of Technology, 1985.
19. Molloy, J., and R. Moeckel. Improving Destination Choice Modeling Using Location-Based Big Data. *International Journal of Geo-Information*, Vol. 6, No. 9, 2017, p. 291. https://doi.org/10.3390/ijgi6090291
20. Molloy, J. *Development of a Destination Choice Model for Ontario*. Technical University of Munich, 2017.
21. Ji, J. Y. *Long-Distance Mode Choice Modeling of Ontario Province*. M.Sc. thesis. Technical University of Munich, 2017.

The Standing Committee on Traveler Behavior and Values (ADB10) peer-reviewed this paper (18-02359).

TRR
JOURNAL OF THE TRANSPORTATION RESEARCH BOARD

Transportation Research Record
2018, Vol. 2672(47) 114–124

DOI: 10.1177/0361198118772954
journals.sagepub.com/home/trr
SAGE

How Should Travel Demand and Supply Models Be Jointly Calibrated?

Ali Najmi[1], Melissa Duell[1], Milad Ghasri[1], Taha Hossein Rashidi[1], and S. Travis Waller[1]

Abstract

Calibration is a critical aspect of model development that has long been recognized by researchers as a challenging issue. In particular, difficulties arise when the observed data used for calibration do not match the model output, which is the case in the majority of transport planning models. In the traditional calibration process, the origin–destination (OD) matrices are the key interface between demand and supply models, which could lead to issues when observed traffic link counts are used to update the OD matrix, causing a loss of key demand characteristics in the process. Developing a unified structure for modeling both demand and supply requires a calibration process that meets the requirements of both types of models, a serious issue which has received less attention in the literature. In this paper, the existing processes of developing and integrating demand and supply models are discussed and then examined using a case study in the Melbourne area. The numerical results show that the standard OD calibration procedure causes unrealistic changes in the OD matrix. Finally, some possible solutions to address the current limitations in development of a unified structure are discussed.

The calibration of transportation planning models plays a crucial role in model application and decision-making. It is generally considered that the effectiveness of a model lies in its ability to accurately replicate actual conditions, which will then be used to produce estimates regarding future situations. Thus, the usefulness of the models and their ability to reliably analyze policy scenarios depends on the quality of calibration. Inadequate calibration can result in significant issues regarding stability, error propagation, and the transferability capacity of the model. For example, Bloomberg et al. conducted a study on six different software programs and found that the calibration differences of 13% in the predicted freeway speeds for existing conditions increased to differences of 69% in the forecasted freeway speeds for future conditions (*1*).

Traditional planning models capture the interaction of two components, demand and supply, in a way that is often considered to be completely independent from one another, as a result of differences in their data resources and modeling techniques. On the demand side, transport demand modelers estimate the myriad complexities of travel demand based on individual decisions on day-to-day activities, changes in land use, and so forth. On the supply side, transport network modelers predict network conditions that will result from a predetermined demand, usually based on an equilibrium assumption (*2*). Despite the interdependence between demand and supply (i.e., network conditions are both determined by and influence travel decisions), traditionally the calibration of each remains largely separated. Moreover, the calibration of both demand and supply is further confused by questions regarding both the calibration procedures and their practical application. From a technical perspective one of the issues is, given a set of macroscopic observations such as travel survey data and road traffic counts, which macroscopic or microscopic rules should be applied to move the simulation or assignment closer to the observations? Typically traffic counts for areas of the network and trip diaries for a part of population are available.

Traditionally, the calibration process of a planning transport model includes separate exercises to estimate parameters of demand models that generate the origin–destination (OD) matrix and to estimate the network parameters (i.e., link capacities). Once both demand and supply models are calibrated, the structure can be further enhanced through an iterative process (using a feedback loop from supply to demand to update network

[1]Research Center for Integrated Transport Innovation, School of Civil and Environmental Engineering, The University of New South Wales, Sydney, Australia

Corresponding Author:
Address correspondence to Ali Najmi: a.najmi@unsw.edu.au

attributes such as travel time), such that the difference between the observed counts and the simulation results is judged to be within an acceptable gap.

From the view of demand modelers, the OD matrix is not calibrated while it is the output of the calibrated demand models. If the luxury of having an observed OD matrix exists, it is then used to assess the precision of the calibrated demand models. On the other hand, from the perspective of supply modelers, an OD is considered properly calibrated if it can optimally reproduce the observed traffic counts. Numerous network modeling-based methods have been proposed in the literature for OD calibration (*3–7*). The central part of all the OD calibration methods is the target/initial OD. In many of the methods, the weighted distance between the updated OD matrix and the target matrix provided from demand models, and also the weighted distance between simulated link flows and observed flows, are minimized subject to flow constraints (*8*).

The motivation for this paper stems from a central research question: Do traditional transport models where the demand and supply is primarily connected via OD matrices produce biased results? Even though the established OD updating methodologies seem to give satisfactory output, the disadvantage may be that they lose critical information contained in demand-side models and their forecasting capabilities. Therefore, the contribution of this paper can be summarized as follows:

- We discuss transportation model calibration from the perspectives of demand, supply, and the integration of the demand and supply;
- Based on a large-scale, realistic case study of Melbourne, Australia, we illustrate the potential downfalls of traditional OD calibration techniques and discuss their implications for models in practice;
- We suggest some possibilities from the demand modeling perspective for the unified calibration of demand and network approaches.

Background

Travel Demand Model Development

Traditional travel demand models are developed separately for system-level and user-level properties in which the individual trips are usually the units of analysis. They include four sequential steps of trip generation, trip distribution, mode choice, and assignment. Demand models are generally developed using travel diary data at the disaggregate level of individuals/households. These models are mainly based on the utilitarian concept that travelers either maximize their utility (random utility maximization [RUM] [*9*]), or minimize their regret (random regret minimization [RRM] [*10*]). In developing each of the demand sub-models, goodness-of-fit tests are usually employed to assess the effectiveness of the model estimations.

In the more recent version of demand models like tour-based and activity-based models (ABMs) (they fall under the domain of agent-based modeling), the travel attributes replication (such as mode choice, destination choice, and trip purpose) plays a critical role in demand calibration. The criterion has been employed in two directions, forward and backward. In the forward approach, the model parameters of the demand models are modified in such a way that the model replicates the observed travel attributes in the household travel survey. On the other hand, in the backward approach, some variables of the demand models are modified based on some information provided by the network such as path flows and travel times. For more information on calibrating four-step and ABMs, the reader is referred to other research (*11–15*).

Network Models Development

The network model is the component of the transportation planning process that outputs link flows as an estimation of the network conditions that will result from the modeled travel demand. The network model represents the supply side, or the providers of road services, by assigning the routes between OD trips. Most often the travel demand is in the form of the OD matrix. Equilibrium is typically sought when the travel times on all routes between an origin and destination are equal (within an acceptable gap function) and minimal (*16*). Although numerous updated approaches exist to expand the traditional equilibrium assumptions, most notably stochastic user equilibrium to address the assumption of perfect information (*17*) and dynamic traffic assignment to capture the time-dependent nature of many traffic flow phenomena (*18*), the traditional static approach continues to be far more prevalent especially in practice, for reasons such as solution uniqueness and stability.

Traditionally, the travel demand and network modeling components of the planning process have been quite separate, a fact which is reflected in both the research communities and the relevant literature. Of course, this is likely a result of the complexity of each modeling aspect; however, this work seeks to make the case for how and why this fundamental disconnect may bias the model results in terms of both the demand and the supply outputs.

The Connection between the Demand and Network Models

In many applications of transport planning models, the policy options that are being examined require a

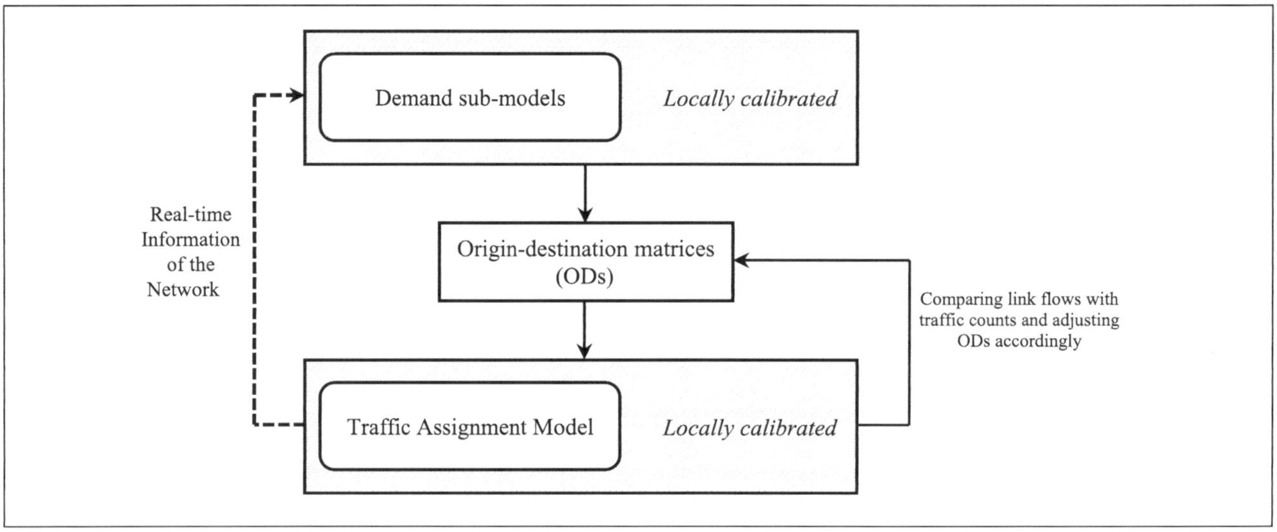

Figure 1. Framework of individually calibrated demand and supply models.

connection between the network and demand models (i.e., demand and supply cannot be treated as completely independent). In trip-based, tour-based and many ABMs, the linkage between demand and supply is the OD matrix. Previous works have found that OD matrices generated on the basis of surveyed data are more accurate in comparison with OD matrices that are simulated (*19*).

As Figure 1 illustrates, there are two separate parts to the calibration of four-step models. First, there is a calibration process for individual trips generated by the calibrated demand sub-models (such as trip generation, trip distribution, and mode choice). The output of this process is the OD matrix, which is input to the assignment model. However, the link flows that result when an OD matrix is applied with the assignment model can be substantially different from the observed traffic count data. This happens for a variety of reasons, including the differences in time scale between the demand data and the observed counts (*20*). Therefore, a second calibration process based on the assignment model and the OD matrix is often employed, as shown in the figure.

OD matrix updating (also called estimation or calibration) is the common approach to minimize the discrepancy between the modeled and observed link flows, because updating of the survey data underlying the travel demand model is difficult and costly. OD matrix updating is usually based on current traffic count data collected on a set of links. The established approaches can be categorized into four modeling approaches including maximum entropy, maximum likelihood estimator, Bayesian inference, and generalized least square estimator (*21*). However, the accuracy of the estimated matrices depends on the updating model used, the input data errors, and the set of links with collected traffic counts (*22*). The OD matrix updating problem has many solutions because in any practically sized network, the number of OD pairs will by far exceed the number of links with observed traffic counts, and thus the equation system for estimating the OD matrices is underspecified (*8*). Therefore, to estimate the proper OD matrices, in the literature, two indices of the distance between the updated OD and target OD (or the observed OD if available) and the distance between the estimated link flows and observed link flows are applied.

More advanced four-step models include an iterative process in which the calibrated demand and supply models communicate. As can be seen in the Figure 1 using a dotted line, in each iteration, the network-level attributes such as path travel times are considered exogenous information to be used in the demand model. The updated demand model feeds a new OD matrix to the traffic assignment model, which produces new network attribute data (i.e., a new set of path travel times or link cost). Including a feedback loop creates a stronger linkage between the demand and supply but still faces limitations because the connection is only through updating variables like travel time, not related to the calibration of either model component.

The process of calibration for ABMs is similar to four-step models. For calibration of ABMs, demand modelers use a travel diary of a sample of travelers to develop a model replicating the behavior of people. In a technical way, the likelihood of modeling the observed decisions is maximized given the sample. The sample is then expanded to replicate the population assuming that the parameters of the model developed on the sample are unbiased estimators of the true parameters of the population. Then trips (tours or daily activity itinerary) are

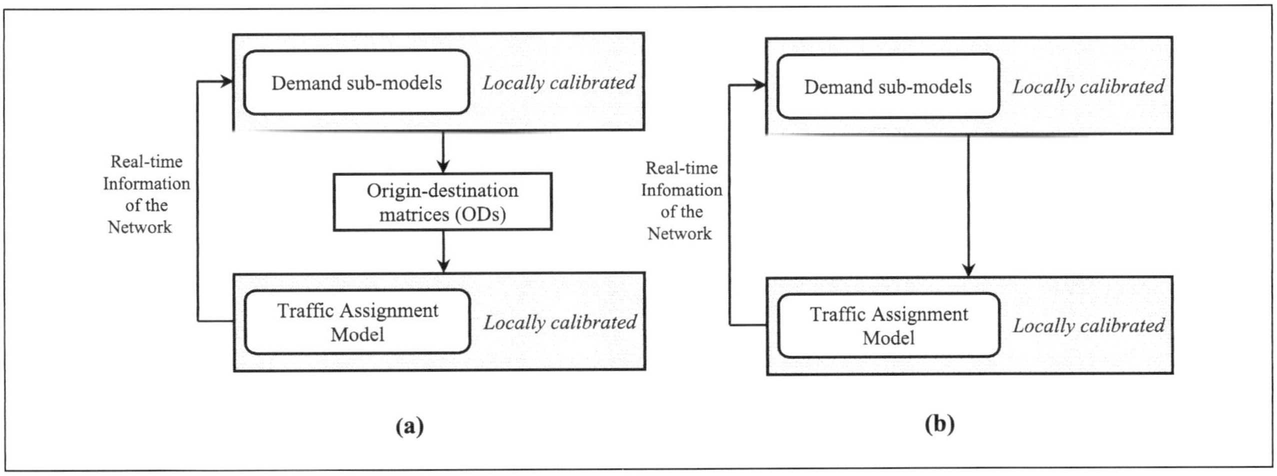

Figure 2. The linkage of ABM and TA in simulation; (*a*) OD-based models, (*b*) non-OD-based models.

inputted to the traffic assignment model. The network modelers calibrate the traffic assignment model using the link flows. In the so-called joint structures proposed in the literature (the difference between this so-called integrated structure and our proposed integrated structure will be elaborated in detail later), travel demand and traffic assignment models, a feedback loop is again proposed (*23,24*). Similar to Figure 1, the role of the feedback loop is to provide network-level information such as path travel times to the already calibrated travel demand models (for example to the activity purpose, mode choice, and destination choice models) in which travel time plays a significant role. As shown in Figure 2*a*, the feedback loop is mainly used during the simulation, while during the calibration exercise the role of the feedback loop is, like before, to update *variables* used in the demand model. In line with using the feedback loop in four-step models, Auld and Mohammadian (*23*) and Lin et al. (*24*) suggest feedback mechanisms between ABM and traffic assignment models. In these models, the time-dependent ABM is fed into a traffic assignment model at specific time intervals. The predicted travel times resulted from running traffic assignment is then returned to the ABM in order to be used for rescheduling activities. This process is repeated until a state of convergence is reached.

As shown in Figure 2*b*, the recent direction of demand and supply models such as ABMs and dynamic traffic assignment models provides an agent-based alternative to the OD-based approach. Nonetheless, similar to Figure 1, the calibration process is still done in an independent manner for both demand and supply models. As Figure 2*b* shows, each model would still be separately calibrated and then joined in a sequential manner, similar to what happened during the simulation process.

In practice, the majority of models that are applied by real transport planning agencies around the world are OD based, and the demand and supply models are calibrated separately. Moreover, although there is a feedback loop between demand and supply models in some models, it is almost universally used in the simulation phase. The next section explores the possible conflicting results from disjointly calibrating the demand and supply models.

Case Study

In order to investigate the impact of network-based OD updating on the OD matrix itself, we use a case study of the city of Melbourne. In the case study, we want to compare the changes in an OD matrix that result from two common approaches. The first approach is based on a demand model estimation, while the second method uses a network-based OD updating approach. In order to explore the impact of both approaches, we perform a relative comparison on how each method changes the OD matrix. This section explains the details of this experiment and discusses the findings.

Data

Household Travel Survey. The main data source for demand in this study is the household travel surveys collected in the Victorian Integrated Survey of Travel and Activities (VISTA) in 2007, from which we generated several different OD matrices, as explained in the following section. The zoning system in this matrix is defined based on the census collection districts (CCD) in Melbourne. However, to expand the OD matrix to the population level, a larger zoning system of statistical local areas is used.

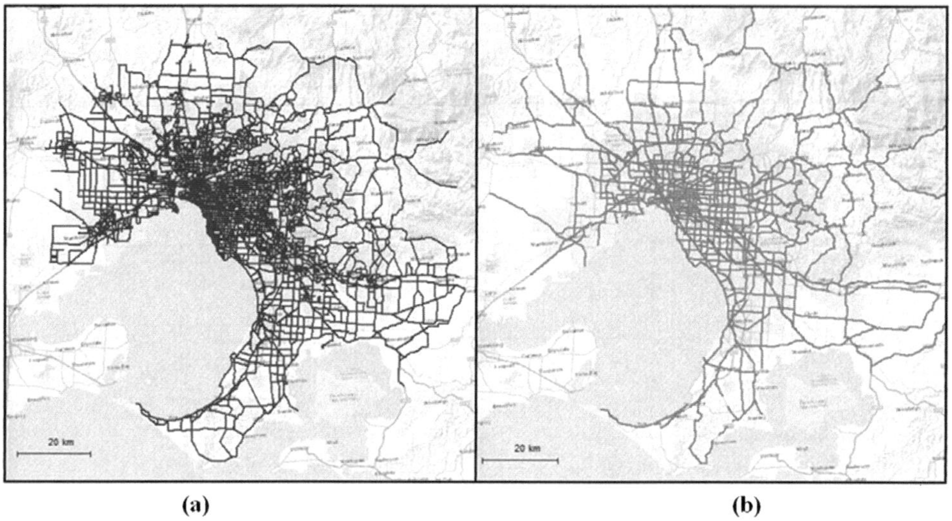

Figure 3. Melbourne network; (*a*) Melbourne network links, (*b*) publicly available traffic estimates.

Network and Traffic Counts. This case study focuses on the greater area of Melbourne, Australia illustrated in Figure 3. The network consisted of approximately 44,000 links and 16,000 nodes, which would be considered a very large-scale regional network. The original network data were obtained courtesy of CityX at the Monash University (*25*), which included the link parameters for the conventional BPR travel cost function (*20*). The data were then processed to extract the vehicle-only network. The traffic counts are based on annual SCATS signal averages and are publically available from VicRoads Open Data Portal (*26*). The demand centroids were based on CCDs as obtained from the Australia Bureau of Statistics.

Two map-matching processes were necessary to complete the dataset. First, the CCD centroids were created and spatially matched to the nearest network node via the creation of a centroid connector, using ArcGIS. Second, the traffic counts were matched to the network links based on spatial location and their geometric angle. As a result of various disparities between the datasets (the roads in the traffic dataset covered greater distances than network links but could only be matched to a single network link), the final processed dataset resulted in 3,655 network links with counts to use as benchmarks of observed link flows.

Methodology of the Case Study

For this case study, we are interested in exploring what is a reasonable change in the OD matrix and then we want to examine the changes that result from a common OD matrix updating approach.

First, we need to generate a hypothetical forecasted OD matrix. We used the household travel data from VISTA and an approach introduced by Ghasrikhouzani et al. (*27*), where trip purpose, travel mode, time of day, commute distance, and attributes of trip destination are jointly modeled using a modified decision tree, which was previously shown to produce the best aggregate fit to data. Using this demand model, we produce two matrices for the target 2009. The first is for normal conditions, which we call OD^P and the second for an extreme condition, which we call OD^E.

The extreme condition is presented as an upper bound for the reasonable change in an OD matrix that would be produced by a demand model. For the extreme condition, we assume that the entire population comprises full-time workers. This is not intended to be a realistic scenario; instead, the extreme assumption creates a suitable benchmark for measuring the impact of OD calibration/updating method on an OD matrix. We define the difference between OD^P and OD^E as a matrix of changes p'. The matrix of changes p' is an indicator of the largest potential change under an extreme condition of a demand scenario. According to the analysis with the case study dataset, the daily total number of trips under the extreme condition has increased 7% compared with the estimate under the normal conditions.

In the next step, we want to explore the impact of a traditional OD matrix updating method. We define the base case (generated from the VISTA dataset) as OD^0, which will be updated using the set of observed flow counts. We solve the equilibrium assignment for the OD matrix using the traffic assignment by paired alternative segments (TAPAS) solution approach (*28*) and then compare the observed flows with the model output. As expected, equilibrium flows from the traffic assignment model do not match the observed link flows and calibration is necessary.

We use the OD updating method by Juarez and Chavez (5), the modified version of the method in Spiess (3), to calibrate the input OD^0. This method updates the demand of OD pairs in order to match the count flows. The updating process is conducted in a bi-level optimization process. The sub-problem is the user equilibrium problem which guarantees the user equilibrium condition for the assigned flows to the network (solved using TAPAS in the current implementation). The upper level minimizes an objective function (Equation 1) with two parts: (i) the difference between count flows and assigned flows (Δv) and (ii) the difference between target and updated OD demands (Δg). The combination of these two terms in the objective function is weighted using the parameter α. By definition, larger values of α would result in higher weight for the difference between count flows and assigned flows. For $\alpha = 1$, the updating method is exactly the same Spiess method which is widely used in the literature.

$$\min Z(g) = \frac{\alpha}{2}\sum (\Delta v)^2 - \frac{1-\alpha}{2}\sum (\Delta g)^2 \quad 0 \leq \alpha \leq 1 \qquad (1)$$

The output of the updating method is an updated OD matrix OD^U. At this stage, a comparison between the elements of OD^O and OD^U would result in a matrix of changes of p which represent the magnitude of changes from the OD updating process. The case study analysis compares the results for p and p' in the next section.

Effectiveness of the OD Calibration Method

Intuitively, p values are expected to be considerably smaller than p' values because the magnitude of changes from the calibration process is expected to be lower than the changes resulting from a change as extreme as the assumption of full employment. However, the results suggest otherwise. In this study, when $\alpha = 1$, nearly 30% of the p values are larger than their corresponding p' values. This means, for 30 percent of OD pairs, the calibration procedure has a higher impact compared with the extreme condition of full employment.

Table 1 presents a cross-classification of p versus p' for $\alpha = 1$, $\alpha = 0.9$, and $\alpha = 0.5$. Assigning lower values to α, the calibration algorithm would consider higher weights for differences between target and updated OD. In contrast, the higher weights for α would be considered for matching the count flows.

Ideally, a transport modeler does not expect the OD calibration to modify the initial values significantly because there are characteristics in the target/initial OD matrix that are important for any policy analysis. Therefore, all of the elements of p matrix should ideally fall into the category of $-25\% \leq p < 25\%$ in Table 1. However, Table 1 shows that this is not the case. It should be noted that the limits in the table are chosen intentionally for better elaboration of the changes in p and p'. In Table 1, for $\alpha = 0.5$ the number of elements in the favorable category of p is higher than those for $\alpha = 0.9$ and $\alpha = 1$. This is because when $\alpha = 0.5$, changes in the OD matrix are more penalized in the calibration process.

The most problematic cases are bolded in Table 1. These cases represent the situations where the calibration process has generated considerably larger changes in demand compared with the changes that would occur from an extreme change in the demand model itself (as represented by OD^P and OD^E). When $\alpha = 1$, in 497 out of 2,248 cases, while the extreme condition cause a minor change in demand (changes less than 25%), the updating procedure produce a change of more than 25%. Furthermore, the updating process changes 159 of the OD pairs more than 100%, 29 of which are even changed more than 500%. However, this condition would be better when the results of $\alpha = 0.9$ and $\alpha = 0.5$ are considered. For these cases, the number of OD pairs with minor changes in extreme demand condition and major changes in calibration process would reach 144 and 37, respectively.

Figure 4 visualizes the relationship between p and p' for the cases of $\alpha = 0.5$ and $\alpha = 1$. Each point in this plot corresponds to one of the elements in the OD table. The horizontal axis represents p values and the vertical axis represents p' values. Ideally, one would expect the updating not to make major changes to OD elements, which would mean the points would be scattered close to the vertical axis in this plot. Note that there would be no limitation on the p' values, as the impact of the extreme scenario on travel demand is unknown. Figure 4 shows a noticeable number of points far from the vertical axis and very close to the horizontal axis. These points represent a major change in the OD calibration process for the OD pairs with minor changes under the extreme demand condition.

Therefore, while OD updating plays a critical role in connecting the demand-side and supply-side models in the literature, it cannot transfer the OD demand truly and completely. This leads to an interruption in the linkage between these two sorts of models.

Discussion

While previous efforts have been made to join the demand and network models in some advanced integrated transport models (see Figure 2), the linkage is

Table 1. Percentage of Changes in OD pair—Updating Method versus Extreme Scenario

Change in p	$-100\% \le p' < -75\%$	$-75\% \le p' < -50\%$	$-50\% \le p' < -25\%$	$-25\% \le p' < 25\%$	$25\% \le p' < 50\%$	$50\% \le p' < 75\%$	$75\% \le p' < 100\%$	$100\% \le p' < 200\%$	$200\% \le p' < 500\%$	$500\% \le p'$	Sum
Alpha = 0.5											
$-100\% \leqslant p < -75\%$	0	***0***	***0***	***0***	***0***	***0***	0	0	0	0	0
$-75\% \leqslant p < -50\%$	0	0	***0***	***0***	***0***	1	0	0	0	1	2
$-50\% \leqslant p < -25\%$	2	4	4	***4***	1	5	0	1	5	1	27
$-25\% \leqslant p < 25\%$	133	401	296	703	88	161	58	130	125	42	2137
$25\% \leqslant p < 50\%$	8	9	3	***23***	3	4	3	5	2	2	62
$50\% \leqslant p < 75\%$	1	2	***1***	***6***	***0***	3	1	1	0	2	17
$75\% \leqslant p < 100\%$	0	***0***	***0***	***0***	***1***	***0***	0	0	0	0	1
$100\% \leqslant p < 200\%$	***0***	***0***	***0***	***2***	***0***	***0***	***0***	0	0	0	2
$200\% \leqslant p < 500\%$	***0***	***0***	***0***	***0***	***0***	***0***	***0***	***0***	0	0	0
$500\% \leqslant p$	***0***	***0***	***0***	***0***	***0***	***0***	***0***	***0***	***0***	0	0
Sum	144	416	304	738	93	174	62	137	132	48	2248
Alpha = 0.9											
$-100\% \leqslant p < -75\%$	0	***0***	***1***	***0***	***0***	***3***	0	0	0	1	3
$-75\% \leqslant p < -50\%$	1	3	***3***	***4***	***0***	3	0	0	2	1	17
$-50\% \leqslant p < -25\%$	6	17	8	***19***	4	2	0	8	6	0	70
$-25\% \leqslant p < 25\%$	119	359	282	644	80	151	55	118	119	37	1964
$25\% \leqslant p < 50\%$	10	27	8	***44***	6	11	3	6	4	5	124
$50\% \leqslant p < 75\%$	6	4	***0***	***14***	***2***	1	2	2	0	2	33
$75\% \leqslant p < 100\%$	2	***4***	***1***	***8***	***0***	***2***	2	2	1	2	24
$100\% \leqslant p < 200\%$	***0***	***2***	***1***	***4***	***1***	***1***	***0***	1	0	0	10
$200\% \leqslant p < 500\%$	***0***	***0***	***0***	***1***	***0***	***0***	***0***	***0***	0	0	1
$500\% \leqslant p$	***0***	***0***	***0***	***0***	***0***	***0***	***0***	***0***	***0***	0	0
Sum	144	416	304	738	93	174	62	137	132	48	2248
Alpha = 1 (Spiess (3))											
$-100\% \leqslant p < -75\%$	12	***21***	***8***	***46***	***7***	***12***	4	4	4	9	127
$-75\% \leqslant p < -50\%$	9	25	***18***	***46***	***2***	11	4	13	5	3	136
$-50\% \leqslant p < -25\%$	25	53	29	***91***	8	34	10	15	23	7	295
$-25\% \leqslant p < 25\%$	74	241	200	439	57	87	32	72	63	20	1285
$25\% \leqslant p < 50\%$	8	21	17	***43***	7	5	4	11	19	2	137
$50\% \leqslant p < 75\%$	3	10	***6***	***15***	***1***	7	4	6	4	2	58
$75\% \leqslant p < 100\%$	2	***5***	***4***	***10***	***1***	***2***	1	4	2	1	32
$100\% \leqslant p < 200\%$	***5***	***23***	***8***	***34***	***7***	***4***	***1***	5	4	2	93
$200\% \leqslant p < 500\%$	***4***	***14***	***7***	***9***	***1***	***6***	***2***	***5***	6	1	55
$500\% \leqslant p$	***2***	***3***	***7***	***5***	***2***	***6***	***0***	***2***	***2***	1	30
Sum	144	416	304	738	93	174	62	137	132	48	2248

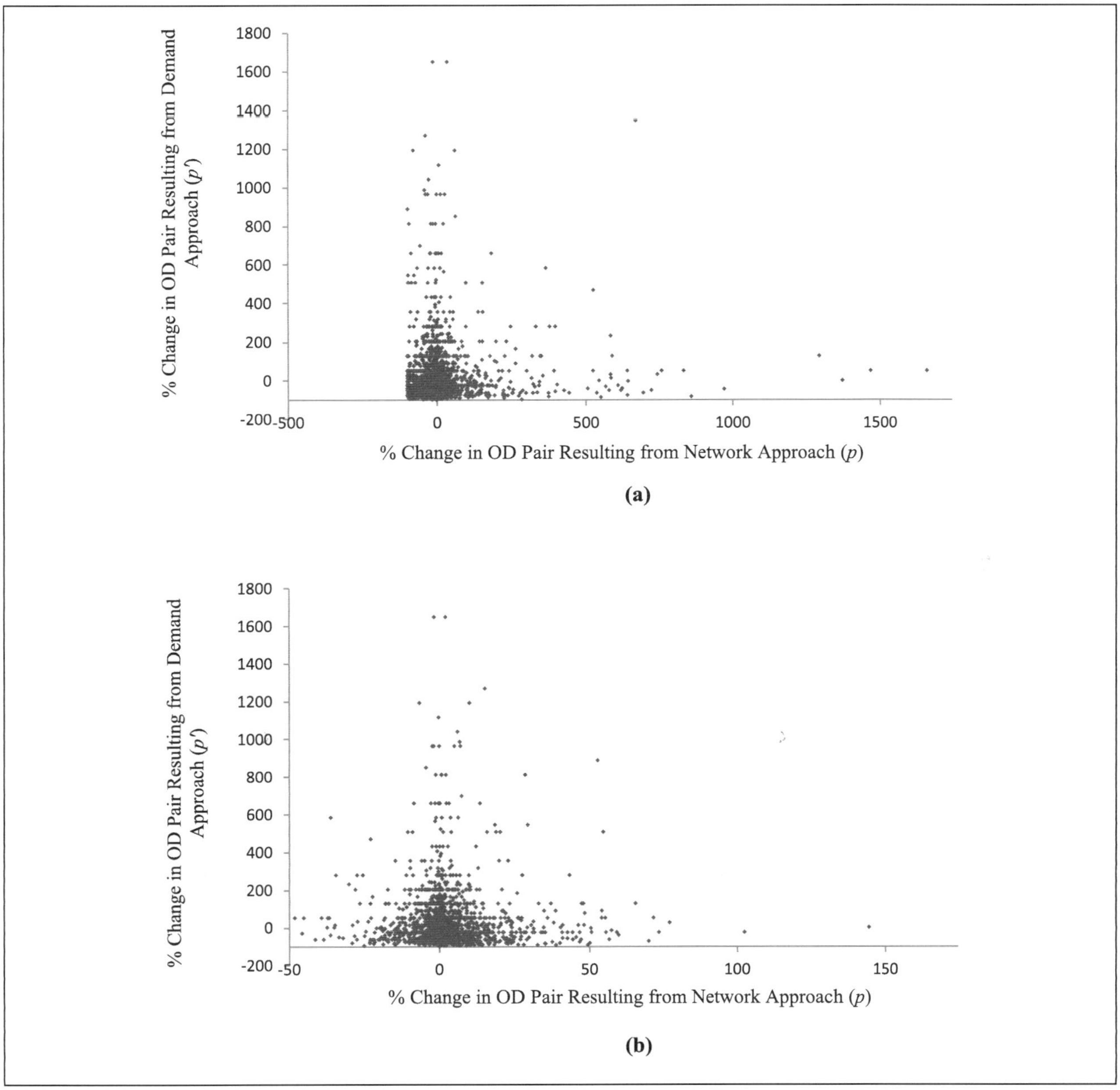

Figure 4. The percent of the changes in OD pair by updating method versus extreme scenario; (*a*) in the case of $\alpha = 1$(*b*) in the case of $\alpha = 0.5$.

primarily limited to a feedback loop where a simulation-based process passes exogenous travel times to the demand from the network. This study demonstrated that established calibration methods may significantly change the OD demands. This approach could sacrifice the accuracy and predictive power of the demand model and also lose the reciprocal relationship with network model.

This situation can be made worse when the OD matrix is considered as the sole connection between supply and demand. When tours/trips are summarized into the OD matrix, variables used in the process of generating the OD matrix that may represent important characteristics such as stochasticity and adaptivity are no longer present. As a result, some information about the demand is lost if an OD matrix represents demand. However, because of lack of a better solution, OD matrices remain the most common representative of demand to be used in traffic assignment models.

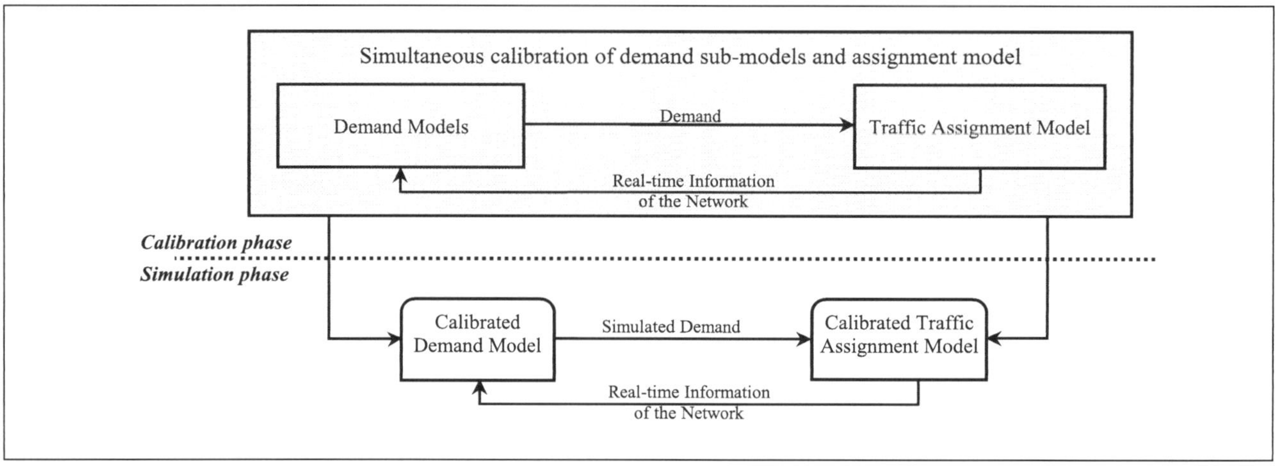

Figure 5. The desirable structure in developing a fully integrated transportation model.

From the network perspective, it is necessary to update the OD matrix as a result of the temporal difference between when the survey data (for the demand) was collected and the time the link counts were observed, which creates conflicting output between the assignment model and the observed link counts. However, when the OD matrix is changed, important characteristics that were directly represented in the demand model but no longer modeled in the OD matrix will be lost.

It is expected that a joint model will generate more consistent results when demand and supply are jointly calibrated. Figure 5 depicts a desirable structure for an integrated demand and supply model. As discussed, the OD updating/calibrating method, as an integral part of OD-based models development, significantly affects demand outputs. Therefore, a sequential model calibration process which directly links demand and supply models is expected to produce more reasonable results. This is the situation where the models are simultaneously calibrated and fully integrated so that they can be used for running scenarios for forecasting, as the demand model is completely sensitive to changes in input variables and is capable of directly information the supply model about consequences of these changes.

When trying to unify demand and supply models, some modifications to the demand model will be necessary, although as previously discussed, these modifications should not be applied (or at least with very limited magnitude applied) to the OD matrix but to the structure of the demand model. The first option is to change the parameters which result from fitting the demand sub-models on surveyed data so that they can reproduce the observed traffic counts. Updating the parameters can include adjusting the constants in the demand sub-models and adjusting the sensitivity of each of the sub-models in choosing an alternative to some variables. As an example, a modeler can adjust the auto availability constants (*14*) and also adjust the sensitivity of choosing a destination for doing a recreational activity to the distance between origin and destination zones, such that the downstream assignment model provides better fit to the observed link flows. It is important to note here that by doing so, the fit to the observed travel behavior data is decreasing at the cost of having better fit to the observed network data. One may also argue that after updating a parameter the model is no more behavioral than it was when it was calibrated by only looking at the demand data. A solution to this problem is to update the underlying assumptions about the error term components reflecting unobserved factors not captured in the parameters of the demand model. In this way the behavioral feature of the demand model is not touched, and the unobserved components are expected to reflect some features of the network. Although updating the error term assumptions seems effective in developing a fully integrated transportation model, it is very complicated.

It is also possible, although less desirable, to weight the agents and the activities of travelers in the joint calibration of demand and supply models. This can procure a high resemblance between the link flows and observed traffic counts (*15*). Examples of the possible changes include changing the trip generation rates for different individuals. Changing these weights is similar to changing data. Since the surveyed data are the most reliable data available, weighting the activities and agents is not recommended, unless there is a valid reason. A valid reason, as an example, could be the change in the attractiveness of a zone because of the opening of a shopping center after the year the surveyed data are collected. Higher weights can be given to the agents in the zone and its surrounding area.

Conclusion

Despite the interdependent relationship between demand and supply, in transport planning models the two aspects are often treated separately. The result may be unrealistic and unpredictable changes in the travel demand, which could result in biased policy predictions. Although the literature shows a growing recognition of the importance of integrating the demand and supply models in the simulation phase, there are fewer proposals regarding integration during the calibration process itself. Certainly the accuracy of the transport models in forecasting various scenarios is highly affected by the procedures taken in linking the demand and supply models and the calibration afterward.

In this paper, the usual process of linking the demand and supply models in the literature is presented, and then the potential impact of an adverse effect of OD updating as the main tool for the linkage is discussed, using the Melbourne area case study. The results revealed that the impact of OD updating in changing a considerable number of the OD pairs is higher compared with the extreme condition of full employment. Therefore, it is important to continue research in other approaches for calibrating the unified structure of demand and supply models, with more focus on demand components instead of on ODs and network attributes. Finally, some possible solutions to improve the quality of the calibration, focusing of demand model, were discussed.

Acknowledgments

The authors would like to thank Hillel Bar-Gera for sharing the TAPAS executable to solve the assignment model for the case study. We would also like to thank the Australian Research Council for their support under Linkage Grant # LP160100450.

Author Contributions

The authors confirm contribution to the paper as follows: study conception and design: Ali Najmi, Milad Ghasri, Taha H. Rashidi; data collection: Ali Najmi, Melissa Duell, Milad Ghasri; analysis and interpretation of results: Ali Najmi, Milad Ghasri, Taha H. Rashidi, S. Travis Waller; draft manuscript preparation: Ali Najmi, Melissa Duell, Milad Ghasri, Taha H. Rashidi. All authors reviewed the results and approved the final version of the manuscript.

References

1. Bloomberg, L., M. Swenson, and B. Haldors. *Comparison of Simulation Models and the Highway Capacity Manual.* Presented at 82nd Annual Meeting of the Transportation Research Board, Washington, D.C., 2003.
2. Wardrop, J. Some Theoretical Aspects of Road Traffic Research. *Proceedings of the Institution of Civil Engineers*, Vol. Part II, No. 1, 1952, pp. 325–378.
3. Spiess, H. *A Gradient Approach for the OD Matrix Adjustment Problem.* Publication No. 693. Centre de Recherche sur les Transports, Université de Montréal, 1990.
4. Spiess, H. A Maximum Likelihood Model for Estimating Origin–Destination Matrices. *Transportation Research Part B*, Vol. 21, No. 5, 1987, pp. 395–412.
5. Juarez, L. H., and M. V. Chavez. OD Matrix Adjustment for Transit Networks by Conjugate Gradient Iterations. *Revista Investigacion Operacional*, Vol. 36, No. 2, 2014, pp. 115–126.
6. Cascetta, E., and S. Nguyen. A Unified Framework for Estimating or Updating Origin/Destination Matrices from Traffic Counts. *Transportation Research Part B*, Vol. 22, No.6, 1998, pp. 437–455.
7. Lundgren, J., and A. Peterson. A Heuristic for the Bilevel Origin-Destination-Matrix Estimation Problem. *Transportation Research Part B: Methodological*, Vol. 42, No.4, 2008, pp. 339–354.
8. Abrahamsson, T. *Estimation of Origin-Destination Matrices Using Traffic Counts: A Literature Survey.* IIASA Interim Report IR-98-021/May, 1998.
9. McFadden, D. Econometric Models for Probabilistic Choice Among Products. *Journal of Business*, 1980, pp. S13–S29.
10. Chorus, C. G., T. A. Arentze, and H. J. P. Timmermans. A Random Regret-Minimization Model of Travel Choice. *Transportation Research Part B: Methodological*, Vol. 42, No.1, 2008, pp. 1–18.
11. Schiffer, R. G., and T. F. Rossi. *New Calibration and Validation Standards for Travel Demand Modeling.* Presented at 87th Annual Meeting of the Transportation Research Board, Washington, D.C., 2008.
12. Walker, J. L. Making Household Microsimulation of Travel and Activities Accessible to Planners. *Transportation Research Record: Journal of the Transportation Research Board*, 2005. 1931: 38–48.
13. Cascetta, E., and F. Russo. Calibrating Aggregate Travel Demand Models with Traffic Counts: Estimators and Statistical Performance. *Transportation*, Vol. 24, No.3, 1997, pp. 271–293.
14. Bowman, J. L., M. A. Bradley, and J. Gibb. The Sacramento Activity-Based Travel Demand Model: Estimation and Validation Results. *Proc., European Transport Conference 2006. CD-ROM. Association for European Transport*, Strasbourg, France, 2006.
15. Cools, M., E. Moons, and G. Wets. Calibrating Activity-Based Models with External Origin-Destination Information: Overview of Possibilities. *Transportation Research Record: Journal of the Transportation Research Board*, 2010. 2175: 98–110.
16. Sheffi, Y. *Urban Transportation Networks: Equilibrium Analysis with Mathematical Programming Methods.* Prentice Hall Inc., N.J., 1985.
17. Daganzo, C. F., and Y. Sheffi. On Stochastic Models of Traffic Assignment. *Transportation Science*, Vol. 11, No.3, 1977, pp. 253–274.
18. Peeta, S., and A. K. Ziliaskopoulos. Foundations of Dynamic Traffic Assignment: The Past, The Present and

The Future. *Networks and Spatial Economics*, Vol. 1, No.3–4, pp. 233–265, 2001.
19. Doblas J, and F. G. Benitez. An Approach to Estimating and Updating Origin-Destination Matrices Based Upon Traffic Counts Preserving the Prior Structure of a Survey Matrix. *Transportation Research B: Methodological*, Vol. 39, No.7, 2005, pp. 565–591.
20. Traffic Assignment Manual. Bureau of Public Roads. *U.S. Department of Commerce*, Washington D.C., 1964.
21. Caggiani, L., M. Ottomanelli, and D. Sassanelli. A Fixed Point Approach to Origin–Destination Matrices Estimation Using Uncertain Data and Fuzzy Programming on Congested Networks. *Transportation Research Part C*, Vol. 28, 2013, pp. 130–141.
22. Bera S, and K. V. Krishna-Rao. Estimation of an Origin-Destination Matrix from Traffic Counts: The State of the Art. *European Transport*, Vol. 49, 2011, pp. 3–23.
23. Auld, J., and A. Mohammadian. Framework for the Development of the Agent-based Dynamic Activity Planning and Travel Scheduling (ADAPTS) Model. *Transportation Letters*, Vol. 1, 2009, pp. 245–255.
24. Lin, D. Y., N. Eluru, S. T. Waller, and C. R. Bhat. Integration of Activity-Based Modeling and Dynamic Traffic Assignment. *Transportation Research Record: Journal of the Transportation Research Board*, 2008. 2076: 52–61.
25. CityX. *DynaMel: Dynamic Traffic Assignment Model of Melbourne*. accessed 1 July 2016. http://www.cityxlab.com/dynamel-dynamic-traffic-assignment-model-of-melbourne.html.
26. VicRoads Open Data Portal [Online]. https://vicroads-public.sharepoint.com/InformationAccess/SitePages/Home.aspx. Accessed March 2016.
27. Ghasrikhouzani, M., T. H. Rashidi T, and S. T. Waller. *Novel System of Disaggregate Models for Travel Demand Modeling Using Decision Tree and Random Forest Concepts*. Presented at 94th Annual Meeting of the Transportation Research Board, Washington D.C., 2015.
28. Bar-Gera, H. Traffic Assignment by Paired Alternative Segments. *Transportation Research Part B: Methodological*, Vol. 44, No.8, 2010, pp. 1022–1046.

The Standing Committee on Traveler Behavior and Values (ADB10) peer-reviewed this paper (18-03382).

Article

TRR
JOURNAL OF THE TRANSPORTATION RESEARCH BOARD

Transportation Research Record
2018, Vol. 2672(47) 125–134

DOI: 10.1177/0361198118792324
journals.sagepub.com/home/trr
SAGE

A Modeling Analysis of Impact from E-Hailing Service on Non-Work Travel Mode in Shanghai, China

Jiayu Zhong[1], Xin Ye[1], Ke Wang[1], and Dongjin Li[1]

Abstract

With the rapid development of mobility services, e-hailing service have been highly prevalent and e-hailing travel has become a part of daily life in many cities in China. At the same time, travelers' mode choice behaviors have been influenced to some degree by different factors, and in this paper, a web-based retrospective survey initially conducted in Shanghai, China is used to analyze the extent to which various factors are influencing mode choice behaviors. Then, a multinomial-logit-based mode choice model is developed to incorporate the e-hailing auto mode as a new travel mode for non-work trips. The developed model can help to identify influential factors and quantify their impact on mode choice probabilities. The developed model involves a variety of explanatory variables including e-hailing/taxi fare, bus travel time, rail station access/egress distance, trip distance, car in-vehicle travel time as well as travelers' socioeconomic and demographic characteristics, etc. The model indicates that the e-hailing fare, travel companions and some travelers' characteristics (e.g., age, income, etc.) are significant factors influencing the choice of e-hailing mode. The alternative-specific constant in the e-hailing utility equation is adjusted to match the observed market share of the e-hailing mode. Based on the developed model, elasticities of LOS attributes are computed and discussed. The research methods used in this paper have the potential to be applied to investigate travel behavior changes under the influence of emerging travel modes. The research findings can aid in evaluating policies to manage e-hailing services and improve their levels of services.

Technologies are driving an unprecedented wave of innovations in mobility services (*1*). E-hailing services have been provided by either application-based start-up companies or traditional car service companies to meet diverse travel needs. E-hailing is a process of ordering a car, taxi, limousine, or any other form of transportation pick-up via a mobile device. The "e" stands for electronic and "hail" refers to the traditional process of signaling for an approaching taxicab to stop. Compared with other forms of public transportation, the advantages of taxis include speed, privacy, comfort, no parking fees, and their door-to-door and 24-hour-a-day attributes (*2*). What is more, e-hailing applications can provide benefits to both drivers and customers, such as less searching and waiting times, and paying fares electronically, etc. E-hailing services have developed rapidly and become popular all over the world. Some of the most well-known e-hailing companies are Lyft, Uber, and DiDi. For example, DiDi, which is a leading mobile transport platform in China, offers app-based mobility options for nearly 400 million users (*3*). And its largest number of orders is more than 20 million a day. According to a report from DiDi (*4*), the frequency of using e-hailing in Shanghai is about 4.6 times per capita a month. Travelers do not even need to download and install applications, they can use the service easily via Alipay (the largest mobile payment platform in China) or WeChat (the largest mobile social media platform in China) on their smartphones. E-hailing services have been growing rapidly and e-hailing travel has become a part of daily life in Shanghai.

New technologies have changed the travel environment (*5–7*). The rapid development of e-hailing services has already brought changes in the travel mode choices of urban residents. Especially for non-work trips, e-hailing services may cause travel mode switches or even induce new trips. The focus of this paper is to explore factors influencing the probability that a traveler chooses

[1]Key Laboratory of Road and Traffic Engineering of the Ministry of Education, College of Transportation Engineering, Tongji University, Shanghai, China

Corresponding Author:
Address correspondence to Xin Ye: xye@tongji.edu.cn

e-hailing for non-work trips when alternatives are available. A multinomial logit model is developed to identify factors that are significant in affecting non-work travel mode choices.

This paper is organized as follows. The next section presents a brief literature review on the previous research into e-hailing services, followed by an overview of the retrospective travel survey, the choice-based sampling method and the multinomial logit model structure. Data sources and sample descriptions are then presented. The focus of this paper is to develop a multinomial logit model based on the survey data and zone-to-zone network skims in Shanghai. Efforts are then made to screen factors with potential impacts on mode choices based on statistical inferences of coefficient estimators. Through this process, the factors influencing travelers' mode-choice behaviors are selected and their implications discussed based on model estimation results. And the alternative-specific constant is then adjusted in the e-hailing utility equation to match the population market share. Based on the developed model, elasticities of LOS attributes are computed and discussed. Finally, conclusions are drawn and possibilities for future research are discussed.

Literature Review

Since e-hailing services have emerged only recently, the number of studies on this topic is very limited. Some studies focus on how e-hailing affects system efficiency. Li and Xia analyzed and compared data collected before and after the emergence of taxi-calling mobile apps in Shanghai, and found that taxi-calling mobile apps had a positive effect on the overall efficiency of the taxi industry (*8*). Chen and Xu collected the order data of the Didi taxi app from the API interface in Guangzhou, and found that taxi reservations reduced the cost of individual motorized travel but inevitably imposed more pressure on the urban transportation system (*9*).

Another group of studies explored the e-hailing impacts on travelers based on travel surveys. Gao et al. conducted two questionnaire surveys among Beijing's car-sharing and e-hailing users, who were asked questions about both their current and previous travel modes, frequency of use, etc., to estimate changes in residents' travel mode choice through descriptive analysis (*10*). Clewlow and Laberteaux conducted a comprehensive household travel survey in five major U.S. cities including questions on the use of car-sharing and ride-hailing services and showed the impacts of new shared-use mobility services on travel behavior (*11*). The findings suggest that early adopters of ride-hailing services tend to be younger, are more highly educated, have higher incomes, and are more likely to reside in dense urban areas. Rayle et al. in 2014 collected survey data from 380 travelers at three ride-sourcing "hot spots" in San Francisco (*12*). In the survey, respondents were asked if they would still have made a given trip had ride-sourcing services not been available and, if so, how they would have traveled. This survey method is similar to the one we used and is called a "retrospective travel survey" in our study. All the above studies conducted descriptive analysis, rather than modeling analysis, based on survey data.

For modeling analysis, Dias et al. developed a bivariate ordered probit model, based on a survey data derived from 2015 Puget Sound Regional Travel Study, to understand the influence of various exogenous socio-economic and demographic variables on the frequency of use of ride-sourcing and car-sharing services (*13*). Through the literature review, we realize that attempts have not been made to develop a discrete choice model to understand the impact of e-hailing services on travel mode choices. This paper aims to fill this gap by developing a logit-based mode choice model using retrospective travel survey data.

Research Methodology

Retrospective Travel Survey on E-Hailing

The key data used in this study are extracted from a survey designed by the authors and administered by a survey company. It is a web-based retrospective survey on e-hailing taxi/car users, involving a total of 1,190 respondents in April 2017. In this survey, all the respondents were asked to report the information on their most recent e-hailing trip, including origin and destination locations, trip beginning time, trip purpose, fare, etc., with the aid of records in their smartphone applications. Then, assuming that an e-hailing service had not been available for that trip, the respondents were asked what they would have chosen for that trip from the following options in sequence: "change trip destination," "change departure time," "change the number of companions," "change travel mode," and "cancel the trip." Detailed information on the change was then collected as well as respondents' household and personal information. Thus, if the individuals who would have chosen to change travel mode are selected to form a subsample, each individual in the subsample virtually makes two travel mode choices. The first choice must be the e-hailing taxi/car among all the available modes while the second choice will be another travel mode including all the modes including regular taxi hailed in the traditional way but excluding the e-hailing mode. These data of the two choices can be pooled for model development.

Choice-Based Sampling Method

If the researcher is particularly interested in factors that affect the choice probability of a specific mode, the size of a random sample from the entire population should be very large to ensure a reasonable number of observations on that mode. To facilitate data acquisition and guarantee data quality, data can be collected directly from travelers using specific travel mode(s) if a mode choice model needs to be developed. For example, in the study on mode choice for commuting, it is often more convenient to survey transit users at stations and car users at parking lots than to interview commuters in their homes. This method of random sampling within a particular mode is called the choice-based sampling method.

In this study, although e-hailing is very common for the residents in Shanghai, a random sample in most areas would only involve a small number of individuals who exactly use the e-hailing service on the survey day. Thus, each respondent in the sample is asked to report information on his/her most recent e-hailing trip. It is important to note that these individuals should not form a biased sample, since the e-hailing service is highly prevalent in Shanghai and almost everyone uses it in their daily life. Even if some older people do not use smartphones, they still need to use e-hailing services with the help of their younger family members. That's because it is now difficult to call a taxicab without using the e-hailing application in Shanghai.

For model development based on a choice-based sample, Manski and Lerman's (*14*) proposed the exogenous sampling maximum likelihood estimation method for consistent estimators. It allows researchers to estimate logit models using choice-based samples without involving complex estimation procedures. Specifically, if the researcher is using a purely choice-based sample and includes an alternative-specific constant in the representative utility for each alternative, then estimating a logit model, as if the sample were exogenous, produces consistent estimates for all the model parameters except the alternative-specific constants. Furthermore, these constants are biased by a known factor and can therefore be adjusted so that the adjusted constants are consistent (*15*). In particular, the expectation of the estimated constant for alternative j, labeled $\hat{\alpha}_j$, is related to the true constant $\alpha^*{}_j$ by

$$E(\hat{\alpha}_j) = \alpha^*{}_j - ln(A_j/S_j) \quad (1)$$

where A_j is the share of decision-makers in the population who chose alternative j, and S_j is the share in the choice-based sample who chose alternative j. Consequently, if A_j is known (that is, if population shares are known for each alternative), then a consistent estimate of the alternative-specific constant is the constant $\hat{\alpha}_j$ that is estimated on the choice-based sample plus the log of the ratio of the population share to the sample share. In this study, the alternative-specific constant can be adjusted based on relevant data of market shares.

Multinomial Logit Model and Choice Set Restriction

A discrete choice approach will be used to model travelers' mode choice behaviors. The MNL model is widely applied in transportation. In this study, the universal choice set is defined to contain the following five main modes, including driving, e-hailing taxi/car, rail, bus, non-motorized mode in the first choice scenario. In the second choice scenario, there are almost the same five modes, except that e-hailing taxi/car is replaced with regular taxi. Since not all travelers can choose a mode from the universal choice set, the choice set needs to be differentiated for each traveler. The choice set for each traveler is developed in the following ways. Driving mode was not available if a traveler did not have a private car. If the rail/bus in-vehicle time from the zone-to-zone network skim takes a missing value because the trip origin or destination does not have good access to metro/bus station, the traveler could not ride the metro/bus for this trip. Taxi and non-motorized modes were available for all the travelers.

The multinomial logit model can be formulated as follows. A decision-maker, labeled i, faces a choice set C_i containing j alternatives. The logit model is obtained by assuming that the decision maker "i" chooses the alternative from C_i with the maximum utility. And the representative utility is usually specified to be linear in parameters: $V_j = X_j\beta_j$, where X_j is a vector of observed variables relating to alternative j. With this specification, the logit probabilities can be calculated as:

$$P(y_{ik} = 1) = \frac{exp(X_{ik}\beta_k)}{\sum_{j \in C_i} exp(X_{ij}\beta_j)} \quad (2)$$

where y_{ik} is a dummy variable indicating alternative "k" is chosen.

The parameter vectors β can be estimated by maximizing the log-likelihood function:

$$LL(\beta) = \sum_{i=1}^{n} \sum_{k \in C_i} y_{ik} ln \left[\frac{\exp(X_{ik}\beta_k)}{\sum_{j \in C_i} \exp(X_{ij}\beta_j)} \right] \quad (3)$$

where $y_{ik} = \begin{cases} 1 \text{ if the } i^{th} \text{ individual chooses alternative } k (U_{ik} = \max\{U_{ij} | j \in C_i\}) \\ 0 \quad \text{otherwise} \end{cases}$

Since e-hailing taxi/car is just like another form of regular taxi, they have similar characteristics. In fact, many cities such as New York in the United States and most cities in China, e-hailing applications have been dominantly adopted by drivers of regular taxis (*2*). The utility function of a taxi can be treated the same as that of e-hailing except for the alternative-specific constant, which quantifies the difference or benefit from the e-hailing service relative to the regular taxi service.

Through the above data preparations, the choice model can be estimated based on the pooled data by employing the Nlogit Software.

Data Sources and Sample Description

The dataset used for analysis and model estimation is extracted from an e-hailing travel survey and zone-to-zone skims. Besides, market share data were used to adjust the alternative-specific constant in the e-hailing utility equation.

Market Share Estimate

The population market share data are derived from the comprehensive travel survey data in Shanghai. According to a report on the comprehensive survey (*16*), the daily trip rate in Shanghai is 2.16. As mentioned before, people use e-hailing service for travel about 4.6 times each month per capita in Shanghai (*4*), which means about 0.1533 times a day. This number is the average use for the entire population in Shanghai. As mentioned in the text, WeChat and Alipay platforms, which are used by most people, are easy to use to hail a car in Shanghai. Even if some older people do not use smartphones, they still need to use e-hailing services with the help of their younger family members. In such situations, we assumed that every person considers e-hailing as one of their choices. Thus, the market share of e-hailing taxi/car mode can be estimated as 0.1533/2.16 = 7%. Before the emergence of e-hailing services, the market share of taxis was 5%, as per the comprehensive travel survey data. Thus, the e-hailing service contributes to a 2% increase in market share.

Zone-To-Zone Skim Matrices

Zone-to-zone skim matrices of different travel models are important for mode choice model development. The rail and bus networks were built up with TransCAD based on GIS information of roads, metro/bus lines, metro/bus stops. The transit and road network map of Shanghai is shown in Figure 1. In the entire city, a total of 5,432 residents' committees are considered to be traffic analysis zones. Residents' committees are called "Juweihui" in Chinese, which is the basic administrative unit of residents in Shanghai. Traffic speeds on roads are estimated using the floating taxi GPS data by time-of-day periods. The zone-to-zone network skims were generated by three time-of-day periods (morning peak: 7:00 to 9:00 a.m.; evening peak: 5:00 to 7:00 p.m.; and flat period: the rest

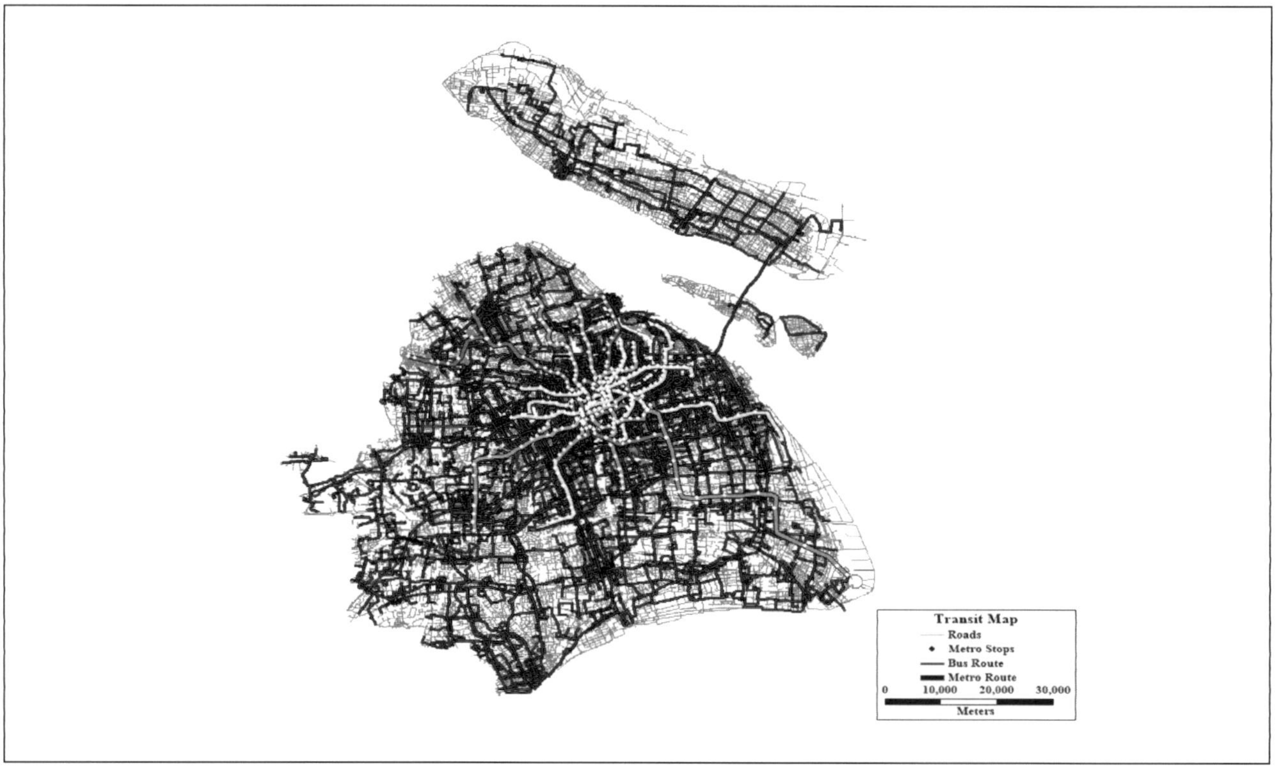

Figure 1. Shanghai transit and road network.

of the day), including car in-vehicle travel time, rail/bus in-vehicle time, rail fare, station access/egress walk time and distance, etc. The level-of-service data were then merged into mode choice data by zone ID and trip beginning time period.

Sample Description

The entire survey sample consists of 1,190 respondents who reported 1,190 e-hailing trips in total. Of those, 25.7% were work or work-related trips, 20% were trips returning home, while the remaining 54.3% were non-work trips. In the subsample of non-work trips, 12.8% of respondents would have chosen to cancel the trip, 19.0% would have chosen to change trip departure time, only 4.8% would have chosen to change trip destination, when faced with the scenario that the e-hailing service was not available. The statistics imply that e-hailing services have become a force to induce new trips while some of them choose to change departure times possibly due to the concerns on the time budget allocation but few of them would change trip destination. Since this study is focused on non-work travel mode choices, only 507 respondents from the survey reporting a non-work trip and intending to only change travel mode, accounting for 78.6% of observations in the non-work trip sample, is selected to form a subsample for mode choice model development.

The personal and household characteristics of those 507 respondents in the subsample are shown in Table 1. Females account for 53.7%, slightly more than males. The average age was 34 years old, while it can be seen that the majority of users are between 20 and 49 years old. These young and middle-aged people are the main force of e-hailing users. Their monthly income ranges from less than 2,000 to more than 30,000 RMB Yuan. Most people are well educated, with an associate degree or above. The job category included full-time job, part-time job, full-time student and part-time student, and so forth. Most people in the sample had full-time jobs. At the same time, occupations were classified in different groups, for example,financial industry, construction industry, computer services and software industry, and so forth. The computer services and software industry accounted for 11.2%, retail industry 11.8%, while some others are distributed sporadically. The sample covers almost all types of occupation in the city. Most respondents were living with families (87%) and some were living in dormitories. The survey collected information on the age distribution of residence members, as shown in Table 1.

The information on the latest e-hailing trip was critical for our study. Trip information, includes trip origin and destination locations, trip beginning times, trip purpose, the number of companions, fare, and so forth. In Figure 2, one can see the distribution of trip origin and

Table 1. Descriptive Statistics of Personal Characteristics (N = 507)

Personal characteristics	Type	Percent	Personal characteristics	Type	Percent
Age (years)	<20	2.76	Gender	Female	53.70
	20–30	34.71		Male	46.30
	31–40	42.80	Job category	Full-time job	82.20
	41–50	16.77		Part-time job	2.00
	51–60	2.17		Full-time student	5.30
	>60	0.79		Part-time student	0.20
Income (RMB Yuan)	<2 k	5.72		Self-employed	4.90
	2–4.5 k	12.03		Housewife	3.20
	4.5–6 k	16.37		Retiree	0.80
	6–8 k	20.12		Unemployed	0.80
	8–10 k	16.57		Others	0.60
	10–15 k	16.96	Residential type	Family	87.00
	15–20 k	5.52		Dormitory	11.60
	20–30 k	5.33		Other	1.40
	>30 k	1.38	Residence members	Infant (0–3 years)	17.20
Education	Elementary school	0.39		Preschooler (4–6)	17.00
	Junior school	1.18		School child (7–11)	20.50
	High school	6.51		Teenager (12–18)	13.80
	Technical school	1.73		60–70 years old	23.90
	Associate degree	22.88		71–80 years old	5.90
	Bachelor degree	51.68		Over 80 years old	2.40
	Master degree	10.06		Other	39.30
	Doctoral degree	2.57			

Note: 1 USD = 6.7 RMB Yuan.

Table 2. Descriptive Statistics of Trip Information (N = 507)

Trip information	Type	Percent	Trip information	Type	Percent
Mode shift	Taxi	48.10	Trip purpose	Personal business	16.37
	Rail	16.00		Studying	6.51
	Bus	14.40		Grocery shopping	0.59
	Non-motor	5.10		Shopping	6.51
	Driving	16.40		Eating	10.85
				Entertaining	17.16
				Daily life	2.37
				Exercising	0.59
				Visiting	13.81
				Medical treatment	8.09
				Picking someone up	7.30
				Others	9.85

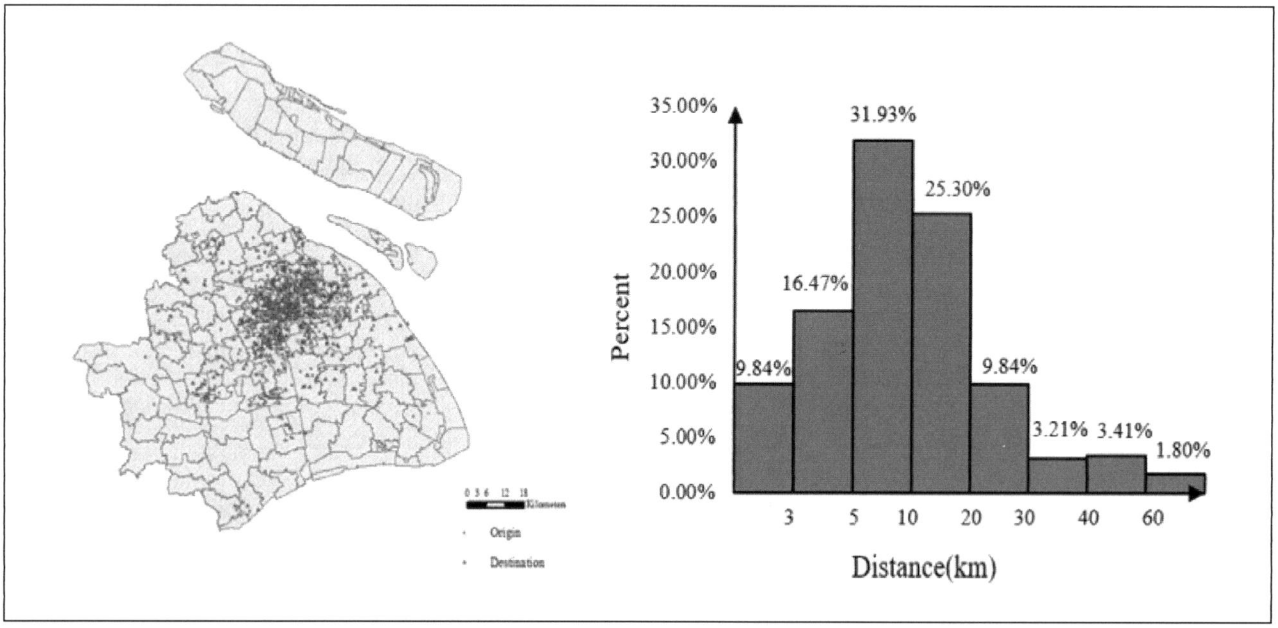

Figure 2. Spatial distributions of trip origins and destinations.

destination locations, which are spread widely across the entire city and reflect well the spatial distribution of the population in downtown and suburban areas. Meanwhile, trip distance distribution ranges widely from less than 3 km to more than 60 km but is mainly in the range of 3–20 km, as in Figure 2. The statistics of non-work trip purpose are shown in Table 2. The important information being collected is how travelers would deal with that trip if the e-hailing service was not available. The subsample only contains non-work trips with only potential mode shift. As shown, the majority of them would chose to travel by traditional taxi instead of e-hailing, accounting for 48.1%, while 16.0%, 14.4%, 5.1% and 16.4% would choose metro, bus, non-motorized and driving mode instead. The statistics indicate the mode shift to the e-hailing taxi/car mode does not only occur from traditional taxi mode but also from other public transit and non-motorized modes. Thus, it will be valuable to develop a mode choice model to better understanding how e-hailing services affect non-work travel market shares.

The level-of-service attributes of different modes and the number of companions have great impact on travel mode choices. The average number of companions is 0.92, which is equivalent to 1.92 for the customer occupancy rate. This number is close to that in a former report on the high occupancy rate (about 1.8) of taxi use in Shanghai (*17*). The descriptive statistics of LOS attributes can be found in Table 3. The Shanghai taxicab service price structure is given as follows: during the period

Table 3. Descriptive Statistics of LOS Attributes (N = 507)

LOS attributes	Mean	S.E.	Maximum	Minimum
Fare of taxi (RMB Yuan)	45.68	51.00	452.41	14.00
Fare of e-hailing (RMB Yuan)	39.48	35.80	280.00	8.00
Rail access distance (km)	1.07	0.77	3.60	0.02
Rail egress distance (km)	0.80	0.60	3.44	0.11
Rail initial waiting time (min)	2.29	0.49	5.50	2.00
Rail in-vehicle time (min)	23.88	18.64	105.00	2.00
Bus initial waiting time (min)	4.14	2.94	60.00	2.00
Bus in-vehicle time (min)	44.23	37.78	252.07	1.49
Bus access distance (km)	0.41	0.31	2.34	0.02
Bus egress distance (km)	0.42	0.33	2.34	0.05
Trip distance (km)	13.20	14.72	127.11	0.58
Car in-vehicle time (min)	23.47	15.06	143.90	2.33

Note: 1 USD = 6.7 RMB Yuan.

between 5 and 11 p.m., taxi drivers charge a flat rate of 14 Yuan for the first 3 kilometers plus a charge of 2.4 Yuan for each additional kilometer over 3 kilometers and 3.6 Yuan for each additional kilometer over 10 kilometers. Thus, the monetary cost on taxi can be calculated for every trip as per skim distance and charge rate. Since the fare is observed for each e-hailing trip, it is directly used for model estimation. The differences between taxi and e-hailing fares in Table 3 is caused by this treatment since e-hailing users in real situations can take advantage of some discounts if using electronic coupons offered by the e-hailing platform.

Model Estimation Results

The model is estimated based on the pooled sample, where each individual makes a mode choice twice under two different scenarios or cases. The universal choice set in both cases contains five alternatives. In Case 1, e-hailing is an alternative mode and has already been chosen. In Case 2, e-hailing is assumed to be unavailable and traditional taxi becomes a substitute. E-hailing and traditional taxi modes will be assumed to share the same utility function except the alternative-specific constant. The main differences between E-hailing and traditional taxi modes include less waiting time and more convenient payment methods. It is challenging to isolate and quantify those factors and therefore alterative-specific constants are simply specified to capture the advantage of e-hailing over traditional taxi mode. The model estimation is completed by using Nlogit. The details of estimation results are shown in Table 4, where all the selected independent variables have significant and expected impacts on mode choice probabilities. Otherwise, we checked the correlations of all the variables specified in the model. Most correlation ratios are near zero, indicating weak correlation among explanatory variables.

The positive or negative factors in the utility functions of different alternatives are reflected by the sign of coefficients. The age terms suggest that the probability of choosing e-hailing increases with age but peaks at about 40 years old. Middle-aged people favor taking an e-hailing or regular taxi since taxi provides the door-to-door and high-quality service that they prefer. The probability is declining when the age is greater than 40. The elderly may be sensitive to price or need help from others to reserve e-hailing taxis. People with a high income are more likely to travel by taxi since they are less concerned about the cost. Travelers working in public management occupations show less tendency to use e-hailing or regular taxis possibly because they usually have more environmental awareness. The number of companions positively affects the e-hailing or regular taxi choice probability. When people travel with their friends or family members, they can share the cost and save travel time. The fare takes a negative factor indicating that the cost of e-hailing or a regular taxi is users' concerns.

In the rail utility function, some level-of-service attributes significantly affect the rail choice probability. Accessibility to a rail station is a significant contributor to rail choice probability and long access/egress distance to/from a station discourages people from choosing rail. The age of travelers is a positive contributor to rail choice probability. Elderly people can take advantage of public transport policies, such as fare discounts or waivers. Meanwhile, since there is a traditional virtue in offering seats to elderly people in China, they can usually get a seat on the metro. And it is found that travelers who engaged in the culture, sports, and entertainment industry or were full-time students are more likely to travel by rail. Since these people are more energetic and like to travel around, their trip destinations are scattered in every corner of the city. Rail can meet their needs for long-distance trips at a low cost.

Table 4. Model Estimation Results

Mode	Variable	Coefficient-statistic	t-Statistic
Taxi	Constant of taxi	−6.2909	−10.40
	The traveler's income is between 15k and 20k (RMB Yuan) a month	0.8833	1.98
	The traveler engages in public management industry	−1.0085	−2.04
	The number of companions	0.1300	1.60
	Fare of taxi (RMB Yuan)	−0.0517	−1.74
	Age of traveler (Years)	0.3087	7.31
	Age squared	−0.0038	−6.15
Rail	Constant of rail	−1.7786	−2.22
	Rail access distance (km)	−0.4000	−1.75
	Rail egress distance (km)	−0.4276	−1.55
	The traveler engages in culture, sport or entertainment industry	2.6222	3.97
	The traveler is a full-time student	0.7125	1.60
	Age of traveler (Years)	0.0262	1.54
Bus	Constant of bus	−1.2456	−2.37
	Bus in-vehicle time (min)	−0.0200	−3.71
	The traveler engages in transportation industry	1.1686	2.53
	The traveler engages in science & technology industry	1.4678	1.93
	The traveler is a full-time student	0.7125	1.60
	Age dummy (41–50 years old)	0.7954	2.20
Non-motor	Trip distance (km)	−0.1909	−3.54
	The traveler lives with her/his families	−1.5862	−3.50
Driving	Constant of driving	−1.5144	−2.17
	The traveler has a driving license	1.4654	2.89
	The traveler engages in energy industry	2.1559	1.98
	The traveler's income is between 2k and 4.5k (RMB Yuan) a month	−1.7935	−2.26
	The traveler has a part-time job	1.8396	2.43
	Car in-vehicle time (min)	−0.0262	−2.21
E-hailing	The traveler's income is between 15k and 20k (RMB Yuan) a month	0.8833	1.98
	The traveler engages in public management industry	−1.0085	−2.04
	The number of companions	0.1300	1.60
	Fare of taxi (RMB Yuan)	−0.0517	−1.74
	Age of traveler (Years)	0.3087	7.31
	Age squared	−0.0038	−6.15
Summary Statistics			
Number of observations: 507 × 2 = 1,014			
LL ($\hat{\beta}$): −560.259			
$\rho^2(0)$: 0.656			
$\hat{\rho}^2(0)$: 0.578			

Note: 1 USD = 6.7 RMB Yuan.

Bus in-vehicle time is a significant factor affecting bus choice probability. The longer the bus in-vehicle time, the smaller the probability. For a short distance trip, taking a bus may be convenient and quick. But if a trip distance is long, boring and involves a long in-vehicle time this leads to a poor travel experience. It is interesting to find that travelers between 41 and 50 years old are more likely to use the bus instead, possibly because travelers in this age group undertake heavy family burdens and become sensitive to higher prices. It is found that travelers who work in the transportation and science & technology industries or were full-time students were more likely to choose the bus. The bus's low fares appeal to full-time students. Companies in the transportation and science & technology industries may have subsidies for bus travel.

It is confirmed that the trip distance is a determinant of non-motorized mode choice probability. Long distances discourage people from traveling by non-motorized modes such as walking or cycling. In particular, people who live with family members are less likely to favor using this mode since people usually go out with their family members for non-work purposes. Non-motorized modes are not suitable for people traveling together, except for short-distance trips.

In driving mode utility, it is undoubtedly the case that having a driver's license increases the driving probability. Driving choice probability decreases with in-vehicle time since driving a car requires a driver's full attention which, coupled with road congestion in peak hours, can leave drivers feeling bored and anxious. People with lower

Table 5. Constant Adjustment Results

Case	Constant	Market share
1	–9.1861	7.0%
2	–9.5777	5.0%

incomes are less likely to drive since it is costly to use and park a car. Travelers working in the energy sector show a greater tendency to drive since they may develop a preference for driving to work. It is interesting to find that a traveler with a part-time job is more likely to drive. It is possibly because they need to undertake a wide variety of daily activities and need a high degree of mobility during the day.

According to Equation 1, the constants in taxi and e-hailing utility equations can be adjusted to match respective market shares (approximately 5% and 7%, as mentioned before) in the population. The constants after the adjustment are listed in Table 5. One can see that the constant in Case 1 exceeds the constant in Case 2 by 0.3916. This difference can be understood as the contribution from the e-hailing service to the overall utility value of traditional taxi mode, incorporating all the benefits from short searching and waiting times, less uncertainty, convenient payment method, etc.

Based on the developed models, both direct and cross elasticities of LOS attributes are computed and all the results are shown in Table 6. One can see that the fare elasticity of e-hailing/taxi is –0.220/–0.231. It means if e-hailing/taxi fares go up by 10%, users will decrease by 2.20%/2.31%. It shows that the passengers of e-hailing/taxis are not very sensitive to trip expenditures. Since the taxi fare is much higher than that of other forms of public transport for the same trip, low-income people may frequently not consider taking a taxi. On some special occasions, such as going to a hospital or having an emergency to handle, people would not be too concerned about fares. It helps to explain the low fare elasticity. If rail access/egress distance becomes longer, travelers will be less likely to use rail but they do not seem highly sensitive to the access distance and egress distance. The absolute values of direct elasticities are less than 0.2 while those of most cross elasticities are less than 0.1. The bus in-vehicle time's direct elasticity is about –0.44 while cross-elasticities range from 0.097 to + 0.275. The most elastic attribute is the non-motorized trip distance, which has a direct elasticity of –0.79 on the non-motorized mode but it has much less cross elasticities on all other modes. The car in-vehicle time's direct elasticity is about –0.24 but the cross elasticity is less. In general, most LOS attributes arc not considered of high elasticity.

Conclusions and Discussions

This paper is an initial attempt to explore the impact of e-hailing services on non-work travel mode choices based on retrospective travel survey data. Multinomial logit models are developed for both traditional taxi and e-hauling taxi/car modes, where multiple level-of-service attributes and travelers' household and personal characteristics are specified as explanatory variables into the utility functions for alternative travel modes. Alternative-specific constants in traditional taxi and e-hauling taxi/car utility functions are adjusted to match the overall market shares observed from the population. Then, the additional benefit contributed from the e-hailing service to taxi mode can be quantified by the difference between alternative-specific constants in respective utility functions.

Utilities of alternative modes are compared in order to identify influential factors behind observed choice behaviors. The model estimation results indicate that high incomers, the middle-aged and people with companions are more likely to take a taxi or use an e-hailing service. People working in public management show less tendency to use a taxi or an e-hailing taxi/car. Full-time students are more likely to take a bus or train. A traveler in a part-time job is more likely to drive. Significant LOS attributes include transit/e-hailing fare, rail access/egress distance, bus in-vehicle time, car in-vehicle time and non-motorized trip distance. Elasticity analysis shows

Table 6. Elasticity of LOS Attributes

	E-hailing /Taxi		Rail		Bus		Non-motorized		Driving	
LOS attributes	Case1	Case2	Case1	Case2	Case1	Case2	Case1	Case2	Case1	Case2
Fare of taxi/e-hailing	–0.220	–0.231	+ 0.020	+ 0.014	+ 0.018	+ 0.015	+ 0.008	+ 0.006	+ 0.014	+ 0.010
Rail access distance	+ 0.110	+ 0.113	–0.174	–0.170	+ 0.097	+ 0.100	+ 0.054	+ 0.055	+ 0.056	+ 0.057
Rail egress distance	+ 0.100	+ 0.102	–0.159	–0.156	+ 0.085	+ 0.087	+ 0.047	+ 0.048	+ 0.057	+ 0.058
Bus in-vehicle time	+ 0.246	+ 0.275	+ 0.206	+ 0.212	–0.442	–0.436	+ 0.124	+ 0.127	+ 0.097	+ 0.099
Non-motorized trip distance	+ 0.116	+ 0.120	+ 0.091	+ 0.093	+ 0.139	+ 0.143	–0.787	–0.780	+ 0.049	+ 0.050
Car in-vehicle time	+ 0.113	+ 0.112	+ 0.099	+ 0.100	+ 0.069	+ 0.070	+ 0.036	+ 0.036	–0.244	–0.238

that most LOS attributes are not quite elastic enough to change the market shares. However, the emerging e-hailing service has increased the market share of taxis from 5% to 7% in Shanghai, which is equivalent to a relative change of 1.4. This technological innovation should be considered of high elasticity to the market change.

The retrospective travel survey and choice modeling methods used in this paper have the potential to be applied to investigate travel behavior changes under the influence of emerging travel modes, such as shared cars, shared bicycles, and so forth, in the mobile internet environment. Further study will make attempts to apply big data. Traditional travel survey method is still widely used currently because researchers can design survey questions to collect information on individual personal and socioeconomic characteristics, although it is limited by the sampling process and potential discrepancies between respondents' actual behavior and their responses. If one can combine e-hailing big data and survey data, more relevant variables may be used in a mode choice modeling process. A more sophisticated method may be developed for data collection and model estimation, through which the change of travel behavior can be understood better in the new environment.

Acknowledgments

This research is supported by the National Natural Science Foundation of China. The project numbers are No. 71671129 and No. 71734004.

Author Contributions

The authors confirm contribution to the paper as follows: study conception and design: Jiayu Zhong, Xin Ye, Ke Wang; data collection: Jiayu Zhong; analysis and interpretation of results: Jiayu Zhong, Xin Ye; draft manuscript preparation: Jiayu Zhong, Xin Ye, Dongjin Li. All authors reviewed the results and approved the final version of the manuscript.

References

1. Peng, C., and N. Yu. Connecting E-hailing to Mass Transit Platform: Analysis of Relative Spatial Position. *Transportation Research Part C: Emerging Technologies*, Vol. 77, 2017, pp. 444–461.
2. He, F., and Z. J. M. Shen. Modeling Taxi Services with Smartphone-Based E-Hailing Applications. *Transportation Research Part C Emerging Technologies*, Vol. 58, 2015, pp. 93–106.
3. DiDi. *Overview*. www.didichuxing.com/en/aboutus/overview. Accessed June 18, 2017.
4. DiDi and CBNData. *China's Smart Mobility Data Report in 2015*. person.zju.edu.cn/attachments/2016-04/07-1461299406-696460.pdf. Accessed June 18, 2017.
5. Tang, J., S. Zhang, X. Chen, F. Liu, and Y. Zou. Taxi Trips Distribution Modeling Based on Entropy-Maximizing Theory: A Case Study in Harbin City—China. *Physica A Statistical Mechanics & Its Applications*, Vol. 493, 2018, pp. 430–443.
6. Tang, J., F. Liu, W. Zhang, R. Ke, and Y. Zou. Lane-Changes Prediction Based on Adaptive Fuzzy Neural Network. *Expert Systems with Applications*, Vol. 91, 2017, pp. 452–463.
7. Tang, J., F. Liu, Y. Zou, W. Zhang, and Y. Wang. An Improved Fuzzy Neural Network for Traffic Speed Prediction Considering Periodic Characteristic. *IEEE Transactions on Intelligent Transportation Systems*, Vol. 18, No. 9, 2017, pp. 1–11.
8. Ye, L., and T. Xia. Evaluation of Efficiency of Taxi Industry: Effects of Taxi-Calling Mobile Apps. Presented at 94th Annual Meeting of the Transportation Research Board, Washington, D.C., 2015.
9. Chen, J., and L. Xu. The Compact of Reserve Rent-a-car to Urban Transportation. *Traffic & Transportation*, No. A02, 2016, pp. 162–165.
10. Gao, Y., A. Jian, and Y. Quan. The Impact of APP-Based Car Sharing on Travel Mode Shift and Transportation Operation Performance. *Urban Transport of China*, Vol. 14, No. 5, 2016, pp. 1–8.
11. Clewlow, R., and K. Laberteaux. Shared-Use Mobility in the United States: Current Adoption and Potential Impacts on Travel Behavior. Presented at 95th Annual Meeting of the Transportation Research Board, Washington, D.C., 2016.
12. Rayle, L., S. Shaheen, N. Chan, D. Dai, and R. Cervero. App-Based, on-Demand Ride Services: Comparing Taxi and Ridesourcing Trips and User Characteristics in San Francisco. Presented at 93rd Annual Meeting of the Transportation Research Board, Washington, D.C., 2014.
13. Dias, F. F., P. S. Lavieri, V. M. Garikapati, S. Astroza, R. M. Pendyala, and C. R. Bhat. A Behavioral Choice Model of the Use of Car-Sharing and Ride-Sourcing Services. *Transportation*, Vol. 44, No. 6, 2017, pp. 1307–1323.
14. Manski, C. F., and S. R. Lerman. The Estimation of Choice Probabilities from Choice Based Samples. *Econometrica*, Vol. 45, No. 8, 1977, pp. 1977–1988.
15. Train, K. E. *Discrete Choice Methods with Simulation*. Cambridge University Press, 2009, pp. 66–67.
16. Shanghai Urban-Rural Construction and Transportation Development Research Institute. The Main Results of the Fifth Comprehensive Traffic Survey in Shanghai. *Traffic & Transportation*, Vol. 31, No. 6, 2015, pp. 15–18.
17. Mao, B., Y. Yin, and W. Guan. Countermeasures and Suggestions for the Development and Management of Urban Taxi in China. *Comprehensive Transportation*, No. 9, 2013, pp. 35–38.

The Standing Committee on Traveler Behavior and Values (ADB10) peer-reviewed this paper (18-04087).

Article

TRR
JOURNAL OF THE TRANSPORTATION RESEARCH BOARD

Transportation Research Record
2018, Vol. 2672(47) 135–145

DOI: 10.1177/0361198118782760
journals.sagepub.com/home/trr
SAGE

Alternative Activity Pattern Generation for Stated Preference Surveys

He He[1], Bilge Atasoy[1], J. Cressica Brazier[1], P. Christopher Zegras[1], and Moshe Ben-Akiva[1]

Abstract

We present a systematic method for generating activity-driven, multi-day alternative activity patterns that form choice sets for stated preference surveys. An activity pattern consists of information about an individual's activity agenda, travel modes between activity episodes, and the location and duration of each episode. The proposed method adjusts an individual's observed activity pattern using a hill-climbing algorithm, an iterative algorithm that finds local optima, to search for the best response to hypothetical system changes. The multi-day approach allows for flexibility to reschedule activities on different days and thus presents a more complete view of demand for activity participation, as these demands are rarely confined to a single day in reality. As a proof-of-concept, we apply the method to a multi-day activity-travel survey in Singapore and consider the hypothetical implementation of an on-demand autonomous vehicles service. The demonstration shows promising results, with the algorithm exhibiting overall desirable behavior with reasonable responses. In addition to representing the individual's direct response, the use of observed patterns also reveals the propagation of impacts, that is, indirect effects, across the multi-day activity pattern.

Individuals generally do not make decisions to travel in isolation. Rather, they make these choices simultaneously with, or as a result of, decisions to participate in activities. Accounting for the motivations underlying individuals' activity-travel behavior has become increasingly recognized as essential to understanding today's dynamic urban systems. This activity-based lens is particularly useful for examining complex behavior, such as trip chaining and discretionary travel, as well as the impacts of the range of "soft" measures associated with mobility management. Considering the growing popularity of flexible work hours and shared on-demand modes, and the potentially disruptive changes associated with autonomous vehicles (AVs) and mobility as a service, we need better ways to understand how such innovations will influence individuals' demand for activities and, ultimately, the performance of the mobility and broader urban system.

To understand the possible implications of new products and services unfamiliar to a consumer, researchers have long used the stated preference (SP) survey. Travel behavior analysis often uses SP surveys, for example, to understand how users might respond to a new mode, such as high speed rail, or infrastructure, such as bike lanes. Such analyses, however, tend to be trip-focused—conditional on a trip being made. New mobility innovations might, however, impact individuals' activity implementation plans or schedules by, for example, allowing more time to partake in discretionary activities in a new location. Thus, understanding the process by which individuals and households schedule their activities remains a core challenge in inferring responses to new mobility or activity options. Even though activity scheduling is something that nearly everyone actually does, to some degree, on a daily basis, formalizing and generalizing it analytically is difficult. Activity scheduling is a multidimensional task that requires the decision maker to choose a number of activities to complete, the sequence of these activities, the time, duration, and location of each activity, and, when relevant, the mode of travel and route between activity locations. Even assuming a reasonable discretization of time and space, the resulting number of possible combinations is both computationally demanding and too large for any person to feasibly consider, cognitively. Generating alternative activity patterns, thus, must aim to be computationally efficient,

[1]Massachusetts Institute of Technology, Cambridge, MA

Corresponding Author:
Address correspondence to He He: thehe@mit.edu

while mimicking people's actual decision-making processes.

In this paper, we propose a systematic method for generating alternative activity patterns that will form choice sets for SP surveys within an activity-based modeling framework. The method takes advantage of detailed multi-day activity diaries collected using Future Mobility Sensing, a smartphone-based activity-travel survey platform (*1*). We start with individuals' revealed activity-travel patterns as the base for the new activity patterns. Then, using the proposed method, we adjust the patterns in response to hypothetical scenarios. In addition to representing the individual's direct response, the use of observed patterns also reveals the propagation of impacts across a multi-day activity pattern. Here, we review existing work addressing the challenges of generating alternative activity patterns. We then present the new method and the issues we considered in its development, and finally we show the results of a proof-of-concept application of the proposed method using an activity-travel survey conducted in Singapore.

Background

Since the inception of the activity-based approach to travel demand modeling, researchers have proposed and operationalized numerous activity-based frameworks. Each has taken on the challenge of activity scheduling. Early work established that proposed activity-based models must recognize and take into account the decision-making behavior of individuals and households (*2, 3*). Furthermore, they must be complete in scope, that is, have sufficiently high resolution and flexibility to capture the behavioral changes of interest. Lastly, they must be feasible and practical to operationalize. Early frameworks applied mathematical programming and neural networks in attempts to manage the complexity of the problem (*4, 5*). However, these methods did not simulate how people actually plan their schedules. More recent modeling frameworks have used more sophisticated ways of representing planning and decision-making processes. These approaches have generally fallen on a spectrum between entirely econometrics-based methods, building on discrete choice theory, for example, SimMobility (*6*), and largely rule-based methods that build on decision heuristics, for example, TASHA (*7, 8*) and ALBATROSS (*9*). Most modeling frameworks can be considered hybrids, applying a combination of heuristics and econometrics or at least using the concept of utility maximization to simulate rationality. Pinjari and Bhat provide a comprehensive overview of activity-based modeling frameworks (*10*).

A major challenge to operationalizing activity-based models has to do with realistically representing how individuals schedule activities and, thus, what alternative activity patterns might exist for a given individual. For the purposes of an activity-based SP survey, aimed at understanding how an individual's mode and activity preferences might change in the face of new alternatives, realism would be gained by presenting activity pattern alternatives based on respondents' observed activity patterns. This requires methods that can capitalize on the availability of existing patterns and estimate the utility of activities for a given individual. The AMOS model, a micro-simulator designed to predict behavioral responses to travel demand management (TDM) measures, offers a precedent (*5*). AMOS uses a neural network approach to generate TDM options based on revealed preference and SP survey data. Its focus, however, is on commuters and it uses pre-coded response types. For example, in its application to the Washington DC area, seven response types are possible, including "no change,""change work departure time,""work from home," and four different options for work trip mode changes (*11*). The neural network calculates probabilities for each response type based on the respondent's socio-economic characteristics and existing activity patterns, and responses are selected using Monte Carlo simulation. Afterwards, the model adjusts the rest of the activity pattern to accommodate the response if necessary. However, these adjustments are mostly constraint-based. Furthermore, the prescriptive nature of the pre-coded responses does not capture the diversity of people's potential response strategies (*12*) and largely eliminates the possibility of response discovery.

As micro-simulators have become more advanced, and as technology-based survey methods have allowed for more comprehensive time-use and travel data collection, for example, CHASE (*13*), REACT! (*14*), and UTRACS (*15*), several modeling frameworks have incorporated near-term rescheduling modules. These frameworks adapt agents' planned activity patterns in response to unforeseen circumstances that agents do not take into consideration in the original activity scheduling phase. The components of this problem, that is, an existing activity pattern and a change in circumstances triggering adaptations, are the same as those we are faced with in generating alternative activity patterns for SP surveys. Hence, the methods employed by various activity-based microsimulation frameworks for finding solutions to the rescheduling problem might also apply to our problem. AURORA, a rescheduling module developed as an extension to the ALBATROSS framework, assigns utility to activities based on their duration (*16, 17*). Specifically, the utility function is S-shaped, monotonically increasing with activity duration, and has horizontal asymptotes corresponding to the minimum and maximum utilities that can be derived from an activity. The second derivative of the utility function is positive until it reaches an inflection point, after which it becomes negative. AURORA allows for a diverse combination of responses, including changes to travel mode, episode durations, locations, and sequence, as well as insertion and deletion of episodes. The model chooses a response using an iterative optimization procedure that imitates bounded rationality and accounts for people's resistance to change by introducing a disutility of each change operation. The primary drawbacks to the AURORA model are the difficulty of estimating the large number of parameters of the S-shaped utility function (*18*, *19*) and its lack of multi-day schedule integration (*20*). Another agent-based modeling framework, MATSim, uses a simpler two-parameter utility formulation, in which utility is a logarithmic function of activity duration (*21*). In other words, utility is monotonically increasing with activity duration with an ever-diminishing marginal effect. A rescheduling extension for MATSim now applies an iterative approach similar to that in AURORA (*22*). However, this extension simplifies the procedure considerably in order to reduce computational demand.

Activity Pattern Generation Procedure

We propose a procedure for generating personalized alternative activity patterns triggered by changes in the transportation system or in the characteristics of activities. The procedure builds on the concepts and methods developed in AURORA (*17*), the MATSim rescheduling module (*22*), and others. Specifically, our algorithm is also an iterative optimization procedure that uses elementary operations to search incrementally for better alternatives. We assume the logarithmic formulation for activity utility and make use of the same elementary operations as implemented in MATSim, and we also adopt a notion of *resistance to change* similar to AURORA, however we apply it directly to changes to schedule times rather than the number of elementary operations. Despite the similarities, our objective differs slightly. We are not necessarily interested in the most likely response to system change; rather, we want to explore feasible responses by generating alternative activity patterns that account for activity demands. Furthermore, we examine multi-day activity patterns and allow for flexibility to schedule activities on different days; for example, a disruption on a given day might result in the rescheduling of a subsequent activity to a different day in the multi-day pattern. This approach presents a more complete view of an individual's activity demand, which in reality is rarely confined to a single day. Lastly, the application of the methods to comparatively small samples in SP surveys, as opposed to a synthesized population of agents, somewhat relaxes the constraints on computation time.

For our proposed procedure, an alternative activity pattern is generated using an iterative hill-climbing algorithm to search for a local maximum utility, given a set of preference parameters. The hill-climbing algorithm is a mathematical optimization technique that takes an observed pattern and incrementally improves it by changing one element at a time. The algorithm is complete when the utility of the pattern can no longer improve by changing any one element. Some modelers question the appropriateness of utility maximization from a behavioral standpoint. However, its application here is local in nature: the number of possible activity patterns is severely limited by the procedure of changing single elements combined with the requirement that each iteration of the algorithm must score better than the last, it must "move up the hill." In other words, alternative patterns cannot change drastically by "jumping between hills." The hill-climbing algorithm can be interpreted as a search for the best pattern that is similar to the observed pattern. The sequential approach, together with the non-exhaustive search procedure, considerably reduces computation time compared with simultaneous optimization approaches. This is especially the case as the activity patterns become more complex with the introduction of more activity types, locations, modes, etc. The procedure generates each alternative activity pattern as the "optimal choice' for a given set of preference parameters; thus, we can generate a different alternative pattern by assuming a different set of parameters. In other words, a complete choice set of alternative activity patterns can be generated by varying the individual's assumed preferences, either systematically or randomly. The preference parameters describe an individual's activity demands, modal preferences, value of time, and flexibility.

Figure 1 shows the steps within each iteration of the algorithm. For the first iteration, the *observed activity pattern* serves as the baseline. The final activity pattern output from the procedure is the *alternative activity pattern*. During each iteration of the hill-climbing algorithm, we generate and propose numerous activity patterns, called *proposed activity patterns*. Lastly, within these proposed activity patterns, an *activity agenda* consists of a set of consecutive episodes, for example, home–work–home–shop–home. This activity agenda does not provide information about the timing, duration, or location of episodes, or any information about travel.

Generate Proposed Activity Agendas

Within each iteration the agenda of the baseline pattern serves as a skeleton. We generate every unique proposed agenda that we can possibly create by applying the single elementary operations to the baseline agendas. The number of proposed agendas generated in this step depends

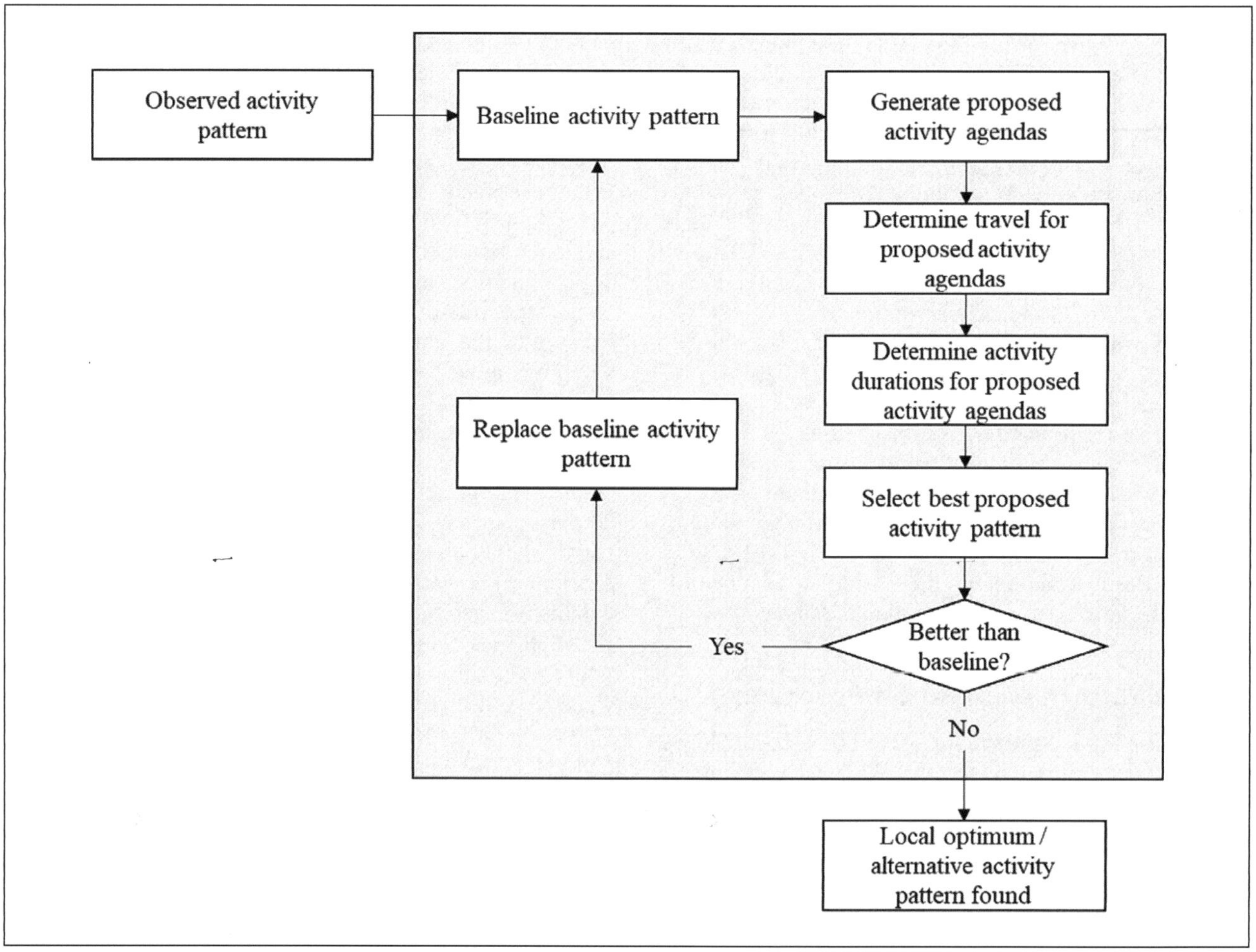

Figure 1. Alternative activity pattern generation procedure.

primarily on the number of episodes in the observed activity pattern and the number of different activity types. Figure 2 illustrates the elementary operations used to modify the baseline agenda. These are the same elementary operations used by Balac and Axhausen (*22*).

First, an attempt is made to insert a new episode between two episodes in the baseline agenda (Figure 2*a*). This operation occurs with every activity type between every pair of consecutive episodes in the baseline agenda. To preserve the "home" episodes as the anchors of the daily agendas, new episodes are not inserted at the beginning or end of the agenda. If the inserted episode is different in type from both the preceding and subsequent adjacent episodes in the agenda, a new agenda is created. Otherwise, the procedure continues on to the next potential slot or activity type.

Next, an attempt is made to remove a single episode at a time (Figure 2*b*), for every episode in the baseline agenda. If the episode is not an overnight "home" episode, it is removed and a new agenda is created. Otherwise, the procedure moves on to the next episode in the agenda. Following the episode removal, if two consecutive episodes are of the same type, one of them is removed.

Finally, two episodes in the baseline agenda are swapped if neither is an overnight "home" episode (Figure 2*c*). This operation is attempted for every possible episode pair combination in the agenda. By allowing this swapping operation, proposed activity patterns with higher total utility can be generated. However, these patterns will likely also bear less resemblance to the individual's observed activity pattern. Whether or not swapping generates reasonable changes in general requires further testing. For our example application in the next section, we do not apply the swapping operation.

Determine Travel for Proposed Activity Agendas

This step determines the mode for every trip in every agenda using the principle of utility maximization. First,

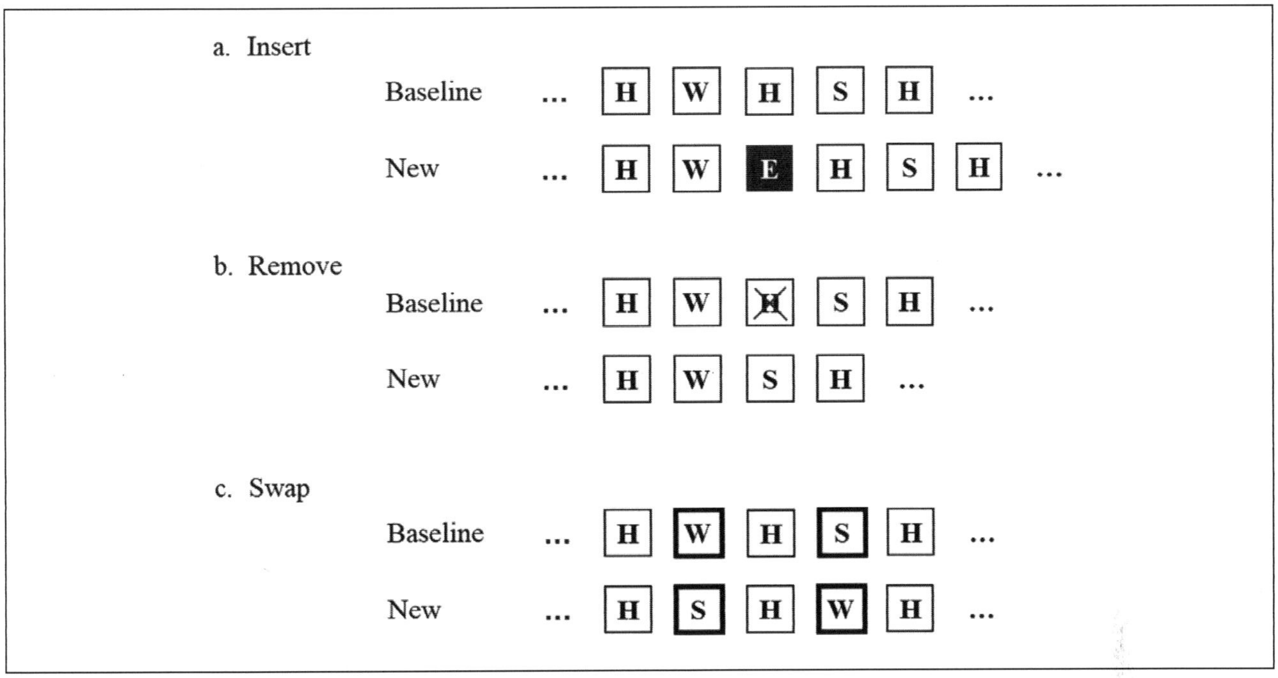

Figure 2. Example of the insert, remove, and swap operations.

each proposed agenda is aligned with the observed agenda and the closest episodes of the same type within the same day are paired across the two agendas. Note that this comparison is with the original, observed agenda, which is generally different from the baseline agenda after the first iteration. Typically, episodes that have undergone the "insert" or "remove" operations cannot be paired, because they do not have a corresponding episode in the observed pattern or proposed agendas, respectively. For paired episodes, the activity location remains the same. For unpaired episodes with multiple possible activity locations (e.g., if the user could complete a shopping activity at similar stores in different locations), we take a myopic approach. Specifically, trips are evaluated sequentially and the mode choice becomes a joint mode–destination choice, because each trip entails the choice between all possible modes to all possible activity locations for the given activity type. The choice of mode–destination for the current trip might have adverse effects on downstream mode–destination choices; however, we do not account for these effects at present. A more nuanced method for selecting activity locations is part of the ongoing work.

For a given origin–destination pair and mode m, we assume that the utility of a trip V_{trip} is a function of the travel time and travel cost:

$$V_{trip} = \beta_m + \beta_{m,T} \cdot T_m + \beta_{m,C} \cdot C_m \qquad (1)$$

where:

- β_m is an alternative specific constant for mode m
- $\beta_{m,T}$ is the parameter for travel time by mode m
- T_m is the travel time by mode m
- $\beta_{m,C}$ is the parameter for travel cost by mode m
- C_m is the travel cost by mode m

The parameters can be varied for this utility function to generate a set of different activity patterns, or the parameters can be estimated from a pre-survey. In addition to the utility of the trip itself, the selection of mode should account for the opportunity cost of travel time, that is, time that could be spent productively on other activities. For example, if it takes a person 20 minutes to drive home from work and 60 minutes to walk, the drive option offers an additional 40 minutes to spend at home or to carry out another activity later on, from which the person can derive additional utility. In other words, the duration of the trip affects utility outside of the trip itself. However, the selection of modes should account for this opportunity cost. Therefore, we estimate the magnitude of the opportunity cost for a trip OC as:

$$OC = -\frac{V_{activity,\,total}}{D} \cdot T_m \qquad (2)$$

where:

- $V_{activity,total}$ is the total utility from all episodes in the observed activity pattern
- D is the total duration spent on activities in the observed activity pattern

The procedure of assigning utility to activities follows in the next section. For each trip, the selected mode and destination, if applicable, are those which yield the maximum total utility, being the minimum sum of trip disutility and the opportunity cost of travel time.

Determine Activity Durations for Proposed Activity Agendas

With travel determined, we now determine the duration of each episode in each activity agenda such that the total utility of each agenda is maximized. We assume that the utility of an activity a is a function of the duration of the activity D_a:

$$V_{activity} = \beta_a \cdot \ln(\alpha_a \cdot D_a) \tag{3}$$

Here β_a and α_a are scale and shape parameters, respectively, for the utility of activity type a. This formulation of utility follows the method that Charypar and Nagel applied to MATSim (*20*). Again, we can derive or estimate the parameters such that the observed activity pattern is optimal without system changes, or we can vary them to generate a set of different activity patterns. For positive parameters, the utility function is monotonically increasing with duration but diminishing at the margin. The diminishing marginal effect can be interpreted as fatigue, boredom, or other forms of efficiency loss as the duration of an activity increases. For some activities in which utility is not necessarily derived intrinsically, such as work or household chores, Equation 3 models the extrinsic value of the activity, such as a salary or a clean house. In this framework, the scheduling process is activity-driven, because travel only occurs when the utility derived from an activity outweighs the disutility of traveling there. Furthermore, if we use reasonable values for the β_a and α_a parameters, the principle of utility maximization ensures a balanced composition of activities, since the diminishing marginal utility, such as experiencing boredom or fatigue, encourages individuals switch to activities that are more pressing or rewarding. We should also note that because the utility function in Equation 3 is non-linear, calculating utility based on duration per episode versus per day or per week will yield different results (Figure 3). Here, three short episodes, each with duration d_a, yield a total of 41.7 utility units if calculated per episode. However, the same three episodes yield only 24.8 utility units if the episode durations accumulate over the analysis period.

Calculating utility on a per episode basis generally results in more frequent-short episodes, because the marginal utility is higher for short episodes. On the other hand, calculating utility on a per week basis results in fewer long episodes in order to minimize the disutility of travel. We handle this discrepancy by distinguishing between mandatory/maintenance activities and discretionary activities. In particular, we accumulate activity durations over the multi-day analysis period for mandatory and maintenance activities, such as work, shopping, and personal errands, while we use daily accumulation for discretionary activities, such as social and recreational.

Beyond the utility of the activity itself, proposed episodes are penalized for changing the individual's schedule, that is, the observed activity pattern. In principle, this is similar to the disutility of change used by Joh et al. (*17*). However, that study applies the change to the operation itself, whereas our disutility of change is a function of differences in episode start times between episodes in the observed pattern and the new pattern. Only

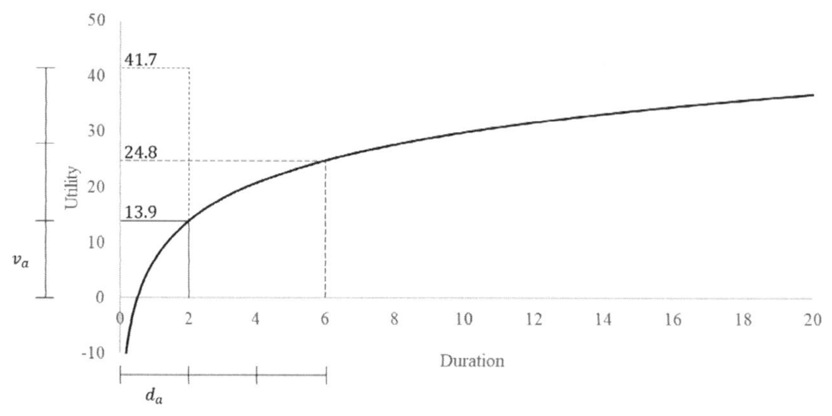

Figure 3. Illustration of activity utility discrepancy for different utility accumulation methods.

proposed episodes that successfully pair with episodes in the observed agenda will produce this disutility of change. In other words, newly inserted episodes and episodes that shifted to different days do not have an associated disutility of change. For an episode of activity type a, Equation 4 gives the disutility of change, V_{change}:

$$V_{change} = -\frac{i_a}{f} \cdot (t_{start,new} - t_{start,obs})^2 \quad (4)$$

where:

- i_a is a parameter representing the inertia of activity type a
- f is the flexibility of the individual
- $t_{start,new}$ is the start time of the episode in the proposed pattern
- $t_{start,obs}$ is the start time of the episode in the observed activity pattern.

Intuitively, activity types that are more difficult to move temporally have higher inertia and produce greater penalties when their start times have to change. On the other hand, flexible individuals experience lower penalties for changes overall. Similar to other parameters in this framework, the activity-specific inertia and flexibility parameters can be varied to generate different activity pattern alternatives, or they can be derived from a pre-survey or a participant's observed pattern. As part of our ongoing work, we are testing different functional forms; the current formulation is only an a priori hypothesis of the disutility of change. Validating these formulations and the parameters remains challenging, however, and the methods to do so empirically are not obvious.

With all travel already fixed, the duration of each episode can be determined by maximizing the utility of the proposed pattern. The optimization problem is formulated as:

$$\max_{d}\left(\sum V_{activity} + \sum V_{change}\right) = \sum_{A} \beta_a \cdot \ln(\alpha_a \cdot D_a) + \sum_{e'_{new}}^{E'_{new}} -\frac{i_{a,e'_{new}}}{f} \cdot \left(t_{start,new,e'_{new}} - t_{start,obs,e'_{new}}\right)^2$$

$$s.t. \sum_{r_{new}}^{R_{new}} T_{r_{new}} + \sum_{e_{new}}^{E_{new}} d_{e_{new}} - \sum_{r_{obs}}^{R_{obs}} T_{r_{obs}} - \sum_{e_{obs}}^{E_{obs}} d_{e_{obs}} = 0 \quad (5)$$

where:

- d_e is the duration of an episode e
- D_a is the sum of durations of episodes e of activity type a that accumulate by day or by analysis period
- A is the set of all activity types a
- E is the set of episodes e within the day being optimized
- E' is the set of episodes e' that pair with episodes in the observed pattern within the day being optimized, i.e., a subset of E
- R is the set of all trips r within the day being optimized
- *new* in the subscript denotes that the value is from the new pattern
- *obs* in the subscript denotes that the value is from the observed pattern
- $t_{start,new,e}$ is the start time of episode e in the new activity pattern, i.e., $t_{start,new,e} = 0$ for $e = 0$, and $t_{start,new,e} = t_{start,new,e-1} + d_{e-1}$ for $e>0$.

Because utility for non-discretionary activities accumulates beyond a single day, the episode durations within the day depend on episode durations on other days within the analysis period, as these episode durations undergo optimization. This procedure can be interpreted as the individual's knowledge of the past and plan for the future. The individual currently bases his or her plan for the future on his or her observed pattern. Consider, for example, work durations over a week. On Monday a person knows how much he should work that week (reflected by the β parameter for work). He also has a plan for how much he will work on each day and will try to work a certain number of hours on Monday, so that the total hours add up to the planned hours. However, if he inserts an additional activity on Monday, he will work fewer hours on Monday than in the observed pattern. Subsequently, on Tuesday, he will know he worked too little the previous day and will try to compensate, so that the total hours once again add up.

Select Best Alternative Activity Pattern

Finally, the total utility of each proposed activity pattern is evaluated:

$$V_{total} = \sum V_{trip} + \sum V_{activity} + \sum V_{change} \quad (6)$$

Note that we do not include the opportunity cost of travel time here, because it is already accounted for in $V_{activity}$.

If the best of the proposed activity patterns has a higher utility than the baseline pattern, another iteration of the algorithm runs using the agenda of the best proposed pattern as the new baseline. Otherwise, the current baseline is selected as the best pattern for the given input parameters.

Proof-of-Concept Applications

We tested this proposed procedure for alternative activity pattern generation on a dataset from a multi-day activity survey conducted in Singapore in 2013 (*1*). This smartphone-based survey collected detailed travel, activity, and location information from consecutive weekdays, for example, Monday through Friday in the same week. Figure 4*a* shows one participant's observed weekly activity pattern. A complete SP experiment would entail collecting multi-day or multi-week activity diaries for each respondent to establish a baseline. The respondent would also complete a pre-survey, which along with the observed activity patterns would inform his or her preference parameters. System variables could then be changed to reflect a hypothetical scenario, for example, increased/decreased travel times by certain modes, restricted mode choices, or a new transportation service, etc. Then, using the proposed method and different sets of input parameters, different alternative activity

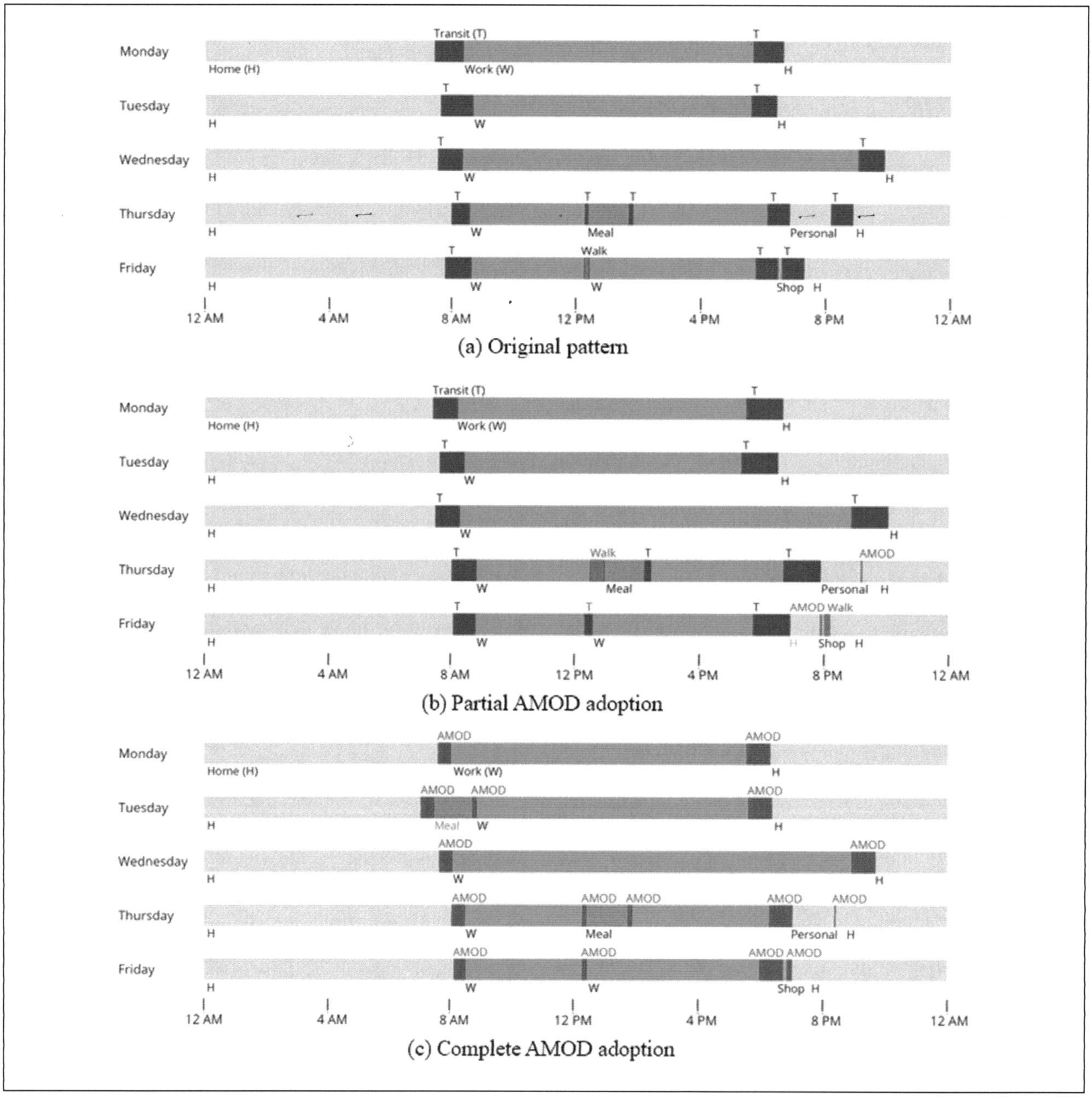

Figure 4. (*a*) An individual's observed weekly activity pattern; (*b*) example alternative activity pattern with partial AMOD adoption; (*c*) complete AMOD adoption.

Table 1. Example User Parameters

Parameter type	Value	Value
Travel mode	$\beta_{m,T}$ [utility/min]	$\beta_{m,C}$ [utility/$]
Walk	−5.5	−4.0
Car	−1.8	−4.0
Transit	−1.2	−4.0
AMOD	−0.8	−4.0
Activity	α_a [1/hour]	β_a [utility]
Home	1.0	12.0
Work	0.3	47.8
Shop	102.1	0.1
Personal	9.1	1.3
Meal	9.1	1.3
Change		
f	150 [no unit]	
i_{home}	10 [utility/hour^2]	
i_{work}	200 [utility/hour^2]	
i_{shop}	20 [utility/hour^2]	
$i_{personal}$	50 [utility/hour^2]	
i_{meal}	20 [utility/hour^2]	

patterns that form a choice set could be generated. In general, input parameters can be derived from the observed patterns or the respondent's pre-survey.

Here, we consider a hypothetical scenario in which an on-demand AV service, Automated Mobility On-Demand (AMOD), is introduced, and we examine examples of alternative activity patterns generated for the abovementioned survey participant. Given that such a service does not yet exist, it is difficult to judge the method's predictive accuracy. However, our objective is only to generate numerous feasible alternatives that form a choice set from which respondents would choose, thus perfect predictions of actual responses are not necessary at this stage. For the purposes of this proof-of-concept demonstration, we assume that AMOD travel times are the same as travel times by automobile and that the cost is double the cost of completing the same trip by transit. Depending on the individual's preferences, adopting AMOD for one or more trips might improve the overall utility of their activity pattern. We select the preference parameters for this demonstration (Table 1) by assuming that the mode-specific constants, β_m, are 0 for all modes and that other preference parameters associated with travel and change fall within a reasonable range. These a priori values were assumed for proof-of-concept demonstration purposes. A different set of parameters would generate different alternative patterns. Lastly, we derive the preference parameters associated with activities from the observed activity pattern, such that activity choices would remain similar if there were no changes to the system.

Work, shopping, and personal errands fall into the category of mandatory/maintenance activities, and thus their durations accumulate over the entire survey period for utility calculation purposes. We categorize home and meal breaks as discretionary activities. Arguably, home can be considered a mandatory/maintenance activity; however, because we never remove overnight home episodes, the remaining home-related decisions are of a discretionary nature, for example, how long to spend at home, whether to return home between activities, or whether to extend the tour.

Figure 4*b* presents the generated alternative activity pattern. In this pattern, the survey participant uses AMOD for two trips, once Thursday night on his way home from the personal errand and once Friday evening for his shopping trip. The convenience of AMOD also introduced a new home episode on Friday, making shopping a separate home-based tour. However, as expected, the relatively high cost of the service results in limited adoption and minimal impact on the observed activity pattern, with the assumed preference parameters.

Figure 4*c* presents a different final alternative activity pattern, in which we explore the impact of complete AMOD adoption. We removed the walk, car, and transit mode options and halved the magnitude of the AMOD parameters, i.e., $\beta_{AMOD,T} = -0.4/min$ and $\beta_{AMOD,C} = -2/\$$. The shorter AMOD travel times allow the survey participant to leave home later in the morning and return home earlier in the evening. For example, on Thursday he would arrive home at 8:26 p.m. as opposed to 8:55 p.m. At the same time, he would add a meal episode on Tuesday, and available work time increases from 47.8 to 48.6 hours.

These examples also highlight the limitations of the proposed method in its current state. The sequential, iterative approach does not account for travel times that vary according to the time of day; thus, we base all travel times in the alternative patterns on evening peak travel conditions. Not only does this potentially affect the selection of mode, it also inaccurately determines the time available for activities. Similarly, it assumes that activity demands are constant throughout the day, a potentially unrealistic property that the additional hour-long meal break before 8 a.m. on Tuesday in the second alternative pattern illustrates.

Conclusion and Future Direction

We present a new method for generating activity-driven, multi-day alternative activity patterns, designed for use in SP surveys. This method generates alternatives using a local optimization algorithm that searches for the best response to a system change that is similar to the survey participant's observed patterns. The alternative patterns also allow us to explore the propagation of indirect effects resulting from the participant's direct response. As a proof-of-concept, we applied the method to data from a multi-day activity-travel survey in Singapore and considered the hypothetical implementation of an on-demand AV service. The demonstration showed promising results with the algorithm exhibiting overall desirable behavior with reasonable responses. However, the examples highlight the method's current limitations, namely inflexibility vis-à-vis activity demand and travel times depending on the time of day. In our continued work, we plan to address these limitations as well as to allow for substitution of activities within activity types, for example, the substitution of office work with teleworking. Incorporating additional ways to constrain alternatives to the observed pattern might also enable more nuance in the generated activity patterns. Currently the measure of change is considered with the difference in start times of activities but the change in the total number of activities is another candidate to impose further constraints.

As we improve the method, we will test the generated activity patterns on respondents to calibrate and ensure that we are generating relevant alternatives. We will soon employ the proposed method in two activity-based experiments focusing on AMOD adoption and carbon emissions feedback. For the first experiment, a SP survey, we will test how to align the experimental treatments, namely the AMOD service, with the activity generation procedure we proposed. For instance, instead of presenting the participants with the full range of AMOD adoption possibilities shown in the proof-of-concept above, we can first gather the participants' stated preferences for their level of adoption of AMOD services or subscription programs. Then, we can use their adoption preferences to tailor the set of alternative activity patterns that we present to the participants in the second part of the SP survey and evaluate whether the participants accept the impact of the AMOD treatment on their overall activity patterns. The proposed method can also be developed into a travel-activity planning advisory application that automatically generates feedback and personalized advice based on users' stated or revealed preferences. In the second experiment, we will test the effectiveness of smartphone-based activity-travel and carbon emissions feedback delivery for encouraging sustainable behavior change. The algorithm's multi-day pattern generation capability can support recommendations for blended travel mode choices over extended study periods. Additionally, by accounting for the multidimensional impacts of changing activities and travel in realistic alternative patterns, this method can help test the effectiveness of an assisted behavioral planning program.

Acknowledgments

The MIT Energy Initiative (MITEI) Seed Grant Fund and the National Research Foundation Singapore, through the FM IRG research group at the Singapore-MIT Alliance for Research and Technology, supported this research.

Author Contributions

The authors confirm contribution to the paper as follows: study conception and design: He, Atasoy, Brazier, Zegras, and Ben-Akiva; analysis and interpretation of results: He; draft manuscript preparation: He, Atasoy, Brazier, and Zegras. All authors reviewed the results and approved the final version of the manuscript.

References

1. Cottrill, C., F. Pereira, F. Zhao, I. Dias, H. B. Lim, M. Ben-Akiva, and C. Zegras. Future Mobility Survey: Experience in Developing a Smart-Phone-Based Travel Survey in Singapore. *Transportation Research Record: Journal of the Transportation Research Board*, 2013. 2354: 59–67.
2. Bowman, J. L., and M. Ben-Akiva. Activity Based Travel Forecasting. *Proc., Activity Based Travel Forecasting Conference*, New Orleans, 1996.
3. Axhausen, K. W., and T. Gärling. Activity-based Approaches to Travel Analysis: Conceptual Frameworks, Models, and Research Problems. *Transport Reviews*, Vol. 12, 1992, pp. 323–341. https://doi.org/10.1080/01441649208716826
4. Recker, W. W. The Household Activity Pattern Problem: General Formulation and Solution. *Transportation Research Part B: Methodological*, Vol. 29, No. 1, 1995, pp. 61–77. http://dx.doi.org/10.1016/0191-2615(94)00023-S
5. Kitamura, R., and S. Fujii. Two Computational Process Models of Activity-Travel Behavior. In T. Garling, T. Laitila, and K. Westin (eds.), *Theoretical Foundations of Travel Choice Modeling*, 1998, pp. 251–279.
6. Lu, Y., M. Adnan, K. Basak, F. C. Pereira, C. Carrion, V. H. Saber, H. Loganathan, and M. E. Ben-Akiva. *SimMobility Mid-Term Simulator: A State of the Art Integrated Agent Based Demand and Supply Model.* Transportation Research Board, Washington, D.C., 2015.
7. Miller, E., and M. Roorda. Prototype Model of Household Activity-Travel Scheduling. *Transportation Research Record: Journal of the Transportation Research Board*, 2003. 1831: 114–121.
8. Roorda, M. J., E. J. Miller, and K. M. N. Habib. Validation of TASHA: A 24-H Activity Scheduling Microsimulation Model. *Transportation Research Part A: Policy and Practice*, Vol. 42, No. 2, 2008, pp. 360–375. http://dx.doi.org/10.1016/j.tra.2007.10.004
9. Arentze, T., and H. Timmermans. *Albatross: A Learning Based Transportation Oriented Simulation System.* Eirass, Eindhoven, 2000.
10. Pinjari, A. R., and C. R. Bhat. Activity-Based Travel Demand Analysis. *A Handbook of Transport Economics*, Vol. 10, 2011, pp. 213–248.
11. RDC. *Activity-Based Modeling System for Travel Demand Forecasting, Metropolitan Washington Council of Governments.* US Department of Transportation, US Environmental Protection Agency, Washington, D.C., 1995.

12. Doherty, S., M. Lee-Gosselin, K. Burns, and J. Andrey. Household Activity Rescheduling in Response to Automobile Reduction Scenarios. *Transportation Research Record: Journal of the Transportation Research Board*, 2002. 1807: 174–181.
13. Doherty, S. T., and E. J. Miller. A Computerized Household Activity Scheduling Survey. *Transportation*, Vol. 27, No. 1, 2000, pp. 75–97.
14. Lee, M. S., and M. G. McNally. On the Structure of Weekly Activity/Travel Patterns. *Transportation Research Part A: Policy and Practice*, Vol. 37, No. 10, 2003, pp. 823–839.
15. Auld, J., M. Z. Frignani, C. Williams, and A. K. Mohammadian. Results of the UTRACS Internet-Based Prompted Recall GPS Activity-Travel Survey for the Chicago Region. *Proc., 12th World Conference in Transportation Research*, Lisbon, Portugal, 2010.
16. Timmermans, H., T. Arentze, and C.-H. Joh. Modeling Effects of Anticipated Time Pressure on Execution of Activity Programs. *Transportation Research Record: Journal of the Transportation Research Board*, 2001. 1752: 8–15.
17. Joh, C.-H., T. Arentze, and H. Timmermans. Modeling Individuals' Activity-Travel Rescheduling Heuristics: Theory and Numerical Experiments. *Transportation Research Record: Journal of the Transportation Research Board*, 2002. 1807: 16–25.
18. Joh, C.-H., T. A. Arentze, and H. J. P. Timmermans. Estimating Non-Linear Utility Functions of Time Use in the Context of an Activity Schedule Adaptation Model. *Proc., 10th International Conference on Travel Behavior Research*, Lucerne, Switzerland, 2003.
19. Nagel, K., K. W. Axhausen, B. Kickhöfer, and A. Horni. Research Avenues. In *The Multi-Agent Transport Simulation MATSim* (Horni, A., K. Nagel, and K. W. Axhausen, eds.), Ubiquity Press, London, 2016, pp. 533–542. http://dx.doi.org/10.5334/baw.97
20. Arentze, T. A., C. Pelizaro, and H. J. P. Timmermans. Implementation of a Model of Dynamic Activity-Travel Rescheduling Decisions: An Agent-Based Micro-Simulation Framework. *Proc., Computers in Urban Planning and Urban Management Conference*, London, Vol. 48, 2005.
21. Nagel, K., B. Kickhöfer, A. Horni, and D. Charypar. A Closer Look at Scoring. In *The Multi-Agent Transport Simulation MATSim* (Horni, A., K. Nagel, and K. W. Axhausen, eds.), Ubiquity Press, London, 2016, pp. 23–34. http://dx.doi.org/10.5334/baw.3
22. Balac, M., and K. W. Axhausen. *Activity Rescheduling Within a Multi-Agent Transport Simulation Framework (MATSim)*. Transportation Research Board, Washington, D.C., 2017.

The Standing Committee on Traveler Behavior and Values (ADB10) peer-reviewed this paper (18-04230).

Article

TRR
JOURNAL OF THE TRANSPORTATION RESEARCH BOARD

Transportation Research Record
2018, Vol. 2672(47) 146–158
© National Academy of Sciences: Transportation Research Board 2018
Article reuse guidelines: sagepub.com/journals-permissions
DOI: 10.1177/0361198118773184
journals.sagepub.com/home/trr
SAGE

Trip Characteristics Analysis of the Effects of a Travel Behavior Change Program

Rosa Arroyo[1], Tomás Ruiz[1], Daniel Casquero[1], and Lidón Mars[2]

Abstract

Travel behavior change programs (TBCP) are measures or actions specifically designed to motivate people to reduce their car use. This study contributes to the literature on rigorous evaluation of the effects of participating in TBCP. In particular, we analyze the effect of TBCP in relation to a potential for a reduction in car use. In contrast to previous studies, the effect of TBCP is analyzed considering two characteristics of the trip: travel companion and time of trip. The TBCP, consisting of three persuasion actions customized to the needs of the participants in the research, was designed and implemented in Valencia (Spain). To evaluate the effects of participating in the TBCP, a two-wave panel survey was carried out, which collected activity-travel scheduling process data from participants before and after the implementation of the TBCP. To properly identify the effects of taking part in the TBCP, participants were divided into two groups: those who were involved in the TBCP, and those who did not participate in any persuasion action (control group, CG). Descriptive and confirmatory analysis, which included the use of sample selection panel data models with treatment effects and random parameters, have been developed. The results suggest that the individuals most affected by participating in TBCP are those carrying out solo and afternoon trips. If traveling with companions, those who do so with household members are more influenced by participating in TBCP. Important research and transportation policy implications are derived from these results.

Travel behavior change programs (TBCP) are measures or actions specifically designed to motivate people to reduce their car use. These programs include personal travel scheduling, travel awareness campaigns, workplace or study place travel plans, and strategies like car sharing (*1–6*).

Typical actions included in TBCP are so-called "soft" measures, such as providing better information on transportation options, appropriate assistance, and motivation or incentives, and disincentive programs for car usage (*2, 6*). The main characteristic of these measures is that they can potentially achieve better cost-effective results in delivering shifts from car to public transportation or non-polluting modes of travel, such as walking or cycling.

Two popular and widespread TBCP are Indimark® and Travel Blending®. Indimark® (*7*) offers individualized marketing of public transportation, walking, and cycling to individuals by providing tailored information. By contrast, Travel Blending® raises people's awareness of their travel patterns and recommends viable changes to reduce possible unnecessary or wasteful travel (*8*).

TBCP have been frequently applied in the past decade. Several large-scale programs of soft measures implemented in the UK are reviewed by Cairns et al. (*5*) and Bamber and Rees (*10*). These sorts of measures were found have the potential to reduce traffic levels significantly, especially in peak periods. Similarly, Friman et al. (*11*) and Skain et al. (*12*) revised several programs using personalized travel planning in Sweden. They also found positive effects of the implemented programs. Other reviews were carried out by Richter et al. (*13*) regarding the implementation of soft measures in several countries, and by Fujii and Taniguchi (*14*) in Japan. They found that in general soft transport policy measures are effective and resulted in significant reductions of car use and CO_2 emissions.

[1]Transport Department, School of Civil Engineering, Universitat Politècnica de València, Valencia, Spain
[2]Social Psychology Department, Faculty of Psychology, Universitat de València, Valencia, Spain

Corresponding Author:
Address correspondence to Rosa Arroyo: maarlo3@cam.upv.es

Different types of soft measures have been tested. For example, travel feedback programs are reviewed by Fujii and Taniguchi (*14*). Examples from Germany and Austria (*15*) focus on workplace initiatives. In Italy, Meloni et al. (*16*) tested personalized travel plans. TBCP based on persuasion strategies have been used in Spain to convince drivers to reduce their car use. Those who participated in the TBCP reduced their car use more than 5%, compared with those who did not participate (*17, 18*).

The majority of soft transport measures have been evaluated shortly after their implementation. Stopher et al. (*19*) described a 5-year study to monitor the effects of a voluntary travel behavior change (VTBC) program. They concluded that the noncoercive policy of VTBC is effective for at least 5 to 8 years after the intervention is undertaken.

However, the methods used to evaluate the effectiveness of TCBP remains a subject of debate among professionals and researchers (*2, 10, 20–28*). Some programs have employed research techniques that did not allow statistical inferences to be drawn from their results. Many program designs still lack adequate CGs to evaluate travel behavior change (*29*). In cases where there is no CG, program designs use pre- and post-tests of participants' behaviors, which cannot exclude the influence of changes in travel behavior that may occur within the general population (*3, 30*). Very frequently, a repeated cross-sectional sample approach is used for the collection of the pre/post data instead of a panel sample approach. This decision may not have a direct biasing impact on the effect estimate itself but the approach makes it much more difficult than, for example, a panel study approach to test the statistical significance of the reported effect (*31*).

In addition, a random assignment of participants to the CG is desirable, so that it is possible to minimize various sources of error variances. However, in practice, it is very difficult to achieve that random assignment, mainly due to attrition in the case of panel-based studies. Therefore, well-designed quasi-experimental studies are acceptable (*10*). Finally, it is also rare to find in the travel behavior literature methods of analysis other than descriptive and explanatory statistics (e.g., *t*-student, Cohen's d, Levene test). Predictive models are required to control the influence of other variables. For example, García-Garcés et al. (*18*) used a predictive (econometric) model that confirmed the significance of participating in TBCP on the changes observed in the time allocated for traveling by private vehicle

Results from the evaluation by Sloman et al. (*32*) and the meta-analysis by Moser and Bamberg (*3*) appear to indicate that TBCP effects persist when self-selection is accounted for, though the effects may be somewhat smaller. These studies found driving trip reductions of 3 to 11% were observed 3 to 24 months after the implementation of TBCP. García-Garcés et al. found a reduction of 5.3% in the proportion of daily travel time that involved a private vehicle (*18*). More recently, Bamber and Rees found an average reduction of the car modal split share of about five percentage points after reviewing 10 studies in the UK (*10*).

The motivation of this study is twofold: first, to contribute to the scarce literature on rigorous evaluation of the effects of participating in TBCP; second, to study the effects of a particular TBCP based on persuasion strategies, which have been customized to the needs of the participants in the research. In particular, the assessment is carried out on observed car and motorbike use by using data from both the first and second waves of a panel survey.

The rest of the paper is organized as follows. A description of the implemented TBCP is presented in the next section. Then, the characteristics of the dataset used are described. This section is followed by an explanation of the methods of analysis used, and empirical results, with conclusions.

Travel Behavior Change Program

As described in Ruiz and García-Garcés (*33*) and García-Garcés et al. (*18*), three different actions based on psychological principles of persuasion (*31*) were designed and implemented in Valencia (Spain). To this end, the answers to a survey that was sent to all participants were considered, in order to identify the internal psychological barriers to the modification of travel behavior. We used these as defined in Ken Wilber's (*32*) four-quadrant structure as a starting point, and focused mainly on individual subjective factors related to travel modes.

The first action was defined to contend with barriers related to perceived behavioral control and instrumental attitudes, by applying persuasion principles of reciprocity and scarcity. Respondents' mobility was analyzed and recurring trips were identified. Once characteristics such as origin, destination, timing, and companion on the most recurring trip were identified, more sustainable travel alternatives were sought. Parameters such as annual savings, the annual reduction of car-related pollution, and travel time were considered. A report with all this information, together with general information concerning transit supply available in their neighborhood, was sent to each respondent by post.

A second action was defined to address barriers related to self-identity and status. Thus, the persuasion principle of authority to influence was applied. Respondents were invited to attend a talk given by a cardiologist and a sports trainer on the relationship between health and physical activity and how walking and riding a bicycle can improve their health.

Finally, a third action was defined to deal with barriers related to affective attitudes. The persuasion principles of social proof and liking were applied. Respondents were invited to watch a video session where people who had recently reduced car use were interviewed on the street about why they had decided to do so.

The number of panelists who participated in the TBCP was 72 out of 118, whereas the rest formed the CG. Each participant in the TBCP received at least two of the treatments described above to motivate them to reduce car use.

Data

Survey Description

As described in García-Garcés et al. (*18*), a two-wave panel survey that collected activity scheduling process data was conducted over a one-year period in the city of Valencia (Spain). The main purpose of this panel survey was to study the potential effect of the TBPC on activity-travel scheduling process decisions. The first and second waves took place during fall 2010 and fall 2011, respectively. Only 62% of the respondents who were surveyed in both survey waves participated in the TBCP, which took place between the two waves.

Both survey waves consisted of three phases. The first phase was a preliminary face-to-face interview where respondents were asked to generate, using paper and pencil, a pre-planned activity-travel agenda for the following week starting the day after the interview. The second phase was developed during the research week, when respondents were using the mobile phone diaries. The third phase consisted of an in-depth telephone interview. Participants were contacted by phone two or three times during the research week and were asked to give reasons for the differences observed between their pre-planned and executed activities and trips.

From another study conducted the year before, 492 individuals agreed to participate in the panel survey. Finally, 165 of them provided complete information in the first wave, being 33.5% of those who initially indicated their willingness to participate. Between the two waves, 47 respondents abandoned the panel due to a change of residence that took them outside the study area, or transfer abroad for work, or they simply opted not to continue. Therefore, 118 individuals participated in both survey waves, so that the final response rate was 24%. One panelist was discarded due to his new employment as a driver in the second survey wave.

Dataset for Empirical Research

Both surveys provided a rich source of detailed information concerning scheduling, rescheduling, and executing daily activities and trips. Due to the purpose of this study, only executed trip decisions collected in both waves are analyzed.

Table 1. Sample Demographic and Socioeconomic Distribution

	CG	TBCP
Gender		
Male	41% (19)	57% (41)
Female	59% (27)	43% (31)
Main occupation		
Employed	65% (30)	74% (53)
Student	30% (14)	24% (17)
Other	4% (2)	3% (2)
Age		
<30	43% (20)	42% (30)
30–50	50% (23)	46% (33)
>50	7% (3)	12% (9)
Car availability		
Low (0–2 days/week)	2% (1)	0% (0)
Medium (3–6 days/week)	9% (4)	3% (2)
High (everyday)	89% (41)	97% (70)

Note: CG = control group; TBCP = travel behavior change programs.

There are 72 participants in the TBCP, the remaining 45 individuals are analyzed as the CG, with the aim of assessing the effects of the program between both waves. The sociodemographic distribution of the sample is shown in Table 1. The sample is mainly composed of habitual drivers, and 94% of the participants declared a high level of car availability.

Analysis and Results

Dependent and Explanatory Variables

The level of use of private vehicles (PV) in both waves has been measured by considering the daily time allocated to PV use in relation to the total daily time allocated to travel per person. This was the dependent variable that provided best results in the analysis. Consequently, the dependent variable is the proportion:

$$\frac{\textit{Daily time allocated to travel in private vehicle}(PV)}{\textit{Daily total time allocated to travel}(Total)}$$

Thus, the dependent variable is a continuous variable which always takes values between 0 and 1.

The unit of analysis is person-days for each survey wave. Person-days with duration lower than 1 minute were discarded. Person-days were also rejected if their duration was higher than a threshold defined as $q_3 + 3(q_3 - q_1)$, where q_i is the quartile i of the frequency distribution of trips for each mode. Finally, a total of 738 person-days are available in each wave. In the first wave there are 3,315 trips, and in the second wave 3,294 trips.

Table 2. Number of Trips and Daily Travel Time

		Wave 1		Wave 2	
		CG	TBCP	CG	TBCP
	Person-days	284	454	284	454
	Trips	1263	2052	1197	2097
	Trips/Person-day	4.45	4.52	4.22	4.62
Daily travel time (min)	Total	83.6	89.3	83.7	86.9
	In private vehicle (PV)	69.5	76.6	69.9	68.4
	PV/Total	0.83	0.86	0.84	0.79

Note: CG = control group; TBCP = travel behavior change programs.

The distribution referencing participants in CG and in TBCP groups is shown in Table 2.

Most of the observed trips per day were carried out by PV as all the participants were habitual drivers. The reduction of total daily travel time for those who participated in the TBCP between wave 2 and wave 1 was 2.7%. The reduction of travel time by PV for those who participated in the TBCP between wave 2 and wave 1 was much higher: 10.7%. Therefore, the decrease of the ratio PV/Total from 0.86 to 0.79 for those who participated can be explained by the reduction of the daily travel time by PV.

Explanatory variables include trip attributes, demographic variables (at individual and household level), and socioeconomic characteristics (Table 3). Willingness to reduce car use is also included as an explanatory variable. Willingness to reduce car use was not a condition for respondents to be accepted as participants in the research.

Descriptive Analysis

The proportion PV/Total slightly increases between survey waves for those who did not participate in the TBCP, no matter if it is a solo or a joint trip (Figure 1). Conversely, those who participated in the TBCP reduced their PV/Total in wave 2. This reduction is significantly higher for solo trips.

The proportion PV/Total increases between survey waves for those who did not participate in the TBCP and are accompanied by household members when they travel (Figure 2). The ratio PV/Total hardly changes between waves for those who did not participate in the TBCP and are accompanied by non-household members. In contrast, those who participated in the TBCP reduce their ratio PV/Total in wave 2, no matter the type of companion.

Table 4 shows, for each characteristic of the trip, the ratio PV/Total both for respondents in the CG and for participants in the TBCP. Independent samples t-test was used to compare the average proportion values for participants in the TBCP and non-participants. The objective is to test the null hypothesis that the difference between those two related means is 0. The t-test assumes that the variability of each group is approximately equal. Without that assumption, a special form of the t-test should be used, so Levene's test for equality of variances is employed (*36*).

Results of Levene's test suggest a possible influence of participating in the TBCP regardless of the number of trips taken per day. Taking into account the total travel time per day in minutes, instead of the number of trips, the results highlight a possible influence of TBCP only on respondents who spend more than 65 minutes per day traveling.

With reference to the influence of companions, results of Levene's test suggest that participating in the TBCP has a positive influence on reducing PV use in trips carried out with companions from outside the household of the individual. However, no evidence is found with regard to the effect of travel with companions compared with trips carried out alone.

The timing of the trip might also be relevant in explaining differences in the ratio PV/Total between waves. Finally, differences in the ratio might be important only for trips carried out on working days.

Empirical Models

As explained previously, this study uses data from a before-and-after panel survey of habitual drivers. Some of the panelists participated in a TBCP, and the rest formed the CG.

The attrition of panelists might prevent the participants in the TBCP and CG groups from being similar in relation to basic demographics and travel mode split. In order to correct this potential sample selection bias, the Heckman two-step procedure (*37*) is used.

In its first stage (Equation 1), a probit model is fit to estimate the probability of being in the control (CG) or treatment (TBCP) group and the results are stored for use by the selection model. In the second stage (Equation 2), a

Table 3. Dependent and Explanatory Variables

Variable	Definition
Dependent variable	
PV/Total	(time allocated to PV per day) divided by (total time allocated to travel per day)
Explanatory variables	
TBCP	1 = Respondent participated in TBCP; 0=otherwise
Trip attributes	
Totaltime	Total time allocated to travel episodes per day (in min)
Privehtime	Time allocated to travel in PV per day (in min)
Totaltrips	Total number of trips per day
Privehtrips	Number of trips in PV per day
Day	1 = if trip is carried out on a weekday; 0 = otherwise
Acomp	1 = if it is a joint trip; 0 = otherwise
Acomp_time	Total time traveled with companions per day (min)
Acomp_time_p	Percentage of time traveled with companions per day
Acomp_HH_time_p	Percentage of time traveled with household companions per day
Acomp_other_time_p	Percentage of time traveled with other companions per day
Morning_trips_p	Percentage of daily trips during the morning
Afternoon_trips_p	Percentage of daily trips during the afternoon
Evening_trips_p	Percentage of daily trips during the evening
Night_trips_p	Percentage of daily trips during the night
Demographic attributes	
Age	0 = Respondent age is between 18 and 30 1 = Respondent age is between 31 and 50 2 = Respondent age is over 50
Gender	1 = Female; 0 = Male
Marital status	
Married	1 = Respondent is married; 0 = otherwise
Divorced	1 = Respondent is divorced; 0 = otherwise
Couple	1 = Respondent lives in couple but not married (co-habiting couple); 0 = otherwise
Single	1 = Respondent is single; 0 = otherwise
Education	0 = Primary school is the highest level of education for respondent 1 = Secondary school is the highest level of education for respondent 2 = Respondent has studied non-university higher education/professional school/further education 3 = University studies is the highest level of education for respondent 4 = Bachelor degree is the highest level of education for respondent 5 = PhD is the highest level of education for respondent
Household attributes	
HHMembers	Number of members at home (including respondent)
Household position	
Head	1 = Respondent lives alone, lives only with his/her partner, or is one of the parents in case the household is a family; 0 = otherwise
Share	1 = Respondent lives in a shared house; 0 = otherwise
Son/Daughter	1 = Respondent lives with his/her parents; 0 = otherwise
FAM_WD	1 = Respondent lives with his/her parents during weekdays; 0 = otherwise
FAM_WK	1 = Respondent lives with his/her parents during weekends; 0 = otherwise
Socioeconomic attributes	
Student	1 = Respondent is studying; 0 = otherwise
Employed	1 = Respondent is employed; 0 = otherwise
Retired	1 = Respondent is retired; 0 = otherwise
Car availability	2 = High car availability (evey day of the week); 1 = Medium car availability (3–6 days per week); 0 = Low car availability (2 days per week or less)
Willingness to change	
Willing	1 = Respondent is willing to change his/her mobility towards a decrease of car use; 0 = otherwise

regression with selection is fit conditionally on the first step estimation. Treatment effects are considered by inserting the estimated propensity to participate in the TBCP (first step of the model) as an explanatory variable of the estimated ratio PV/Total (second step), and using all observations to fit this later model. The following modeling framework for a sample selection panel data model with treatment effects and random parameters is used (*37*):

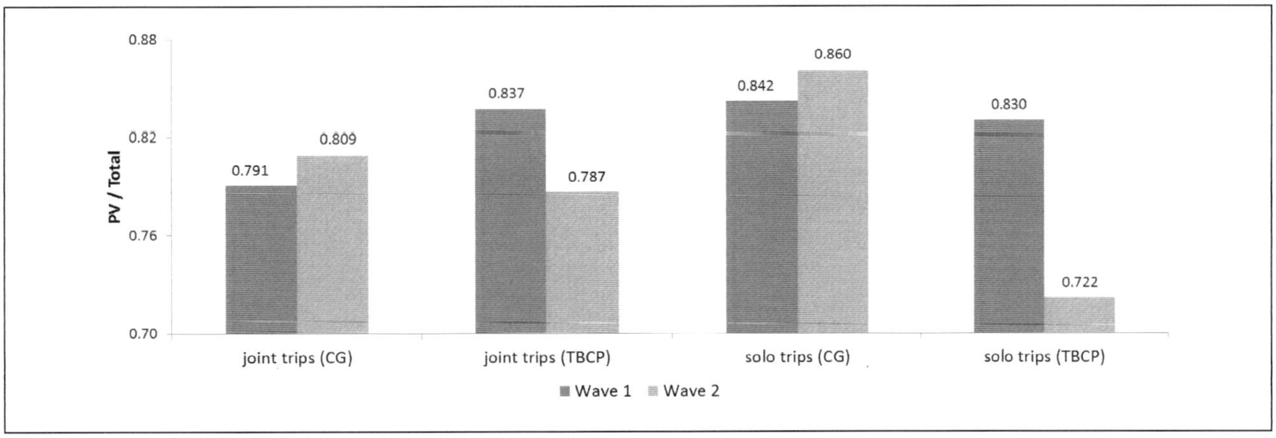

Figure 1. PV/Total between survey waves and participation in TBCP. Joint/solo trips.

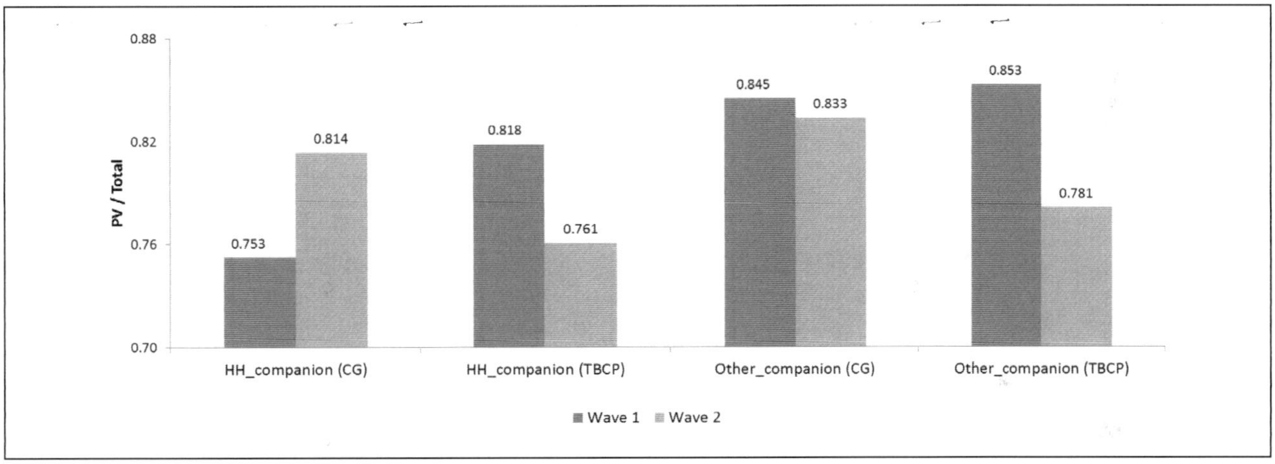

Figure 2. PV/Total between survey waves and participation in TBCP. Type of companion.

$$z_{it}^* = \alpha_i' w_{it} + u_{it}\ z_{it} = 1(z_{it}^*>0), u_{it} \sim N[0,1] \quad (1)$$

$$y_{it} = \beta_i' x_{it} + \gamma_i\ z_{it} + \varepsilon_{it}\varepsilon_{it} \sim N[0, \sigma 2] \quad (2)$$

$$(\varepsilon_{it}, u_{it}) \sim N\left[(0,0), (\sigma^2, 1, \rho\sigma)\right], \mathrm{Corr}[\varepsilon_{it}, u_{it}] = \rho$$

Random parameters are employed in the first and second steps of the models. The underlying motivation of the random parameters is to accommodate individual heterogeneity in the parameters of the models. For example, the general form of α_i as a random parameter is:

$$\alpha_i = \alpha^0 + \sigma\ v_i$$

where α_0 is the mean of the parameter, which is fixed and common to all respondents, while σ is the standard deviation of the parameter, and v_i a normally distributed term. Note that non-random parameters have $\sigma = 0$, so their formulation is $\alpha_i = \alpha^0$.

Following the difference-in-differences approach, a period x treatment group interaction captures the change in the proportion PV/Total of those who participated in the TBCP. To evaluate the effects of participating in the TBCP, we extend Equation 2 to the following specification:

$$y_{it} = \beta_i' x_{it} + \gamma_i z_{it} + m_1 a_i + m_2 w_t + h(aw)_{it} + \varepsilon_{it} \quad (3)$$

where the value of a_i indicates those who participated in the TBCP and w_t indicates the period in which the TBCP were implemented (wave 2). The values of μ_1 and μ_2 denote the coefficients associated with the main effects of the treatment and period variable, while the value η refers to the estimated coefficient of the interaction term between these two variables (*aw*), which captures the effect of participating in the TBCP.

Finally, to assess whether effects vary among characteristics of the trips, we extend our initial Equation 3 into the following specification:

Table 4. PV/Total. Independent Samples *t*-Tests

Variable		Time in PV / Total travel time		Levene's test for equality of variances		*t*-Test for equality of means		
		Wave 1	Wave 2	F	Sig	t	df	Sig(2-tailed)
Number of daily trips	Low_CG (3 or fewer trips per day)	0.850	0.860	0.5	0.46*	−0.2	229	0.82
	Low_TBCP (3 or fewer trips per day)	0.844	0.777	9.3	0.00**	1.6	302	0.11
	Medium_CG (4–5 trips per day)	0.804	0.829	0.5	0.50*	−0.6	213	0.54
	Medium_TBCP (4–5 trips per day)	0.831	0.754	7.4	0.01**	2.4	360	0.02
	High_CG (6 or more trips per day)	0.751	0.759	0.1	0.79*	−0.2	153	0.87
	High_TBCP (6 or more trips per day)	0.833	0.786	14.8	0.00**	1.7	289	0.10
Type of trip companion	Companion_trips_CG	0.791	0.809	0.9	0.34*	−0.6	418	0.55
	Companion_trips_TBCP	0.837	0.787	12.9	0.00**	2.4	724	0.02
	No_companion_trips_CG	0.842	0.860	0.2	0.64*	−0.4	179	0.69
	No_companion_trips_TBCP	0.830	0.722	20.7	0.00**	2.2	229	0.03
	Household_companion_trips_CG	0.753	0.8135	5.8	0.02**	−1.5	266	0.13
	Household_companion_trips_TBCP	0.818	0.7607	10.2	0.00**	2.1	475	0.04
	Other_companion_trips_CG	0.845	0.8334	0.6	0.45*	0.4	331	0.73
	Other_companion_trips_TBCP	0.853	0.7812	17.6	0.00**	2.5	478	0.01
Time of day (>=mean)	Morning_CG	0.797	0.794	1.0	0.31*	−0.7	599	0.50
	Morning_TBCP	0.846	0.756	27.9	0.00**	3.3	955	0.00
	Afternoon_CG	0.792	0.863	8.3	0.00**	−2.1	327	0.03
	Afternoon_TBCP	0.848	0.767	18.4	0.00**	3.1	510	0.00
	Evening_CG	0.823	0.828	0.5	0.49*	−0.1	289	0.90
	Evening_TBCP	0.839	0.773	15.8	0.00**	2.5	471	0.01
	Night_CG	0.829	0.843	0.6	0.45*	−0.3	124	0.79
	Night_TBCP	0.808	0.815	1.2	0.27*	−0.1	192	0.88
Type of day	Working_day_CG	0.792	0.813	0.4	0.55*	−0.7	433	0.48
	Working_day_TBCP	0.850	0.771	35.9	0.00**	3.6	703	0.00
	Non_working_day_CG	0.844	0.852	0.8	0.38*	−0.2	164	0.87
	Non_working_day_TBCP	0.796	0.772	0.6	0.44*	0.6	250	0.57
Total travel time	Low_CG (<=65 minutes)	0.685	0.784	5.2	0.02**	−1.8	201	0.08
	Low_TBCP (<=65 minutes)	0.784	0.715	6.2	0.01**	1.7	325	0.10
	Medium_CG (65–100 minutes)	0.882	0.851	1.3	0.26*	0.9	205	0.37
	Medium_TBCP (65–100 minutes)	0.856	0.804	4.7	0.03**	1.6	292	0.10
	High_CG (>100 minutes)	0.852	0.838	1.0	0.31*	0.4	189	0.68
	High_TBCP (>100 minutes)	0.861	0.805	14.0	0.00**	2.0	334	0.05

Note: *Equal variances assumed / **Equal variances not assumed. Levene's test results that are different for the same variable on CG and TBCP groups are shaded.

Table 5. First Step. Random Parameters Probit Panel Model

	Coefficient	Coef/SE	Coefficient	Coef/SE
Constant	227.175	0.77	17.1412***	3.00
Day	11.6271***	2.81	0.22234	0.61
Trips	0.70325***	2.73	2.75417***	3.05
Age	25.9581***	2.99	3.89751***	2.59
Gender	−35.5314***	−2.98	5.73108***	2.83
HHmember	7.82954***	2.95	59.3898***	3.00
Single	14.8495***	2.80	5.47581**	2.31
Total no. of observations	738			
Pseudo R^2	0.45714			

Note: ***, **, *: significance at 1%, 5% and 10% level.

$$y_{it} = \beta_i' x_{it} + \gamma_i z_{it} + m_1 a_i + m_2 w_t + h_1 (aw)_{it} + h_2 (aT)_{it} + h_3 (wT)_{it} + h_4 (awT)_{it} + \varepsilon_{it} \quad (4)$$

where *T* refers to the characteristic of the trips considered, the values of η_1, η_2, and η_3 denote the lower order effects, while η_4 denotes the coefficient associated with the three-way interaction between the treatment, period, and the characteristic considered (*awT*). In doing so, we estimate whether participating in the TBCP has affected individuals differently considering whether they travel alone or not, the type of companion, the total travel time per day, and the most frequent timing of their trips.

Results

The models are estimated by maximum likelihood using Nlogit software (*38*). The critical level of significance chosen was 0.10 and variables with statistically insignificant estimated coefficients are excluded from the model. The only exception to this general rule applies to the second step of the model; coefficients of the main and two-way interaction effects (treatment and period variables × T) are kept in the models because they are needed to correctly estimate the coefficient associated with the three-way interaction variable.

Results obtained by the two-step sample selection panel data model with treatment effects and random parameters are summarized in Tables 5 and 6. As can be seen in Table 6, the ρ coefficients are significantly different from zero, validating the methodological approach used in this study. Thus, the sample selection models are appropriate to avoid the biased estimation of the coefficients that might result if independent models were to be implemented for the second step. The negative sign of the ρ coefficients suggests that unobserved factors, which increase the probability of being included in the TBCP group, also increase the propensity for a reduction of the PV use ratio. Additionally, at least one variable (DAY) appears with non-zero coefficient in the selection equation but does not appear in the equation of interest, which assures correction for sample selectivity (*39*).

First Step

Results of the random parameters probit in the first step of the model are presented in Table 5. Positive signs of the estimated coefficients are associated with an increased likelihood for the respondent to participate in the TBCP and provide information in both survey waves. All individual coefficient estimates are highly significant. Estimated scale parameters are the standard deviations because parameters are normally distributed. Almost all standard deviations are significant, indicating that unobserved heterogeneity is well captured. The signs of the estimated coefficients indicate where the potential bias caused by differences in participation in TBCP and CG groups can concentrate. Individuals more likely to participate in the TBCP are those who make trips on weekends, more daily trips, are older, male, have more household members, and/or are single.

Second Step

Regression estimates that allow for differences in other demographics and trip characteristics, and correct for sample selection bias, are reported in models 1 to 5 in Table 6. It can be observed that *R*-squared values are low, which is common when studying individual travel behavior.

In the models in Table 6, the variables of interest are the interaction variable between the period and the treatment status, which indicates the global effect on those who participated in the TBCP, and the three-way interaction variable between the period, the treatment, and the trip characteristic considered. Estimated scale parameters are the standard deviations because parameters are normally distributed. Almost all standard deviations of the estimated coefficients of the main and two-way

Table 6. Second Step. Random Parameters Selection Panel Model

Variable	Model 1 T = COM_TRIP Coefficient	Model 1 z	Model 2 T = P_OTHER Coefficient	Model 2 z	Model 3 T = AFTERNOON_TRIPS_P Coefficient	Model 3 z	Model 4 T = EVENING_TRIPS_P Coefficient	Model 4 z	Model 5 T = NIGHT_TRIPS_P Coefficient	Model 5 z
					Means for random parameters					
Constant	.56877***	21.10	.61069***	21.75	.63207***	24.00	.66033***	15.50	.58461***	25.21
Wave	.06508**	2.58	.06103***	2.93	−.03975*	−1.79	.07133***	2.79	.03691*	1.71
TBCP	.60196***	19.56	.58336***	22.99	.56089***	18.89	.55684***	17.22	.55823***	18.08
T	.00078*	1.88	.03545	1.49	−0.04374	−1.16	.03287	.68	.14294**	2.07
Wave*TBCP	**−.11789*****	**−3.37**	**−.11580*****	**−3.92**	**0.01626**	**0.48**	**−.12925*****	**−3.77**	**−.09993*****	**−3.39**
Wave*T	−.00093	−1.59	−.09405***	−2.70	.21273***	3.70	−.14576**	−2.26	−0.08249	−0.88
TBCP*T	−.00128**	−2.43	−.06875*	−1.91	0.02723	0.46	−.10897	−1.64	−.20344**	−2.03
Wave*TBCP*T	**.00140***	**1.83**	**.14025*****	**2.73**	**−.25400*****	**−2.99**	**.17369***	**1.78**	**.23211***	**1.70**
Trips	−.04080***	−9.28	−.04194***	−10.56	−.04234***	−10.89				
Trip_tim	.00240***	10.26	.00255***	12.69	.00256***	12.91	.00112***	7.27		
Acomp_HH_time_p							−.04245***	−2.94		
Acomp_other_time_p							−.03952**	−2.51		
Morning_trips_p	−.08625***	−3.90	−.09616***	−4.48	−.09629***	−4.38	−.14378***	−4.17		
Afternoon_trips_p							−.06464*	−1.85		
Age	−.01704*	−1.68	−.02736**	−2.44	−.02625**	−2.37	−.02129**	−1.97		
HHMember							−.01249**	−2.04		
Divorced	.06327*	1.90							.06749**	1.98
Couple			−.03686***	−2.58	−.03476**	−2.46	−.04056**	−2.38		
Single	.03037**	2.13							0.02171	1.46
Education	−.01900***	−5.22	−.02021***	−5.62	−.02056***	−5.72	−.01667***	−4.44	−.01900***	−5.00
Student			−.03259*	−1.72	−.03473*	−1.91				
Retired			−.17626***	−3.66	−.17056***	−3.60			−.17728***	−3.14

Note: ***, **, * ==> Significance at 1%, 5%, 10% level.

Table 6. Second Step. Random Parameters Selection Panel Model (Cont.)

	T = COM_TRIP		T = P_OTHER		T = AFTERNOON_TRIPS_P		T = EVENING_TRIPS_P		T = NIGHT_TRIPS_P	
Variable	Coefficient	z	Coefficient	z	Coefficient	z	Coefficient	z	Coefficient	z
					Scale parameters for dists. of random parameters					
Constant	.12874***	21.72	.00534	.00	.13978***	26.2	.08853***	15.04	.13588***	21.53
Wave	.04756***	5.46	.00382	.49	0.0052	0.67	.08912***	10.13	.06212***	7.05
TBCP	.01457*	1.68	.02599***	3.47	.02510***	3.41	.00282	.34	.02359***	2.59
T	.00043	.33	.00444	.44	0.00993	0.73	.13842***	7.44	.16048***	5.01
Wave*TBCP	.00938	.84	.00890	.86	0.00913	0.89	.00145	.13	.02188*	1.81
Wave*T	.00060	.31	.00010	.01	0.02143	1.05	.02621	.97	0.04751	1.2
TBCP*T	.00012	.74	.01411	1.07	0.0119	0.65	.00892	.40	0.02318	0.52
Wave*TBCP*T	.00023	.09	.00277	.14	0.03111	1.23	.00451	.13	0.06	1.13
Trips	.00133	1.04	.00021	.18	0.00077	0.67	.00020***	2.87		
Trip_tim	.00017	.25	.00060	.93	0.00046	0.71				
Acomp_HH_time_p							.00279	.30		
Acomp_other_time_p							.03557***	3.65		
Morning_trips_p	.12224***	9.33	.18008***	14.39	.17310***	14.04	.23897***	17.00		
Afternoon_trips_p							.09627***	6.19		
Age	.06996***	11.07	.00230	.39	0.00399	0.69	.03186***	5.13		
HHMember							.00340*	1.95		
Divorced	.01882	.56							0.0128	0.39
Couple			.13699***	13.17	.13466***	13.08	.04030***	3.84		
Single	.02682**	2.35							.04804***	3.77
Education	.00966***	6.14	.00972***	6.30	.00926***	6.1	.01283***	8.09	.02042***	12.08
Student			.03175**	2.56	.02471**	2.03				
Retired			.04799	1.06	0.05123	1.14			0.09605	1.58
Sigma	2.78308***	293.35	2.84461***	331.20	2.80746***	331.51	2.89662***	292.43	2.82958***	273.45
Rho	−.91331***	−78.79	−.93493***	−106.39	−.94414***	−121.14	−.88936***	−56.16	−.87391***	−49.78
Total No. of Observations	738									
Pseudo R^2	0.08130		0.09730		0.09876		0.06240		0.04219	

interaction effects are not significant, indicating absence of heterogeneity. This is not the case for the three-way interaction effect.

The results obtained hold a number of interesting implications. First, the estimated coefficients of the interaction variable between the period variable and the treatment status, which indicates the global effect, are negative in all models. Therefore, participating in the TBCP causes a reduction in the relative use of a PV, which confirms the result of García-Garcés et al. (*18*) who used only data from the second survey wave in their analysis. The estimated reduction of the ratio PV/Total is around 10% in all models.

From model 1, the estimated coefficient for the three-way interaction variable that includes whether or not it is a joint trip is positive, although the absolute value is low. This result suggests that individuals making solo trips are more likely to be positively influenced by participating in the TBCP compared with those carrying out joint trips. Therefore we can conclude that companions during trips increase the rigidity of travel mode decisions.

In model 2, the estimated coefficient for the three-way interaction variable that includes whether or not the companion is a non-household member is positive.This result suggests that individuals making trips with household companions tend to be more influenced by participating in the TBCP than those accompanied by non-household companions. Considering the relative rigidity that companions add to travel mode decisions, if those companions are relatives or housemates, that rigidity is lower when compared with colleagues or acquaintances.

Finally, from models 3, 4, and 5, the estimated coefficients for the three-way interaction variables that include whether the individual carries out more trips during the afternoon, evening, or night, are negative, positive, and positive, respectively. These results suggest that individuals who make most of their trips during the afternoon tend to be more influenced by participating in TBCP compared with those who are morning travelers. In contrast, the more trips made during the evening or night, the less likely it is that participating in the TBCP has the expected effect. Afternoon activities are usually more flexible than morning activities, which may explain the first result. On the other hand, when car use is associated with an evening or night activity, it is more difficult to change to an alternative travel mode, due to reduced availability.

Conclusions

The results of this study confirm that participation in a TBCP has a different effect depending on with whom the individual usually travels, and the most frequent timing of the trips. The importance of the companion in the effect of the TBCP justifies the use of methods based on social network analysis to properly design and implement TBCPs, and to evaluate their efficacy.

The results of this study justify the use of persuasion techniques in the context of TBCP. The main advantage of this kind of strategy is that they can be adapted to overcome different types of individual travel behavior barriers. Therefore, the use of persuasion strategies can be recommended in sustainable urban mobility plans.

A number of transport policy implications can be derived from the results of this study:

- Specific actions should be designed to persuade people traveling in groups to reconsider car use. These new actions could introduce negotiation techniques, so that individuals can find common alternatives to their activity-travel agendas (e.g., a change of activity location that facilitates walking), which permit them to reduce their car use.
- Travel mode alternatives to cars should be widely available during evenings and nights. On-demand public transportation is a system that may be better suited to those trips.

Limitations of the present study are related to the size and characteristics of the sample, and the number of observations after the implementation of the TBCP. It is desirable to repeat the present study recruiting a bigger sample size, which would allow the evaluation of the effect of each individual persuasion action. In addition, it would also be appropriate to have a more diverse sample, particularly including individuals aged over 50 years. Also, more survey waves after the implementation of the TBCP would permit the study of medium- and long-term effects of participating in the TBCP.

Future work using the same dataset includes the evaluation of the effect of participating in the TBCP on the characteristics of the executed activities. This exercise will be developed considering demographics and socioeconomics of the participants, and also the characteristics of the activities. Additionally, we plan to take advantage of the rich dataset to carry out a similar study using the pre-planned activities and trips. The results of this study will shed light on the ways in which the TBCP affects the whole activity scheduling process, which will be useful in the improvement of travel demand operational models and the design of more efficient TBCP.

Acknowledgments

This study has been partially funded by the Ministerio de Economía, Industria y Competitividad of Spain, project MINERVA (TRA2015-71184-C2-1-R).

Author Contributions

The authors confirm contribution to the paper as follows: study conception and design: Tomás Ruiz; data collection: Tomás Ruiz; analysis and interpretation of results: Rosa Arroyo, Tomás Ruiz, Daniel Casquero, Lidón Mars; draft manuscript preparation: Rosa Arroyo and Tomás Ruiz. All authors reviewed the results and approved the final version of the manuscript.

References

1. Brög, W., E. Ker, I. Ryle, and R. Wall. Evaluation of Voluntary Travel Behavior Change: Experiences from Three Continents. *Transport Policy*, Vol. 16, No. 6, 2009, pp. 281–292.
2. Chatterjee, K., and P. Bonsall. Special Issue on Evaluation of Programs Promoting Voluntary Change in Travel Behavior. *Transport Policy*, Vol. 16, No. 6, 2009, pp. 279–280.
3. Moser, G., and S. Bamberg. The Effectiveness of Soft Transport Policy Measures: A Critical Assessment and Meta-analysis of Empirical Evidence. *Journal of Environmental Psychology*, Vol. 28, No. 1, 2008, pp. 10–26.
4. Bamberg, S., S. Fujii, M. Friman, and T. Gärling. Behaviour Theory and Soft Transport Policy Measures. *Transport Policy*, Vol. 18, No. 1, 2011, pp. 228–235.
5. Cairns, S., L. Sloman, C. Newson, J. Anable, A. Kirkbride, and P. Goodwin. Smarter Choices: Assessing the Potential to Achieve Traffic Reduction Using "Soft Measures". *Transport Reviews*, Vol. 28, 2008, pp. 593–618.
6. Stopher, P., and P. Bullock. *Travel Behavior Modification: A Critical Appraisal.* Presented for 26th Australasian Transport Research Forum, Wellington, New Zealand, 2003.
7. Brog, W., and M. Schadler. Marketing in Public Transport Is an Investment, Not a Cost. *Proc., 22nd Australian Transport Research Forum*, Sydney, 1998.
8. Rose, G., and E. Ampt. Travel Blending: An Australian Travel Awareness Initiative. *Transportation Research Part D*, Vol. 6, 2001, pp. 95–110.
9. Taylor, M. Voluntary Travel Behaviour Change Programs in Australia: The Carrot Rather Than the Stick in Travel Demand Management. *International Journal of Sustainable Transportation*, Vol. 1, No. 3, 2007, pp. 173–192.
10. Bamberg, S, and J. Rees. The Impact of Voluntary Travel Behavior Change Measures – A Meta-Analytical Comparison of Quasi-Experimental and Experimental Evidence. *Transportation Research Part A*, Vol. 100, 2017, pp. 16–26.
11. Friman, M., L. Larhult, and T. Gärling. An Analysis of Soft Transport Policy Measures Implemented in Sweden to Reduce Private Car Use. *Transportation*, Vol. 40, 2013, pp. 109–129.
12. Skarin, F., L. E. Olsson, I. Roos, and M. Friman. The Household as an Instrumental and Affective Trigger in Intervention Programs for Travel Behavior Change. *Travel Behavior and Society*, Vol. 6, 2017, pp. 83–89.
13. Richter, J., M. Friman, and T. Gärling. Review of Evaluations of Soft Transport Policy Measures. *Transportation: Theory and Application*, Vol. 2, 2010, pp. 5–18.
14. Fujii, S., and A. Taniguchi. Determinants of the Effectiveness of Travel Feedback Programs – A Review of Communicative Mobility Management Measures for Changing Travel Behavior in Japan. *Transport Policy*, Vol. 13, 2006, pp. 339–348.
15. Ker, I. *Travel Demand Management: Public Transport Business Case.* Contract Report RC5051, Department of Infrastructure, Victoria, 2003.
16. Meloni, I., B. Sanjust di Teulada, and E. Spissu. Lessons Learned from a Personalized Travel Planning (Ptp) Research Program to Reduce Car Dependence. *Transportation*, Vol. 44, 2017, pp. 853–870.
17. Ruiz, T, and P. García-Garcés. Measuring the Impact of Travel Behavior Change Programs on the Activity-Travel Scheduling Process. *Transportation Letters*, 2015, pp. 104–113.
18. García-Garcés, P., T. Ruiz, and K. N. Habib. Effect of Travel Behavior Change Programs on Time Allocated to Driving. *Transportmetrica*, 2016, pp. 1–19.
19. Stopher, P., C. J. Moutou, and W. Liu. *Sustainability of Voluntary Travel Behaviour Change Initiatives: A 5-year Study*. ITLS-WP-13-12. 2013.
20. Stopher, P, C. FitzGerald, and M. Xu. Assessing the Accuracy of the Sydney Household Travel Survey with GPS. *Transportation*, Vol. 34, 2007, pp. 723–741.
21. Stopher, P., and S. Greaves. Guidelines for Samplers: Measuring a Change in Behaviour from Before and After Surveys. *Transportation*, Vol. 34, 2007, pp. 1–16.
22. Stopher, P., and N. Swann. A 6-Wave Odometer Panel for the Evaluation of Voluntary Travel Behaviour Change Programs. *Proc., 30th Australasian Transport Research Forum*, Melbourne, Australia, 2007.
23. Taylor, M. Voluntary Travel Behaviour Change Programs in Australia: The Carrot rather than the Stick in Travel Demand Management. *International Journal of Sustainable Transportation*, Vol. 1, 3, 2007, pp. 173–192.
24. Bonsall, P. Do We Know Whether Travel Planning Really Works? *Transport Policy*, Vol. 16, No. 6, 2009, pp. 306–314.
25. Chatterjee, K. A Comparative Evaluation of Large-scale Personal Travel Planning Projects in England. *Transport Policy*, Vol. 16, No. 6, 2009, pp. 293–305.
26. Cohen, T. Evaluating Personal Travel Planning: If it Is Prohibitively Expensive To Get a Robust Answer Then What Should We Do? *Transport Policy*, Vol. 16, No. 6, 2009, pp. 344–347.
27. Philp, M., and M. A. P. Taylor. *Voluntary Travel Behavior Change and its Potential Implications for Climate Change Mitigation and Adaptation*, 2010.
28. Richter, J., M. Friman, and T. Gärling. Soft Transport Policy Measures: Gaps in Knowledge. *International Journal of Sustainable Transportation*, Vol. 5, 2011, pp. 199–215.
29. Stopher, P., E. Clifford, N. Swann, and Y. Zhang. Evaluating of Voluntary Travel Behaviour Change: Suggested Guide Lines and Case Studies. *Transport Policy*, Vol. 16, 2009, pp. 315–324.
30. Fujii, S., S. Bamberg, M. Friman, and T. Gärling. Are Effects of Travel Feedback Programs Correctly Assessed? *Transportmetrica*, Vol. 5, No. 1, 2009, pp. 43–57.
31. Stopher, P. R., and S. P. Greaves. Sample Size Requirements for Measuring a Change in Behaviour. *Proc., 27th*

Australasian Transport Research Forum, ATRF, Adelaide, Australia, 2004.
32. Sloman, L., S. Cairns, C. Newson, J. Anable, A. Pridmore, and P. Goodwin, *The Effects of Smarter Choice Programmes in the Sustainable Travel Towns: Summary Report*, Report to the Department for Transport, London, 2010.
33. Ruiz, T., and P. García-Garcés. Measuring the Impact of Travel Behavior Change Programs on the Activity-Travel Scheduling Process. *Transportation Letters*, Vol. 7, No. 2, 2013, pp. 104–113.
34. Cialdini, R. B. *Influence: How and Why People Agree to Things*. 1st ed. New York: William Morrow and Company, Inc., 1984.
35. Wilber, K. *Integral Psychology*. Shambhala, Boston, 2000.
36. Levene, H. Robust Tests for Equality of Variances, In *Contributions to Probability and Statistics* (I. Olkin, ed.), Stanford University Press, Stanford, CA, 1960, pp. 278–292.
37. Heckman, J. J. Sample Selection Bias as a Specification Error, *Econometrics*, Vol. 47, No. 1, 1979, pp. 153–161.
38. Greene, W. *LIMDEP Version 10. Econometric Modeling Guide*, 2009.
39. Puhani, P.A. The Heckman Correction for Sample Selection and its Critique. *Journal of Economic Surveys*, Vol. 14, 2000, pp. 53–67.

The Standing Committee on Traveler Behavior and Values (ADB10) peer-reviewed this paper (18-04386).

Article

TRR
JOURNAL OF THE TRANSPORTATION RESEARCH BOARD

Transportation Research Record
2018, Vol. 2672(47) 159–170

DOI: 10.1177/0361198118777604
journals.sagepub.com/home/trr
SAGE

Relationships between Consecutive Long-Term and Mid-Term Mobility Decisions over the Life Course: A Bayesian Network Approach

Bobin Wang[1,2], Soora Rasouli[2], Harry Timmermans[2,3], and Chunfu Shao[4]

Abstract

Long-term and mid-term mobility decision processes in different life trajectories generate complex dynamics, in which consecutive life events are interrelated and time dependent. This study uses the Bayesian network approach to study the dynamic relationships among residential events, household structure events, employment/education events, and car ownership events. Using retrospective data obtained from a web-based survey in Beijing, China, first structure learning is used to discover the direct and indirect relationships between these mobility decisions. Parameter learning is then applied to describe the conditional probabilities and predict the direct and indirect effects of actions and policies in the resulting network. The results confirm the interdependencies between these long-term and mid-term mobility decisions, and evidence the reactive and proactive behavior of individuals and households in the context of various life events over the course of their lives. In this regard, it is important to note that an increase in household size has a contemporaneous effect on car acquisition in the future; while residential events have a synergic relationship with employment/education events. Moreover, if people's residential location or workplace/study location moves from an urban district to a suburban or outer suburban district, this has both lagged and concurrent effects on car acquisition.

In China, increasing levels of car ownership lead to many problems, such as congestion, traffic accidents, and air pollution. By the end of 2016, Beijing ranked first in car ownership with 5.48 million vehicles (*1*). The government is trying a variety of policies to decrease car use and stimulate modal shift from cars to public transport and slow transport, such as the odd-and-even license plate rule, fare subsidies for public transport, and congestion charging. In the short term, these policies may change people's travel behavior. However, the effectiveness of policies in the long run will be improved by better understanding the relationships between long-term and mid-term mobility decision processes.

The life course approach provides a rich framework for better understanding these relationships. Van der Waerden et al. (*2*) provided a conceptual framework for understanding the dynamics of activity-travel behavior, and argued that life course events and critical incidents allow people to reconsider and possibly adapt their current activity-travel patterns. Lanzendorf (*3*) defended the mobility biography approach for longitudinal analysis of travel behavior, distinguishing between lifestyle domain, accessibility domain, and mobility domain. This seminal work led to a small but consistent stream of research, culminating in a recent workshop and book (*4*), in which the scope of the life-oriented approach was expanded to include quality of life (*5*). In general, these developments in transportation research pick up concepts and methods related to the life course approach as it was developed earlier and more intensively in social sciences and demography.

Theoretically, the basic idea underlying the life course approach is that human life history is a sequence of

[1]MOE Key Laboratory of Urban Transportation Complex Systems Theory and Technology, School of Traffic and Transportation, Beijing Jiaotong University, Beijing, P. R. China
[2]Urban Planning Group, Department of the Built Environment, Eindhoven University of Technology, MB Eindhoven, the Netherlands
[3]NUAA, Department of Air Transportation Management, Jiangjun Avenue, Jiangning District, Nanjing, P. R. China
[4]Key Laboratory of Transport Industry of Big Data Application Technologies for Comprehensive Transport, Beijing Jiaotong University, Beijing, P. R. China

Corresponding Author:
Address correspondence to Bobin Wang: bobinwang@bjtu.edu.cn

socially defined events and roles enacted over time (*6*). The central concepts are life trajectory, transitions, and events (*7*). Life trajectories describe different domains in life such as residence, health, education, work, leisure and recreation, and finance. Life events are defined as changes in a person's state that trigger a process of reconsideration of various life trajectories (*8*).

In this study, life events are the result of major decisions in a person's life, including a residential move, changes in the number of household members, workplace/study relocation, and changes in car ownership. Some events may change the spatiotemporal context in which activity-travel decisions are made. Some events constrain or expand the individual's choice sets, such as a change of car ownership. Moreover, the time horizon of different events differs. Long-term decisions are related to changes of residential and work location, while mid-term decisions involve car ownership and changes in household structure. Short-term decisions are related to daily mobility, with respect to trip frequency, mode, destination, route, and time of the day (*9*).

Empirically, research in the life course tradition has unfolded along three different lines of analysis. First and foremost, there has been a substantial body of qualitative and quantitative research into the effects of life events on activity-travel behavior. For example, Sharmeen et al. (*10*) reported the effects of several life events on time allocation to different activities and associated travel. Scheiner (*11*) focused on gender to study the effects of key events on changes in time use over the life course. Second, other researchers investigated the duration of particular states; for example, Rashidi (*12*) investigated the timing and reasons for residential relocation. Beige and Axhausen (*8*) utilized the competing risks model to compare different durations of residence, education, employment, and ownership of mobility tools. Third, a more modest stream of research analyzed the interdependencies between life events. Zhang et al. (*13*) used the exhaustive CHAID approach to investigate the two-way relationships between residential location and car ownership biographies in Japan, considering the influence of household structure mobility and employment/education mobility within the framework of a life-oriented approach. Verhoeven et al. (*14*) developed a Bayesian network to capture the dependencies between residential relocation and changes in household size. Oakil et al. (*15*) found complex direct and indirect dependencies between life events and long-term and short-term mobility decisions.

Although this body of research has enhanced our understanding of travel behavior dynamics, it has several limitations. Most studies only focused on a single life trajectory event or one aspect of the relevant causal and temporal relationships. Very few studies integrated these aspects into an overarching model. Moreover, most studies focused on state dependencies between the life domains, but neglected the relationships between mobility decisions (*16*). In the case of the temporal relationships between different life events, most research only considered the sequence of events, while the event occurrence time received less attention.

Elaborating Verhoeven et al. (*14*), this study utilizes a Bayesian network to investigate the dynamic interrelationships between residential events, household structure events, employment/education events, and car ownership events. A web-based survey was used to collect retrospective data of life trajectories.

This paper is organized as follows. Section 2 proposes the analytical framework and describes the modeling approach. Section 3 presents the survey and sample, and variables used for the model. Section 4 presents a detailed discussion of model results. Finally, important findings and recommendations are summarized in Section 5.

Methodology

Analytical Framework

This study focuses on four life trajectories: residential trajectory, household structure trajectory, employment/education trajectory, and car ownership trajectory. Residential events concern moving house (urban district, suburban district, outer suburban district). Household structure events include birth, someone leaving their parents' house, getting married, getting divorced, and passing away. Employment/education events include a change of school, someone finding their first job, and job transfers. Car ownership events involve adding, replacing and disposing of a vehicle.

Dynamics, interdependency, and time dependency are the properties of these life events examined in this study. There are various reasons for this. First, life events focus on the transition from one state to another. Second, mobility decisions are rarely made in isolation, but are strongly interrelated. A change in one life domain may trigger other mobility decisions. Third, these interrelationships may go forward (lagged effect), backward (anticipated effect), or be synchronic (contemporaneous effect). To consider the timing of every life event in the model, "Time ago/future events" are defined as, "how many years ago/in the future a certain event (using some classification of events) occurred/will occur, with respect to current time, *t*."

An example is shown in Figure 1. There are two life trajectories, and each one has a time line. A year is the unit of time, and the bold vertical lines indicate that a certain life event took place at a particular point in time. People of different ages have different life histories. Each observation records the information about both the last

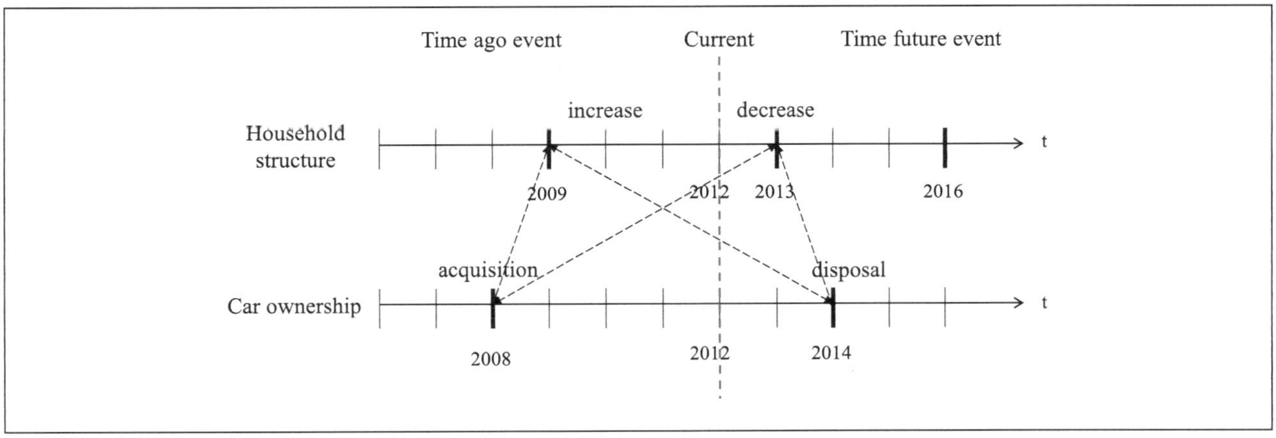

Figure 1. Relationships between household structure events and car ownership events.

event and the next event in different life domains, with respect to the current observation year. For the household structure trajectory in this example, the current observation year is 2012, and the last household structure event happened in 2009, with an increase in household size. The next household structure event occurs in 2013, with a decrease in household size. Thus, the episode duration for the time-ago event is three years, and for the time-future event is one year. The same applies to the car ownership events. Considering the time dependency between these two life events, it becomes clear that the increase in household size has an anticipated effect on car acquisition, while the decrease in household size has a lagged effect on car disposal. It should be noted that time dependencies do not necessarily represent causal relationships between different life events.

Moreover, the external factors influence long-term and mid-term mobility decisions. At the micro level, individual preferences and sociodemographic characteristics determine the order of life events in the life trajectory. At the macro level, economic, social and cultural, and market circumstances constrain the choice set of individuals. Therefore, the formal modeling framework underlying this study is presented in Figure 2.

Modeling Approach

Based on the aforementioned framework, a Bayesian Belief Network (BBN) was used to discover potentially causal relationships in the raw data, and predict the direct and indirect effects of actions and policies given the network structure. The advantages that make BBN attractive are: 1) More complex causal patterns can be included in the model to represent interdependencies between different life events; 2) Model results can directly describe the dynamic relationships between mobility decisions taken at different time points; and 3) Causal relationships can be derived in a flexible way (*14*, *15*).

Hugin software was used for learning the BBN. Two constraints-based algorithms are available for structure learning: the PC algorithm and the NPC algorithm (*17*). The basic mechanism is the same for these algorithms, that is, they are all based on generating a skeleton derived through statistical tests for a set of conditional independence and dependence statements (CID). This study uses the NPC algorithm, as it covers the deficiencies of the PC algorithm and gives a better map of reality relations. The NPC algorithm provides ambiguous regions and interacts with the user to decide the directions for undirected links. Therefore, the use of the NPC algorithm is recommended (*18*).

Although the NPC algorithm is closer to reality, some deterministic links may still be counterintuitive. The reason is that causality is decided on a statistical basis, which may be counterintuitive if not logically possible. Therefore, this study superimposed some constraints (domain knowledge) to fasten the learning, simplify the structure and avoid invalid relationships (*19*). The constraints regarding the causal relationships provide a priori assumptions and conditions about the structure and direct effects for the network. In particular, on theoretical grounds, the following constraints are applied:

1. Age, as a personal attribute, is only allowed to have direct effects on employment/education events and household structure events.
2. Intra-domain relationships between time-ago events and time-future events, and the intra-relationships between different life events in the same life domain, are not considered in the BBN structure.
3. Household structure mobility cannot be decided by workplace/study location, residence location, and car ownership.
4. Car ownership events do not have direct effects on employment/education events and residential

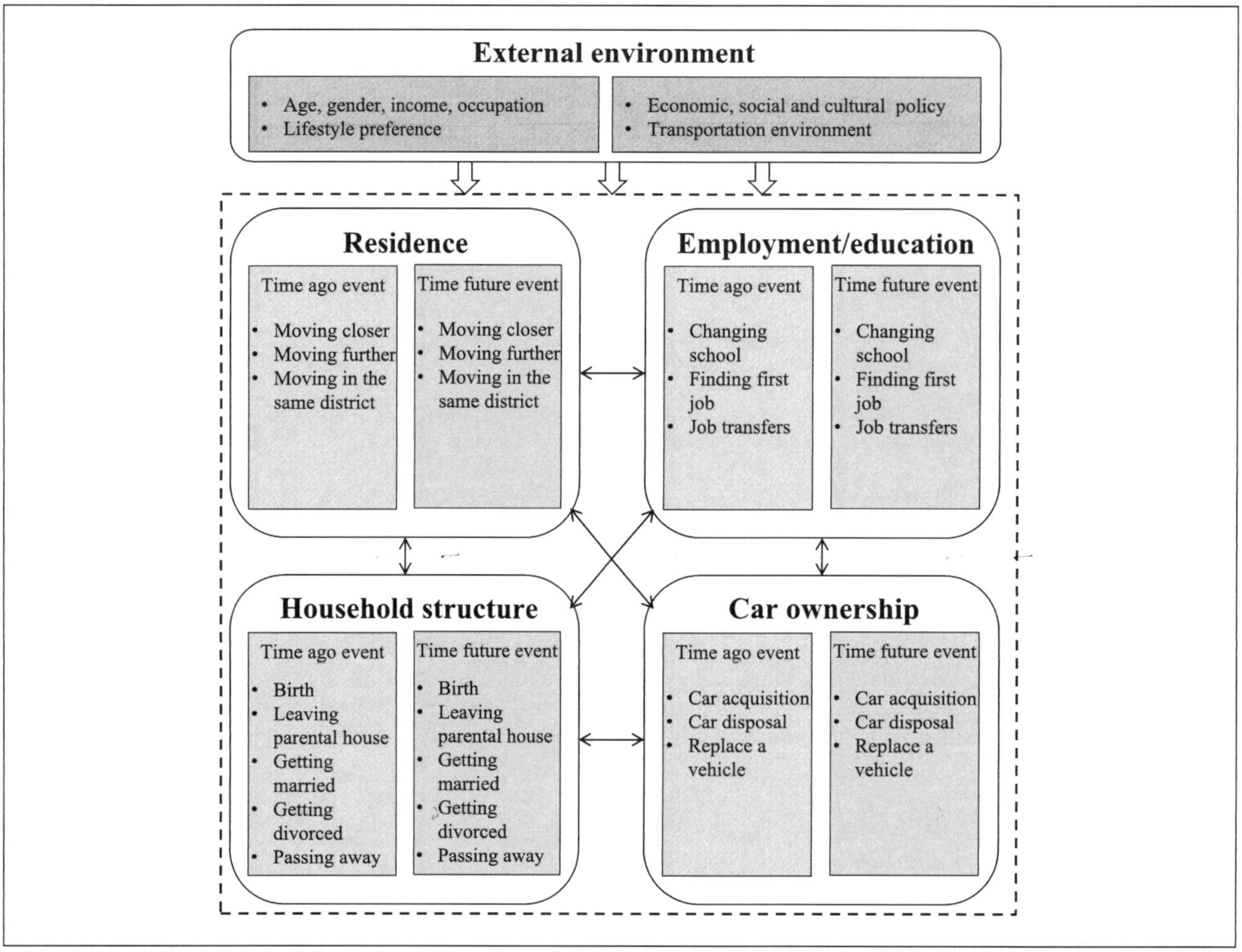

Figure 2. Modeling framework for data analysis.

events. Moreover, residence location changes in the future cannot influence work location changes in the past.

During the parameter learning phase, the CPTs for all nodes in the BBN were specified. As there were some unobserved data, the Expectation-Maximization (EM) algorithm was used to deal with this problem. The EM algorithm tries to estimate the CPTs that maximize the log likelihood of the current joint probability distribution on the case data. The learning process terminates when the difference in log likelihood between two successive iterations reaches a value smaller than the convergence threshold (*20*). The convergence threshold was set to 1.0e-4.

Variables

The selected variables are shown in Table 1. The state "0" means the event happens in the observation year, and this state only belongs to the time ago event. The state "never" means this kind of event has never happened. For example, there is no car in the family at the observation year, so this family cannot add or replace a car in the past, nor dispose of or replace a vehicle in the future.

Survey

Data Collection

To gather information about people's long-term and mid-term mobility decisions over the life course, longitudinal data were necessary. A retrospective approach was used in this study, asking respondents to recall their life events in chronological order. The survey was conducted at the household level, and the questionnaire recorded respondents' life course from when they were 18 years old. If a respondent arrived in Beijing when he/she was older, the respondent was requested to recall the life

Table 1. Explanation of Variables

Life course mobility	Time-ago event	Time-future event	State (years ago/ future)	Description
Employment/education	Workplace/study location changes from *i* to *j* in the past (Ep_*i*_*j*)	Workplace/study location changes from *i* to *j* in the future (Ef_*i*_*j*)	0=0 1=[1–2] 2=[3–4] 3=[5–7] 4=[8–10] 5=>10 Never	*i*, *j* ={urban districts, suburban districts, outer suburban districts}
Household structure	Household size has a change *i* in the past (Hp_*i*)	Household size has a change *i* in the future (Hf_*i*)	0=0 1=[1–2] 2=[3–4] 3=[5–6] 4=>6 Never	*i* ={decrease, increase}
Car ownership	Number of family cars has a change *i* in the past (Cp_*i*)	Number of family cars has a change *i* in the future (Cf_*i*)	0=0 1=[1–2] 2=3 3=4 4=>4 Never	*i* ={decrease (disposal), increase (acquisition), same (replace)}
Residence	Residence location changes from *i* to *j* in the past (Rp_*i*_*j*)	Residence location changes from *i* to *j* in the future (Rf_*i*_*j*)	0=0 1=[1–2] 2=3 3=[4–7] 4=[8–15] 5=>15 Never	*i*, *j* ={urban districts, suburban districts, outer suburban districts}
External factor Age			State (Years old) 1=<20 2=20–35 3=36–50 4=>50	

trajectory events from the time of arrival. The questionnaire content included both the state and life events in every calendar year during the respondent's observation period. Questions related to the four life events, shown in Table 1, include the total number of life events, the time (year) when the events happened, and what exactly was changed by these events. The sample size for present analysis is based on 294 questionnaires with 5,251 observation years.

The design and visual presentation of web-based survey can increase the response efficiency and improve the quality of each answer (*21*). The data were collected in Beijing, China, covering 16 districts of two central urban districts (Xicheng and Dongcheng), four suburban districts (Chaoyang, Fengtai, Shijingshan, and Haidian) and 10 outer suburb districts (Fangshan, Tongzhou, Shunyi, Changping, Daxing, Mentougou, Huairou, Pinggu, Miyun, and Yanqing). Respondents have to be at least 19 years old, and have been settled in Beijing more than one year by the time of the survey.

Age, gender and residential distributions of the sample in that year are summarized in Table 2. It shows that, compared with the Beijing statistics of 2014 (*22*), the sample data are reasonably representative of the population of Beijing, except for the typical bias introduced by web-based samples.

Sample Description

The frequency statistics of the various life events are shown in Table 3. Out of 5,251 observation years, a total of 365 workplace/study location changes occurred. Changes within suburban districts occurred more often (2.9%) than the other places. This suggests that the suburban districts, areas full of job-hopping, provide a lot of job and learning opportunities. Similarly, most residential events happened within the suburban or outer suburban districts. Moreover, car acquisition took place more frequently than car disposal and replacement, and more families increased in size.

Table 2. Sample Distribution

Factor	Level	The census of Beijing in 2014		The sample in 2014	
		N (ten thousand persons)	%	N (person)	%
Gender	Male	1,106.5	51.4	157	53.4
	Female	1,045.1	48.6	137	46.6
	Total	2,151.6	100.0	294	100.0
Residential distribution	Urban district	2,21.3	10.3	20	6.8
	Suburban districts	1,055.0	49.0	175	59.5
	Outer suburb districts	875.3	40.7	99	33.7
	Total	2,151.6	100.0	294	100.0
Age	19			10	
	20–24	223.7	13.1	45	15.9
	25–29	243.6	14.2	55	19.4
	30–39	397.9	23.2	95	33.6
	40–49	354.4	20.7	56	19.8
	50–59	312.8	18.3	24	8.5
	60–69	180.8	10.5	8	2.8
	>70			1	
	total	1,713.2	100.0	283 + 11	100.0

Table 3. Frequency Statistics of the Long-Term and Mid-Term Mobility Decisions

Mobility decisions	Count	%
Employment/education event		
Workplace/study location changes within urban districts	35	0.7
Workplace/study location changes from urban district to suburban district	44	0.8
Workplace/study location changes from urban district to outer suburban district	3	0.1
Workplace/study location changes from suburban district to urban district	34	0.6
Workplace/study location changes within suburban districts	153	2.9
Workplace/study location changes from suburban district to outer suburban district	25	0.5
Workplace/study location changes from outer suburban district to urban district	6	0.1
Workplace/study location changes from outer suburban district to suburban district	21	0.4
Workplace/study location changes within outer suburban districts	44	0.8
Total	365	6.9
Household structure event		
Decrease in household size	67	1.3
Increase in household size	272	5.2
Total	339	6.5
Car ownership event		
Car disposal	6	0.1
Car acquisition	198	3.8
Car replacement	1	0.0
Total	205	3.9
Residential event		
Residence location changes within urban districts	18	0.3
Residence location changes from urban district to suburban district	36	0.7
Residence location changes from urban district to outer suburban district	16	0.3
Residence location changes from suburban district to urban district	26	0.5
Residence location changes within suburban districts	221	4.2
Residence location changes from suburban district to outer suburban district	40	0.8
Residence location changes from outer suburban district to urban district	4	0.1
Residence location changes from outer suburban district to suburban district	28	0.5
Residence location changes within outer suburban districts	79	1.5
Total	468	8.9

In terms of event occurrence in different life trajectories, the frequencies for residential relocation (8.9%), work/education relocation (6.9%), and household member changes (6.6%) were much higher than changes in car ownership (3.9%). Residential relocation happened more frequently than the other mobility aspects,

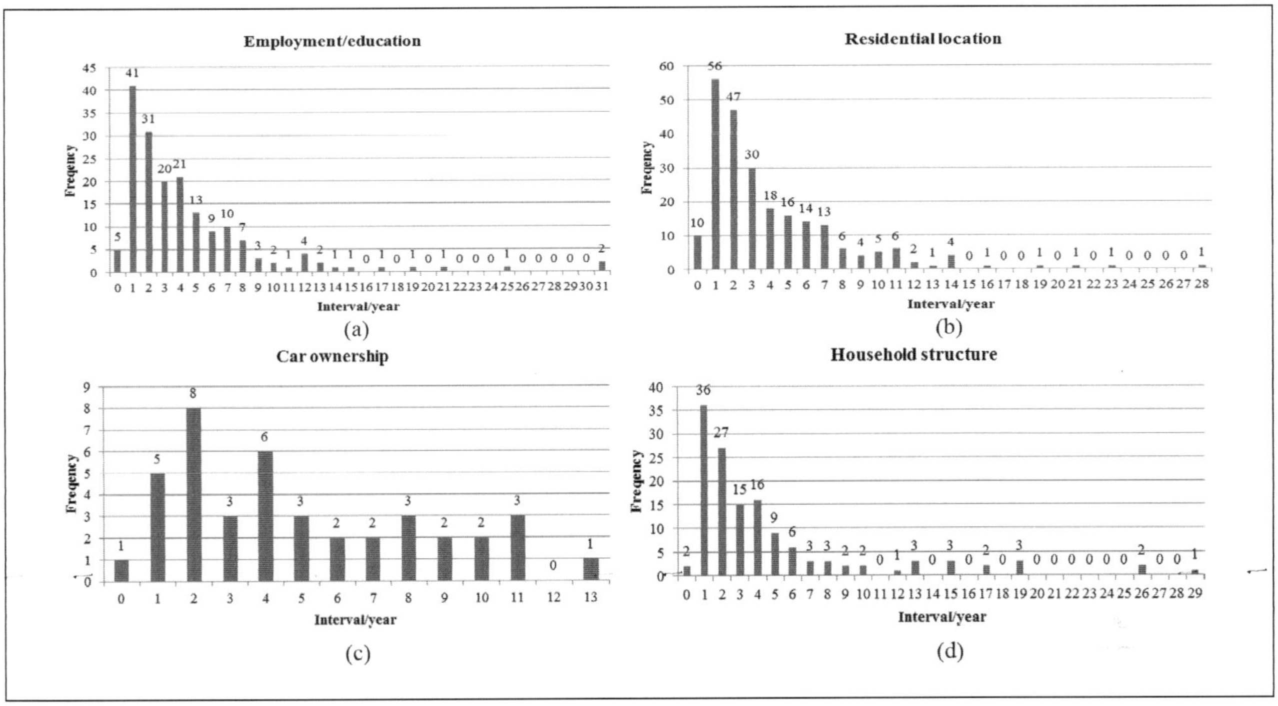

Figure 3. Frequency distribution of the time interval between two consecutive life events.

which is in line with the actual situation of the metropolis in China.

To consider the time dependencies between different life events, the time intervals between two consecutive events in these four life trajectories are shown in Figure 3. It is worth noting that some events may occur in the same year, so the time interval is 0. The results show that most life events in these four life trajectories occurred within two years, and the frequency distributions for these four life trajectories showed similar trends, except car ownership. That is because the changes in car ownership are not as frequent as other mobility aspects, and the respondents rarely had two or more consecutive car ownership changes during the observation period. More importantly, the discrete states of different life events, as shown in Table 1, are decided by the frequency distribution of time intervals.

Results and Analysis

Structure Learning

Hugin software was used to build and estimate the Bayesian Belief Network for life trajectories, using the input data and constraints described earlier. The level of significance was set to the standard value of 0.01 for the learning process. The causal relationships found in the learned network are shown in Figures 4 and 5.

The direct and indirect relationships between life events can be found through these links. Based on the time property of the node, the temporal relationships are distinguished by lagged effects, contemporaneous effects, and anticipated effects. The link directed from the time-ago node to the time-future node refers to a lagged effect, while the reverse direction signifies an anticipated effect. Interrelationships within the same time domain are called the contemporaneous effect. These relationships are discussed below.

Relationships between external factor and life events: "Age" directly links to employment/education events and household structure events, which indicates that the specific mobility occurs at a specific age.

Relationships between car ownership events and other life events: car acquisition in the past was mainly affected by workplace/study location changes and residential location changes in the past. The employment/education event "Ep_outsuburban_suburban" and residential event "Rp_outsuburban_outsuburban" both have a contemporaneous effect on car acquisition in the past. This is understandable as a change in work location or residential location that increases the daily commuting distance leads to the necessity for a car. For car acquisition in the future, work location changes were seen to increase the number of family cars in the future, which can be explained as follows: the increase in salary with job-hopping allows people to earn enough money to buy a

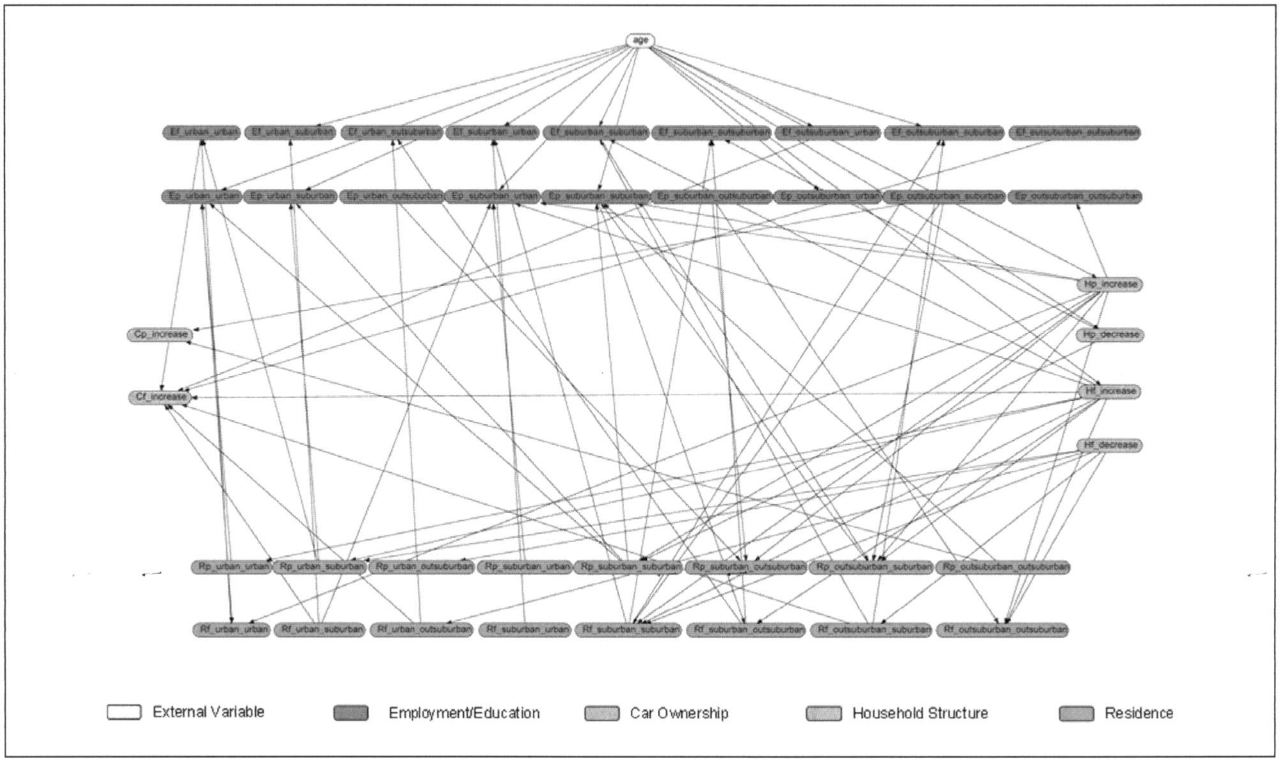

Figure 4. Learned life trajectory network.

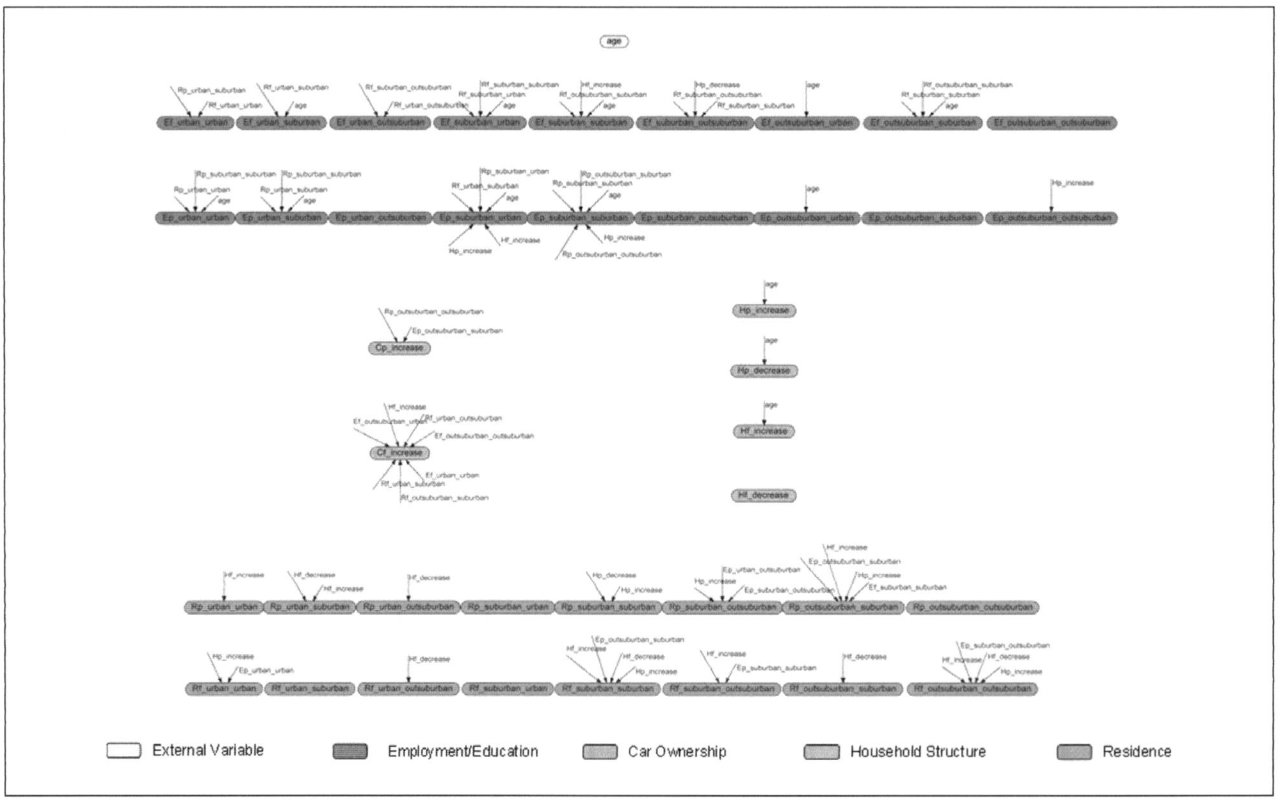

Figure 5. Learned life trajectory network.

car. Residence relocation from urban district to suburban or outer suburban district also has direct effects on car acquisition in the future, because the greater travel distance induces people to consider buying a car in the medium term. Moreover, an increase in household size, such as having a baby, leads to car acquisition in the future.

Relationships between residential events and other life events: residential events seem to be mainly affected by household structure events and employment/education events. Compared with other residential events, "Rp_suburban_suburban", "Rp_suburban_outsuburban", "Rp_outsuburban_suburban", "Rf_suburban_suburban", "Rf_suburban_outsuburban", "Rf_outsuburban_outsuburban" had more links. Work location events had contemporaneous and lagged effects on residence location events, and their locations after change fell into the same area. For example, "Ep_urban_outsuburban" was linked with "Rp_suburban_outsuburban", and "Ep_outsuburban_suburban" was linked with "Rf_suburban_suburban." Meanwhile, household structure events had lagged, contemporaneous, and anticipated effects on residential events.

Relationships between employment/education events and other life events: employment/education events were mainly decided by age, residential, and household structure events. Residential events had lagged and contemporaneous effects on employment/education events. More specifically, there was usually a synergic relationship between them, meaning the variation range for their changed locations is consistent, such as "Rp_urban_suburban" linked to "Ep_urban_suburban," "Rf_urban_outsuburban" linked to "Ef_urban_outsuburban." Moreover, an increase in household size had contemporaneous and anticipated effects on employment/education events. This is plausible in that people usually want to have sufficient economic security before having a baby or getting married.

Relationships between household structure events and other life events: the household structure cannot be influenced by other life events, so it was only affected by age. Based on the above analysis, household structure events had lagged, contemporaneous, and anticipated effects on residential, employment/education, and car-ownership events. Compared with other life events, there are frequent links between household structure events and residential events. This means there is an intimate relationship between these two life trajectories.

Parameter Learning

Based on the network structure, the EM algorithm was used to estimate the CPTs for each node with the observed data. The estimated marginal probability distribution for every node is shown in Figure 6. In the structure learning analysis, the temporal effects can be divided into lagged effects, contemporaneous effects, and anticipated effects. Actually, contemporaneous effects also contain these three kinds of time attributes, if the states of the nodes are taken into account. To differentiate, they are expressed as lag effects, concurrent effects, and lead effects. These specific temporal relationships were not observed in the structure learning, but can only be found in the parameter learning.

One application of this network is to observe the direct and indirect dynamic relationships between the nodes of interest based on the process of simulation. Simulation refers to entering hard evidence into the network and comparing the probabilities of certain nodes of interest with and without evidence. The hard evidence means a certain condition is known and the probability for this state is 100%. When the hard evidence for one or more nodes is entered into the BBN, the probabilities for their related nodes, directly or indirectly, are updated.

In this study, we focused our interest on car acquisition in the future. To solve housing problems among low-income families, the Beijing government annually provides a batch of economically affordable housing to the public, usually located in the outer suburban districts. Structure learning showed that changing residence from urban district to suburban or outer suburban district will lead to car acquisition in the future. So the relationships between remote relocation and car acquisition in the future are analyzed in this study. The relative probability difference (the difference between the updated probability and initial probability divided by the initial value) is used to describe the sign of the evidence effects, and the greater the relative difference, the greater the tendency is. The results are shown in Table 4.

The relevant residential events include three kinds of changes: moving from urban district to suburban district (Rf_urban_suburban), moving from urban district to outer suburban district (Rf_urban_outsuburban), and moving from suburban district to outer suburban district (Rf_suburban_outsuburban). The relative probability differences of car acquisition given evidence of residential events in the future are shown in Table 4. The analysis found that the life event "Rf_urban_suburban" has lag effects on car acquisition in the future. For example, the probability of car acquisition within three years increases by 90.52% if the person decides to relocate from urban district to suburban district in one or two years. Likewise, the residential event "Rf_suburban_outsuburban" had similar effects on car acquisition in the future. For the residential event "Rf_urban_outsuburban,""three years" was a cut-off point. When this residential event happens one or two years later, the probability of car acquisition in four years increases by 78.45%. However, if this residential

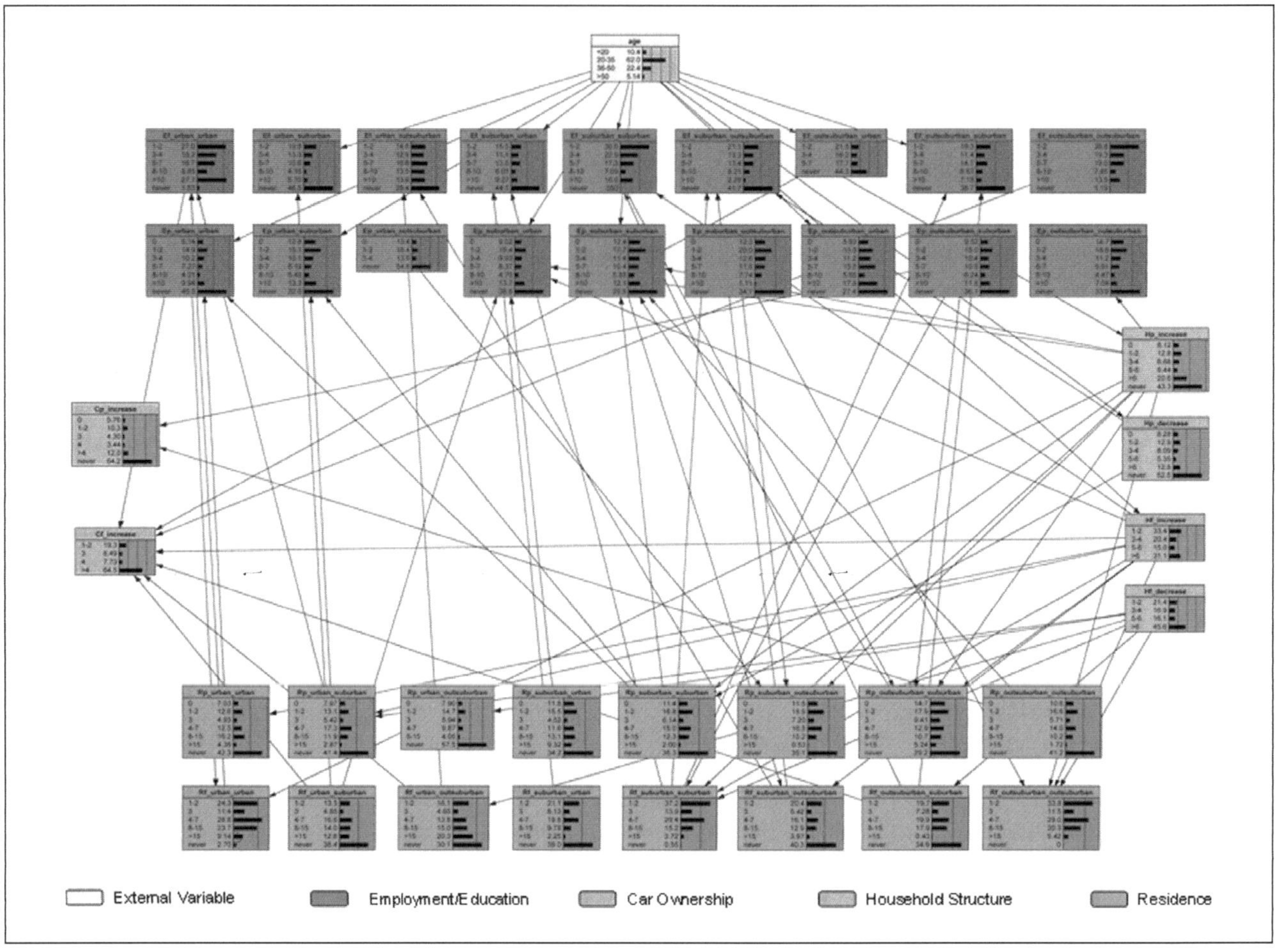

Figure 6. Estimated marginal probability distributions for Bayesian Belief Network.

Table 4. Relative Probability Difference of Car Acquisition Given Evidence of Residential Events in the Future

		Car acquisition in the future			
Evidence of residential events		One or two years later	Three years later	Four years later	More than four years later
No evidence		19.3	8.5	7.7	64.5
Rf_urban_suburban	One or two years later	61.6%	**90.5%**	24.7%	−33.4%
	Three years later	35.1%	−15.0%	**119.3%**	−22.9%
	Four to seven years later	−8.6%	−3.7%	−12.0%	**4.5%**
	Eight to 15 years later	−48.7%	−30.1%	−19.2%	**20.9%**
	More than 15 years later	−54.4%	−38.3%	−37.6%	**25.9%**
Rf_urban_outsuburban	One or two years later	42.6%	69.0%	**78.5%**	−31.3%
	Three years later	−46.6%	**111.5%**	−38.2%	3.9%
	Four to seven years later	−28.5%	1.0%	**25.6%**	5.4%
	Eight to 15 years later	−26.7%	−26.7%	−31.4%	**15.3%**
	More than 15 years later	−65.1%	−52.4%	−56.4%	**33.2%**
Rf_suburban_ outsuburban	One or two years later	8.9%	**13.0%**	10.5%	−5.7%
	Three years later	−4.1%	7.7%	**21.2%**	−2.3%
	Four to seven years later	−31.9%	−28.3%	−10.9%	**14.6%**
	Eight to 15 years later	−60.3%	−51.2%	−56.3%	**31.6%**
	More than 15 years later	−52.4%	−44.8%	−38.3%	**26.2%**

Note: Bold values denote the biggest probability difference between the updated probability and initial probability.

event occurs three or more years later, car purchase is expected to happen simultaneously.

Conclusion

Long-term and mid-term mobility decision processes constitute a complex system of interdependencies, in which life events of different life trajectories are interrelated and time dependent. Changes in one state may trigger various direct and indirect effects through the network. Very few studies have attempted to address the complexity of these decision dynamics. In this study, the Bayesian network approach was used to study the dynamic relationships between consecutive long-term and mid-term mobility decisions over the life course, covering four life trajectories: changes in residence, changes in household structure, changes in employment/education, and changes in car ownership.

The model results provide evidence to confirm the interdependencies between different life events. Considering their causal relationships, the main reasons for car acquisition are the increase of household size, workplace/study location changes, and residential relocation. These are related to the changing of family needs and travel demands. Residential events were closely related to and interacted with employment/education events. Moreover, a synergic relationship was established between these two life trajectories. Household structure events have direct effects on the residential, employment/education, and car ownership events. Compared with other life events, household structure trajectory and residential trajectory have a more intimate relationship with each other.

In terms of temporal relationships, lagged, contemporaneous, and anticipated relationships exist between various life events. Most importantly, changes in household structure have contemporaneous and anticipated effects on residence location events, while an increase in household size has contemporaneous effects on car acquisition in the future. Moreover, more detailed temporal relationships are found from the contemporaneous effects, after considering the timing of the last and the next life events. Residential relocation has a lag effect on car acquisition in the future, if the person expects to move from urban district to suburban district or from suburban district to outer suburban district. This phenomenon is logical: car acquisition and changing house are both big expenditures for a family, which cannot be achieved at the same time. Moreover, when people anticipate a change in workplace/study location from urban district to suburban or outer suburban district in three or more years, it has concurrent effects on car acquisition in the future.

Although this study has established multiple causal and temporal relationships, it still has several limitations and extensions. First, the learned network structure is built on individual life trajectories, without considering interaction between household members. Therefore, it is not guaranteed that the current model will be consistent for members of the same family. Second, the interrelationships between various life events will change with the means of transportation and the spatial and economic context. It would be an interesting extension to study behavioral adaption to new policies or context changes.

Acknowledgments

This work was supported by the National Natural Science Foundation of China (Grant No. 51678044) and Science Fund for Creative Research Groups of the National Natural Science Foundation of China (Grant Number 71621001).

Author Contributions

The authors confirm contribution to the paper as follows: study conception and design: Bobin Wang, Soora Rasouli, Harry Timmermans; data collection: Bobin Wang, Chunfu Shao; analysis and interpretation of results: Bobin Wang, Soora Rasouli, Harry Timmermans; draft manuscript preparation: Bobin Wang, Soora Rasouli, Harry Timmermans, Chunfu Shao. All authors reviewed the results and approved the final version of the manuscript.

References

1. Traffic Management Bureau of the Public Security Ministry. *The Number of National Motor Vehicles and Drivers Kept a Fast Growth in 2016*. Transportation Bureau of the Ministry of Public Security. www.mps.gov.cn/n2255040/n4908728/c5595634/content.html. Accessed January 10, 2017.
2. Van der Waerden, P., and H. J. P. Timmermans. *Key Events and Critical Incidents Influencing Transport Mode Choice Switching Behavior: An Exploratory Study*. Transportation Research Board, Washington, D.C., 2003.
3. Lanzendorf, M. Mobility Biographies: A New Perspective for Understanding Travel Behaviour. Presented at 10th International Conference on Travel Behaviour Research, Lucerne, 2003.
4. Zhang, J. *Life-Oriented Behavioral Research for Urban Policy*. Springer, Japan, 2017.
5. Zhang, J., Y. Tsuchiya, H. Hinohara, and M. Chikaraishi. Citizens' Life Behavior and Quality of Life: Survey and Modeling. Presented at 34th International Association for Time Use Research (IATUR), Matsue City, Japan, 2012.
6. Elder, G. H. Jr., M. K. Johnson, and R. Crosnoe. The Emergence and Development of Life Course Theory. In J. T. Mortimer, and M. J. Shanahan, *Handbook of the Life Course*, Springer, Boston, MA, 2003.
7. Elder, G. H. Jr. *Lifecourse Dynamics: Trajectories and Transitions, 1968-1980*. Cornell University Press, Ithaca, 1985.

8. Beige, S., and K. W. Axhausen. The Dynamics of Commuting over the Life Course: Swiss Experiences. *Transportation Research Part A: Policy and Practice*, Vol. 104, 2017, pp. 179–194. https://doi.org/10.1016/j.tra.2017.01.015.
9. Salomon, I., and M. Ben-Akiva. The Use of the Life-Style Concept in Travel Demand Models. *Environment and Planning A*, Vol. 15, No. 5, 1983, pp. 623–638.
10. Sharmeen, F., and T. Arentze. Incorporating Time Dynamics in Activity-Travel Behaviour Model: A Path Analysis of Changes in Activity and Travel Time Allocation in Response to Life Cycle Events. *Transportation Research Record: Journal of the Transportation Research Board*, 2013. 2382: 54–62.
11. Scheiner, J. Time Use and the Life Course-A Study of Key Events in the Lives of Men and Women Using Panel Data. *European Journal of Transport & Infrastructure Research*, Vol. 16, No. 4, 2016, pp. 638–660.
12. Rashidi, T. H. Dynamic Housing Search Model Incorporating Income Changes, Housing Prices, and Life-Cycle Events. *Journal of Urban Planning and Development*, Vol. 141, No. 4, 2014, p. 04014041.
13. Zhang, J., B. Yu, and M. Chikaraishi. Interdependences between Household Residential and Car Ownership Behavior: A Life History Analysis. *Journal of Transport Geography*, Vol. 2014, No. 34, 2014, pp. 165–174.
14. Verhoeven, M., T. Arentze, H. Timmermans, and P. van der Waerden. *Simulating the Influence of Life Trajectory Events on Transport Mode Behavior in an Agent-Based System*. Intelligent Transportation Systems Conference (ITSC), Seattle, 2007, pp. 107–112.
15. Oakil, A. T. M., D. Ettema, T. Arentze, and H. Timmermans. Longitudinal Model of Longer-Term Mobility Decisions: Framework and First Empirical Tests. *Journal of Urban Planning and Development*, Vol. 137, No. 3, 2011, pp. 220–229.
16. Srinivasan, K. K., and P. Bhargavi. Longer-Term Changes in Mode Choice Decisions in Chennai: A Comparison between Cross-Sectional and Dynamic Models. *Transportation*, Vol. 34, No. 3, 2007, pp. 355–374.
17. Spirtes, P., C. Glymour, and R. Scheines. *Causation, Prediction, and Search*. MIT Press, Cambridge, 2000.
18. Hugin. *Hugin Graphical User Interface Help*. Version 8.4 lite. Hugin Experts A/S, 2016.
19. Cheng, J., D. A. Bell, and W. Liu. An Algorithm for Bayesian Belief Network Construction from Data. *Proceedings of the Sixth International Workshop on Artificial Intelligence & Statistics, Fort Lauderdale*, FL, 1997, pp. 83–90.
20. Lauritzen, S. L. The EM Algorithm for Graphical Association Models with Missing Data. *Computational Statistics & Data Analysis*, Vol. 19, No. 2, 1995, pp. 191–201.
21. Christian, L. M., D. A. Dillman, and J. D. Smyth. Instructing Web and Telephone Respondents to Report Date Answers in a Format Desired by the Surveyor. *Social & Economic Sciences Research Center Technical Report*, Vol. 5, No. 167, 2005, pp. 1–18.
22. Beijing Municipal Bureau of Statistics. *Beijing Statistical Yearbook*. China Statistics Press, Beijing, 2015.

The Standing Committee on Traveler Behavior and Values (ADB10) peer-reviewed this paper (18-04424).

Article

TRR
JOURNAL OF THE TRANSPORTATION RESEARCH BOARD

Transportation Research Record
2018, Vol. 2672(47) 171–181

DOI: 10.1177/0361198118797509
journals.sagepub.com/home/trr
SAGE

Measuring the Impact of Motivations on Travelers' Strategic Decisions in Different Traffic Conditions: Data Collection, Analysis, and Modeling

Dong Pan[1], Samer H. Hamdar[1], John L. Campbell[2], and Amir Farrahi[1]

Abstract

Continued growth in travel demand and the corresponding congestion occurrence accentuates the need for active transportation and demand management (ATDM) for predictive rather than reactive congestion mitigation strategies. These strategies (applications) reduce demand and thus improve the performance of different surface transportation facilities. Earlier research work mainly utilized simulations and field experiments to suggest improvements to ATDM applications associated with pricing and/or provision of information to travelers; however, few such studies considered the dynamics of travelers' motivations as an essential component in the performance of a given ATDM application. In other words, how travelers' motivations change as a function of the ATDM applications' characteristics remains largely unexplored. This study thus examines the theory of planned behavior (TPB) to analyze travelers' motivations associated with different mode choices and captures and measures the corresponding dynamics of motivations when facing different surrounding circumstances and ATDM regimes. Utilizing a survey and an online data collection platform to estimate motivational patterns behind travelers' mode choices, following and extending the TPB paradigm, the study concludes that: (a) travelers' mode choices are primarily determined by intentions (motivations); (b) income and age are two additional characteristics that influence mode choices; (c) travelers' reasoned choices are mainly attitude-oriented; and (d) different attitudinal aspects are accentuated or compromised along with the changes of travel conditions and ATDM applications.

Travelers make various decisions to reach a given goal; such decisions include destination choice, time of day choice, mode choice, route choice, and lane/facility choice (*1*). These decisions are strategic and tactical (and therefore, reasoned) in nature when the traveler's purpose is to optimize the efficiency (e.g., time and cost efficiency) and improve the experience (e.g., safety, convenience, comfort) of their trips based on the information and knowledge available to them. Among the decisions mentioned above, mode choice is discrete, with a manageable set of alternatives (personal vehicle, light rail transit, bicycle, etc., and their combinations). Travelers consider various mode alternatives under different decision-making contexts and eventually select the most favorable one.

Focusing on discrete choice analysis, past studies (especially in the area of economics) estimate the optimal choice or the worthwhileness of a change in choice by the measure of exogenous factors, using maximum utility, transaction cost, etc. (*2*, *3*). These studies can be classified as utility-based studies (i.e., maximum utility theory and its variation including prospect theory) and heuristic-based studies (i.e., information processing heuristics). However, the measures resulting from these studies fail to consider decision-makers' motivations as essential to initiating their decisions (or their changes in decisions) when facing a changing environment. In view of such limitations, this study measures travelers' motivations (mainly attitudes) toward different travel options while adapting to the corresponding travel conditions (e.g., weather and incident conditions) and existing travel demand management strategies (e.g., pricing and available information). The goal of this study is to answer the following two fundamental questions: What kind of

[1]Department of Civil and Environmental Engineering, The George Washington University, Washington, DC
[2]Center for Human Performance and Safety, Battelle Memorial Institute, Seattle, WA

Corresponding Author:
Address correspondence to Dong Pan: pandong@gwmail.gwu.edu

motivations of a traveler are affected by ATDM strategies and surrounding circumstances; How do changes in motivation lead to his/her changes in travel behavior?

Answering these questions requires the framing of travelers' motivations. It is assumed that (mode) choice decisions are reasoned (*4*); in other words, the underlying psychology can be described by reasoned-action models, among which the most prevalent one is Ajzen's theory of planned behavior (TPB) (*5–7*). According to TPB, behavior is primarily determined by intention, a motivational state toward performing a given (physically active) behavior. Intention is further theorized to be a function of three sub-constructs: (a) attitude (i.e., the degree to which performing the behavior is positively valued); (b) subjective norm (i.e., perceived social pressure from significant others to perform the behavior); and (c) perceived behavioral control (i.e., the extent to which performing the behavior is easy and subject to one's will) (*5*, *8*).

Diverse behaviors, such as voting choices, weight loss, attending classes, and committing traffic violations, have been estimated and analyzed through TBP (*5*). In the field of transportation, researchers have utilized TPB as an effective tool to frame travelers' motivational patterns leading to their daily mode choices (such as the choice among car, bus, and bicycle, via a one-off survey design) or the changes of their daily mode choices (such as the shift from driving alone to riding the bus, via a pre-post survey design) (*9*, *10*).

It is argued, however, that though many decisions may have first originated from reasoning, people will not go through a deliberate decisional process every time when a decision is routinized (*11*). As travelers provide feedback for themselves (i.e., contemplating and learning) each time a decision is made, especially when the same (or largely similar) configuration of travel goal/pattern occurs repeatedly, they are likely to take a shortcut in decision making, or the rule of thumb; that is, they link the activation of a typical mode choice directly to the recognition of a travel goal (*12*). Such association is enhanced with time due to the repetition of the goal–act–feedback cycle (*12*, *13*). Thus, strong habits might put a boundary limitation on the applicability of reasoned-action models such as TPB (*13*, *14*). Such limitation may need further investigation through estimating the TBP model using an elaborate experimental setup.

TBP and its limitation in estimating travel behavior (in particular, mode choice) have been examined by many empirical studies. For example, Aarts and Dijksterhuis conducted an experiment at the University of Nijmegen, the Netherlands, to tackle the goal-directed automaticity in two typical modes, namely bicycle and train (*12*). They found that, compared with non-habitual choices (i.e., train), habitual choices (i.e., bicycle) were harder to suppress. This was in line with Ouellette and Wood's conclusion on habits, which claimed that "the direct influence of the past behaviors on the future ones is most pronounced for behaviors that are executed frequently and consistently in a stable context" (*14*). Another experiment was conducted among college students in Germany with a prepaid "semester ticket" on a local bus system (*15*). The results of the experiment (a considerable portion of students who used to drive to campus switched to bus) indicate that even if compelling influence of habits does exist, with sufficient change in decision-making context, reasoned intention will resume the dominant role in decision making, following the description of TPB. These studies, however, failed to look at a more generalized spectrum of travel conditions that include incident and inclement weather conditions. Moreover, these studies did not investigate a specific motivation framing paradigm with the same group of participants over an extended period of time. Such investigation requires a more elaborate data collection platform with more controllability: a virtual travel environment that is based on a real multi-modal network with realistic travel conditions. Such a data collection platform is presented in more details in the following Methodology section.

Following this introductory section, the remainder of this paper is organized as follows. The next section presents the research methodology. The third section presents and analyzes the numerical results. The fourth section summarizes the findings before concluding with some future research directions.

Methodology

This study applies TPB to tackle the motivations that affect travelers' mode decisions with specific focus on one type of trips: morning commute trips. Commuting is the travel from one's place of residence to one's place of work/study, and vice versa. Such behavior is periodic and constitutes the main portion of travel demand during the morning and evening peak hours.

To collect data relevant to this trip type, two types of experiments were conducted: an online survey and an online virtual travel environment. The survey covered a larger sample size to account for the socio-economic characteristics of individuals (in addition to motivations) as attributes to determine people's typical modes for their daily commute; while the virtual environment provided a map-based transportation network with different setups and captured players' in-route mode decisions and the corresponding motivations for an extended period of time.

Survey Design

The survey was designed to study the feasibility of TPB in estimating the mode-choice-correlated behaviors and the underlying motivations. It was divided into the following sections: General Questions (gender, age, education, income, ethnicity, etc.), Social Networks and Information Resources, Intention, Attitude, Subjective Norm, Perceived Behavioral Control, and Environmental/Health Concern. There were 35 single questions and three matrix questions (each matrix question took the quota of five single questions), and for most of the questions related to the TBP paradigm, respondents were provided with options in a seven-point Likert scale to express their levels of agreement/likelihood regarding the given issue/statement.

Sample questions assessing traveler's intention, attitude, subjective norm and perceived control regarding a given mode in the survey are presented below:

- Intention: I plan to take transit for daily commute.
  - Very unlikely.
  - Unlikely.
  - Somewhat unlikely.
  - Neutral.
  - Somewhat likely.
  - Likely.
  - Very likely.
- Attitude: For me to take transit for daily commute is convenient.
- Subjective norm: Most of my family, friends and colleagues (whose opinions I value) think I should take transit for daily commute.
- Perceived control: I am confident that I could take transit for daily commute if I wanted to.

Virtual Travel Environment Design

The virtual travel environment was designed to immerse the participants into a virtual transportation network (Figure 1) to perform their trips, intending to reproduce the morning commute trips that originate in Northern Virginia and end in Washington DC, U.S.A.

Various travel conditions with/without ATDM strategies were manipulated in this study: three travel conditions were classified and simulated, namely, normal commute condition, inclement weather condition, and incident condition. Two types of ATDM strategies were adopted and examined: a pricing strategy (50% discount

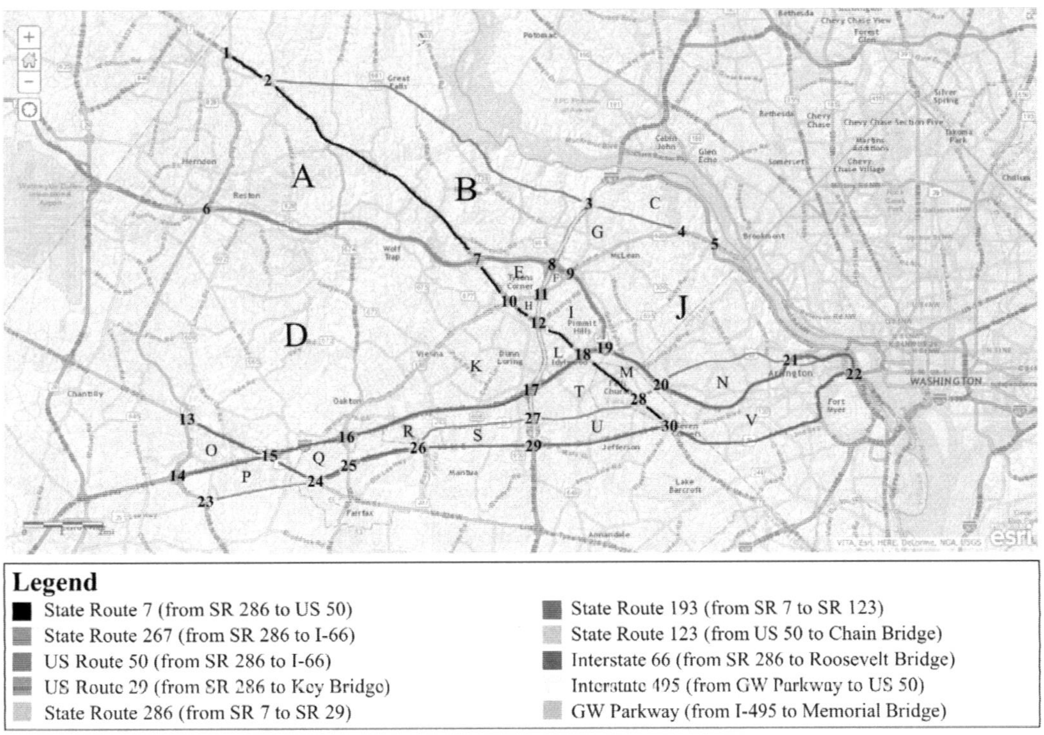

Original Photo: © 2016 Esri®

Figure 1. Roadway network in Northern Virginia translated to the virtual travel environment.

rate on metro) and a traveler information strategy (reminding travelers about the risk of losing income due to the potential late arrival at work).

Experimental Setup. The transportation research team at the George Washington University built a virtual transportation network by imbedding Google® Maps™ with a customized user interface.

A set of origin nodes were pinned in the network, including Ashburn/Leesburg (Node 1 in Figure 1), Centreville/Chantilly (Node 14), Fairfax (Node 24), Falls Church (Node 28), Great Falls (Node 7), Herndon/Reston (Node 6), Tysons (Node 10), Vienna (Node 17). The destination node was Rosslyn (Node 22), the area in Virginia right before entering the national capital, where traffic/transit bottlenecks mostly take place during morning peak hour periods. Participants were randomly assigned to an origin node and "performed" their trips to the destination node.

Participants could choose their preferred travel modes among car (drive-alone), metro, and bicycle. They could also choose their route (if available) under the corresponding mode. Generally, car and bicycle were available on the entire network; however, the availability of metro depended on the location, that is, metro was prohibited if there was no nearby metro station. Participants were provided the opportunities to transfer or reroute; however, once they chose metro or bicycle, car was no longer available.

To help participants make their mode/route decisions, information on the distance, travel time, and cost of using different modes/routes was provided. The "direction" function of Google® Maps™ provides distance and estimated travel time via various modes of transportation without the impact of inclement weather and incident conditions; travel conditions however may affect travel time and bring uncertainty (the variation of travel time). Extensive data mining and overlapping efforts were performed to work out the conditional mean travel time and time variation, as shown in Table 1. The corresponding data libraries that were mined were: traffic data (continuous count stations administered by Virginia Department of Transportation), weather data (The Weather Channel, www.weather.com), and incident data (National Highway Traffic Safety Administration).

The cost of trips varied from mode to mode:

- The cost of driving a car was the gas consumption cost based on mileage ($ 0.16 per mile) plus parking fee, where parking in Rosslyn costs $15 and in-route park-and-ride/bicycle parking costs $5.
- The cost of taking metro was a mileage-based ticket fare ($ 0.32 per mile).
- Riding a bicycle was assumed to be free of charge.

User Interface and Instruction. All participants were required to register with a valid email address and to complete a pre-experiment survey upon their first login. In the virtual travel environment, participants were instructed to arrive at their destinations by 9:00 a.m. They could select their preferred modes/routes and decide their departure time accordingly. A lookup table presented the distance, cost, travel time (mean), and the time variation (minimum and maximum) associated with the different modes for added information; the chosen route option was displayed in the map as well. Participants started their trips by clicking "Go" and they could change their travel choices along several stops on their way; in other words, at different decision points, travelers could, if desired, transfer or reroute by re-deciding their modes/routes and then clicking the "Go" button again (simply clicking "Go" without changing the mode/route choice would maintain the current mode/route). Participants could also review the up-to-present trip summary showing the trip segments that have been experienced (chosen) to help make new decisions for the remainder of their trips. When participants arrived at their destination, they were directed to a post-experiment survey (questions in the pre- and post-experiment surveys

Table 1. Mean Travel Time and Time Variation

Modes	Conditions	Mean travel time	COV	Minimum travel time	Maximum travel time
Car	Normal	1.00*t*	0.14	0.58*t*	1.42*t*
Car	Weather	1.25*t*	0.11	0.84*t*	1.67*t*
Car	Incident	1.92*t*	0.30	0.19*t*	3.64*t*
Metro	All conditions	1.00*t*	NA	1.00*t* −3 min	1.00*t* + 3 min
Bicycle	All conditions	1.00*t*	NA	0.83*t*	1.25*t*

Note: *t* = estimated travel time of corresponding modes acquired from Google Maps; COV = coefficient of variance. Minimum car travel time would be truncated by the corresponding distance over speed limit to avoid over speeding and unrealistically low travel times.

were selected from the first online survey questions) to express their attitudes toward the reliability, economy, and convenience of each mode alternative according to their virtual environment experience.

Back-Office Data. The back-office data could be extracted to generate two output files. One file documents the pre-experiment survey responses and the post-experiment survey responses of each participant; the other file documents the trip information of each participant including origin, destination, and mode choice as well as the corresponding distance, travel time (mean, minimum, and maximum), and cost of available modes.

Numerical Results

This section presents the numerical results and the associated analysis of the two experiments introduced in the Methodology section.

Survey Results

The survey was conducted and responses were collected via the platform of SurveyMonkey® (a paid and customized service for online survey), with two target areas:

1. The Washington Metropolitan Area (WMA), aged between 18 and 64, 120 responses expected with 119 completed responses among 131 attendances;
2. The entire U.S.A., aged between 18 and 64, 180 responses expected with 182 completed responses among 203 attendances.

The key statistical differences between the WMA sample and the U.S. sample are as follows:

1. The average age of the WMA respondents (46.8) is higher than the U.S. (40.5).
2. The WMA respondents are generally better educated; 74.8% of them have received bachelor or higher-level degrees, compared with 54.2% for the U.S. respondents.
3. Respondents from the WMA have a higher average income than the U.S. respondents; that is, $133,000 versus $84,000 in annual household income.
4. A higher percentage of the WMA respondents use metro as their typical commute mode than of the U.S. respondents (13.7% versus 5.9%); correspondingly, a lower percentage of the WMA respondents drive a car (71.0% versus 80.3%).

The survey responses were first digitalized and two statistical analyses were then performed: (a) the use of hypothesis testing to identify whether there is a significant difference between the two target groups; and (b) the use of structural equation analysis to examine TPB, revealing the level of significance between variables in the corresponding hierarchy.

Hypothesis Testing. The research team applied the independent two-sample t-test. The null hypothesis assumes that on the issue asked in a specific survey question, people in WMA do not hold a significantly different opinion from people in the U.S.A.; while the alternative hypothesis assumes that a significant difference does exist. Table 2 presents the t-test results at the confidence level of 95% (i.e., $\alpha = 0.05$), identifying whether to reject the null hypothesis. The purpose of the testing (i.e., the comparison between the two groups) is first to recognize and validate the trends from differences in motivations to differences in mode choices; and second to make sure that responses to the TPB questions can be diverse in statistics before extracting and entering them into the analytical model.

The results are mixed; however, the WMA group has consistently higher mean scores of the intention and control associated with transit (Questions No. 18, 21, 24, 33, 36 in Table 2) than the U.S.A. group and the corresponding *t*-values are higher than those associated with other modes (car and bicycle). This result may indicate that travelers in the WMA are more inclined to take transit for commuting and their inner resistance to transit is lower than the U.S. population in general. This finding, according to TPB, is in line with the fact that a higher percentage of respondents in the WMA sample commute by transit (e.g., metro) compared with the U.S. sample.

Structural Equation Analysis. Structural equation modeling is a multivariate statistical analysis technique that can reveal the correlation between manifest variables and latent variables in structural hierarchy. Utilizing the construct of TPB, the hierarchy of the travelers' determinations of their typical commute modes is illustrated in Figure 2; in addition to intention, personal profile (in line with AASHTO's investigation on the role of gender, age, race, and ethnicity, year of entry, income, vehicle availability, and occupation and industry in commuter mode choice in 2013) (*16*) and concern (about environment and health) (*17*) are also taken into consideration. Table 3 partially presents the coefficients associated with the different suggested travel modes as a result of structural equation analysis (using LISREL®).

Note that in the diagrams in Figure 2, rectangles denote manifest variables (translated from survey responses) and ovals denote latent variables; the parameters between a manifest variable and a latent variable

Table 2. Hypothesis Testing Results

Question no.	No. 17	No. 18	No. 19	No. 20	No. 21	No. 22
Question	Car plan to	Transit plan to	Bike plan to	Car willing to	Transit willing to	Bike willing to
Mean_WMA	5.31	3.06	1.82	5.84	4.18	2.49
StDev_WMA	2.34	2.36	1.57	1.89	2.39	2.05
Mean_US	5.66	2.16	1.96	6.04	3.49	2.72
StDev_US	2.19	1.95	1.63	1.87	2.38	2.18
t-value	1.34	3.58*	0.74	0.92	2.46*	0.93
Result	Fail to reject	Reject	Fail to reject	Fail to reject	Reject	Fail to reject
Question no.	No. 23	No. 24	No. 25	No. 26	No. 26	No. 26
Question	Car try to	Transit try to	Bike try to	Car safe	Car reliable	Car economical
Mean_WMA	5.35	3.63	2.18	5.52	5.89	4.43
StDev_WMA	2.11	2.25	1.82	1.64	1.36	1.95
Mean_US	5.76	2.82	2.35	5.81	6.05	4.83
StDev_US	2.11	2.19	1.87	1.48	1.35	1.96
t-value	1.65	3.12*	0.74	1.61	1.03	1.74
Result	Fail to reject	Reject	Fail to reject	Fail to reject	Fail to reject	Fail to reject
Question no.	No. 26	No. 26	No. 27	No. 27	No. 27	No. 27
Question	Car convenient	Car comfortable	Transit safe	Transit reliable	Transit economical	Transit convenient
Mean_WMA	5.76	5.97	4.48	3.45	4.24	2.84
StDev_WMA	1.80	1.46	1.76	1.58	1.71	1.95
Mean_US	6.15	6.28	4.24	3.77	4.47	2.72
StDev_US	1.52	1.25	1.92	1.89	1.91	1.90
t-value	2.07	2.00	1.11	1.58	1.10	0.53
Result	Reject	Reject	Fail to reject	Fail to reject	Fail to reject	Fail to reject
Question no.	No. 27	No. 28	No. 28	No. 28	No. 28	No. 28
Question	Transit comfortable	Bike safe	Bike reliable	Bike economical	Bike convenient	Bike comfortable
Mean_WMA	3.13	2.31	3.25	4.82	2.24	2.13
StDev_WMA	1.72	1.67	2.05	2.36	1.78	1.46
Mean_US	3.07	2.74	3.89	4.90	2.55	2.55
StDev_US	1.80	1.80	2.13	2.27	1.88	1.86
t-value	0.33	2.09	2.59	0.29	1.41	2.06
Result	Fail to reject	Reject	Reject	Fail to reject	Fail to reject	Reject
Question no.	No. 29	No. 30	No. 31	No. 32	No. 33	No. 34
Question	Car others' opinion	Transit others' opinion	Bike others' opinion	Car easy or difficult	Transit easy or difficult	Bike easy or difficult
Mean_WMA	4.74	2.97	2.16	5.27	2.95	1.83
StDev_WMA	1.97	1.79	1.54	2.07	1.84	1.47
Mean_US	5.16	2.85	2.45	5.57	2.46	2.21
StDev_US	1.95	1.85	1.73	2.10	1.87	1.78
t-value	1.82	0.60	1.49	1.21	2.23*	1.93
Result	Fail to reject	Fail to reject	Fail to reject	Fail to reject	Reject	Fail to reject
Question no.	No. 35	No. 36	No. 37		No. 38	
Question	Car confidence	Transit confidence	Bike confidence		Environmental and health concerns	
Mean_WMA	6.20	4.17	2.50		4.36	
StDev_WMA	1.65	2.38	2.04		1.84	
Mean_US	6.09	3.48	2.86		4.05	
StDev_US	1.81	2.38	2.24		1.98	
t-value	0.53	2.45*	1.44		1.38	
Result	Fail to reject	Reject	Fail to reject		Fail to reject	

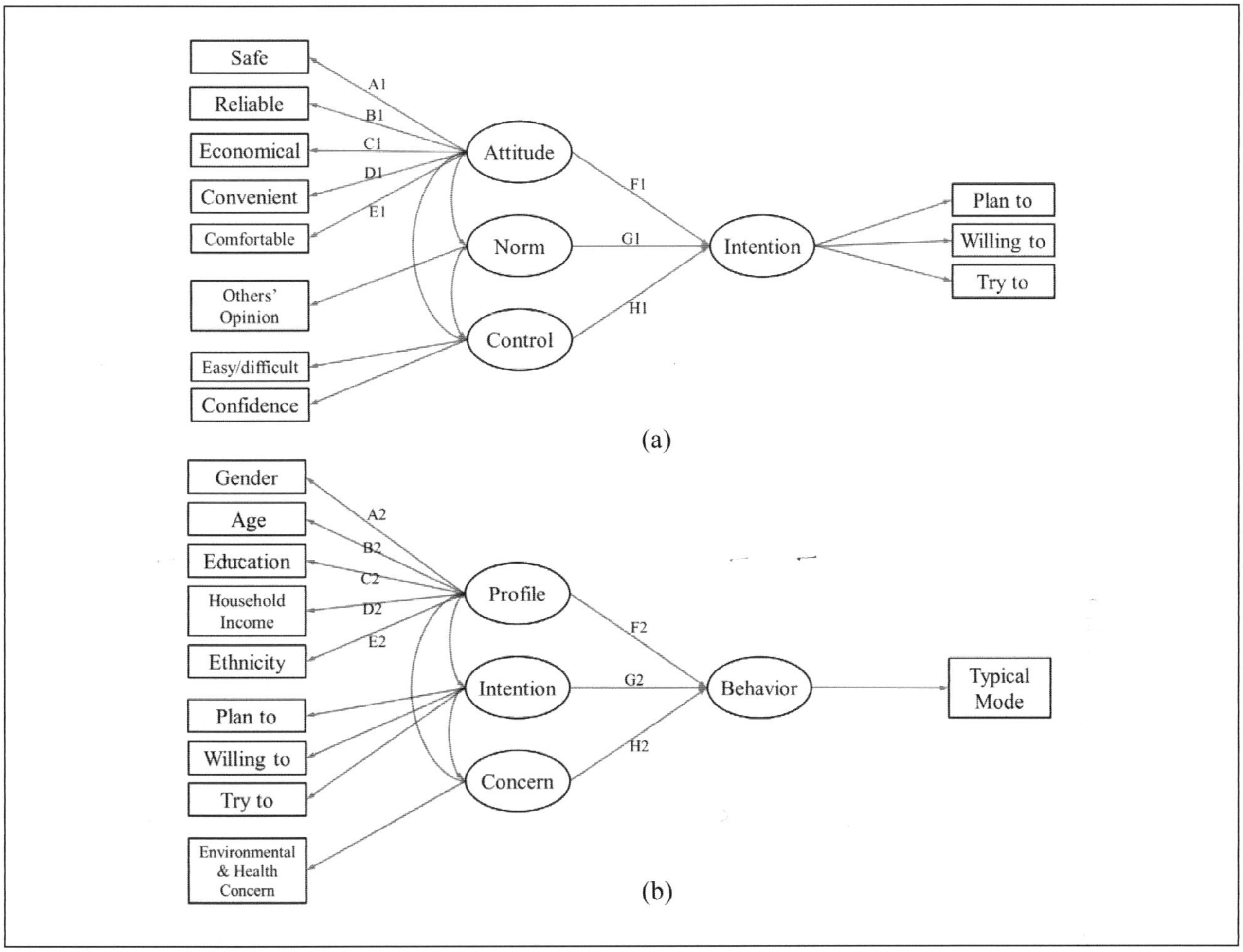

Figure 2. Path diagrams for modeling survey results: (*a*) intention, and (*b*) behavior.

Table 3. Coefficients for Survey Result Modeling

Mode	A1	B1	C1	D1	E1	F1	G1	H1
Car	0.59	0.77	0.42	0.67	0.65	0.73	0.62	0.94
Transit	0.64	0.67	0.55	0.66	0.65	0.46	0.51	0.68
Bike	0.50	0.50	0.26	0.69	0.68	0.83	0.64	0.83
Mode	A2	B2	C2	D2	E2	F2	G2	H2
Car	0.037	0.27	0.13	0.62	0.045	–0.27	0.84	–0.08
Transit	0.043	0.26	0.12	0.65	0.040	0.24	0.61	0.04
Bike	0.037	0.27	0.14	0.61	0.040	0.10	0.48	0.06

are the coefficients of determination (i.e., r^2 or r-squared) and the parameters between two latent variables are the correlation coefficients (i.e., r). Both these two coefficients interpret the likelihood of linearity.

According to coefficients F1, G1, and H1, intention is well correlated to attitude, norm, and control. The best correlation occurs between control and intention when examining car use, emphasizing that commuters drive a car for their daily commute as they feel at ease and confident with that decision; while the least correlation is observed between attitude and intention when examining transit use, indicating that commuters might compromise their attitudes to taking the transit mode for their daily commute.

Table 4. Virtual Environment Scenario Setup

Scenario no.	Conditions	ATDM tools	Number of participations
1	Normal	None	48
2	Normal	Pricing	55
3	Normal	Traveler information	49
4	Weather	None	37
5	Weather	Pricing	34
6	Weather	Traveler information	35
7	Incident	None	43
8	Incident	Pricing	35
9	Incident	Traveler information	35

In addition, coefficients A2 through E2 demonstrate that gender and ethnic identity are not significant to personal profile when predicting commuters' mode decisions; instead, household income is the most critical determinant, followed by age and education. This explains the finding in the survey that the WMA group—with higher age, education, and income level—has a higher percentage of transit users but lower percentage of car users for commuting (for transit, the correlation between profile and mode $r = 0.24>0$, i.e., they are positively correlated; for car $r = -0.27<0$, negatively correlated). In addition, concerns about the environment and health will play a positive role in encouraging transit and bicycle use and suppressing car use. However, profile and concern are both dominated by intention, which is consistent with the TPB construct—behavior (i.e., commute mode choices at the present study) is primarily determined by intention.

Virtual Travel Environment Results

Through targeted emails (list-serve) followed up by in-person group meetings, 25 continuous participants (out of 50 participants contacted) were recruited from the George Washington University. They were on-campus students, faculty members, and staff who commuted to school/work in WMA, who are familiar with the Northern Virginia network, and who volunteered (and were not paid) to participate in the experiments with long-term (six months) commitment. Among these participants, 10 were women (40%); 23 had received bachelor degrees or higher-lever degrees (92%, compared with 74.8% of the WMA sample in the survey experiment); the average age was 32.4 (compared with 46.8 for the WMA sample), and their average reported income was $87,500 (compared with $84,000 for the WMA sample).

Each participant was instructed to go through the virtual experiment at least once but at most three times a week (to attenuate the impacts of learning and habit). The virtual experiments were run on Heroku®, a cloud web host; thus, participants could participate (i.e., "play") in the virtual experiments (i.e., the "game") at their convenience by visiting the designated URL. Tracking and contacting the players was performed via email; for example, the research team sent weekly instructions regarding specific game scenarios and reminders to those who forgot to play the game in the preceding week.

Mode Use Statistics. Table 4 shows the configuration of scenarios (each scenario was announced on a weekly basis). Figure 3 illustrates the corresponding mode use distribution of each scenario.

It can be concluded from Figure 3 that inclement weather conditions boosted the car use; on the other hand, incident conditions suppressed it. In addition, the 50% discount rate on metro (pricing) had caused participants to quit "driving" and to use "metro-dependent" modes (i.e., taking metro or park-and-ride); however, this shift was partially reversed under inclement weather conditions. This result (the reversal of mode shift from car to metro) can be associated with Haustein and Hunecke's concept of perceived mobility necessities: there is an accentuation of the need for mobility (especially for the first and the last mile sections of the trips under inclement weather conditions) leading to the reduction in use of environmentally friendly modes that are not convenient under "abnormal" conditions (*18*).

TPB Modeling. Following the hierarchy described in TPB (Figure 4), structural equation analysis was applied to investigate the correlation between participants' attitudes/norms/controls and their in-route mode decisions in the virtual environment. Participants' norms and controls were acquired from the pre-experiment survey and assumed to remain unchanged; however, their attitudes and in-route mode decisions were collected through the back-office of the virtual environment and updated accordingly after each participation.

Table 5 partially presents the coefficients associated with the different scenarios and modes as a result of structural equation analysis. Note that some TPB

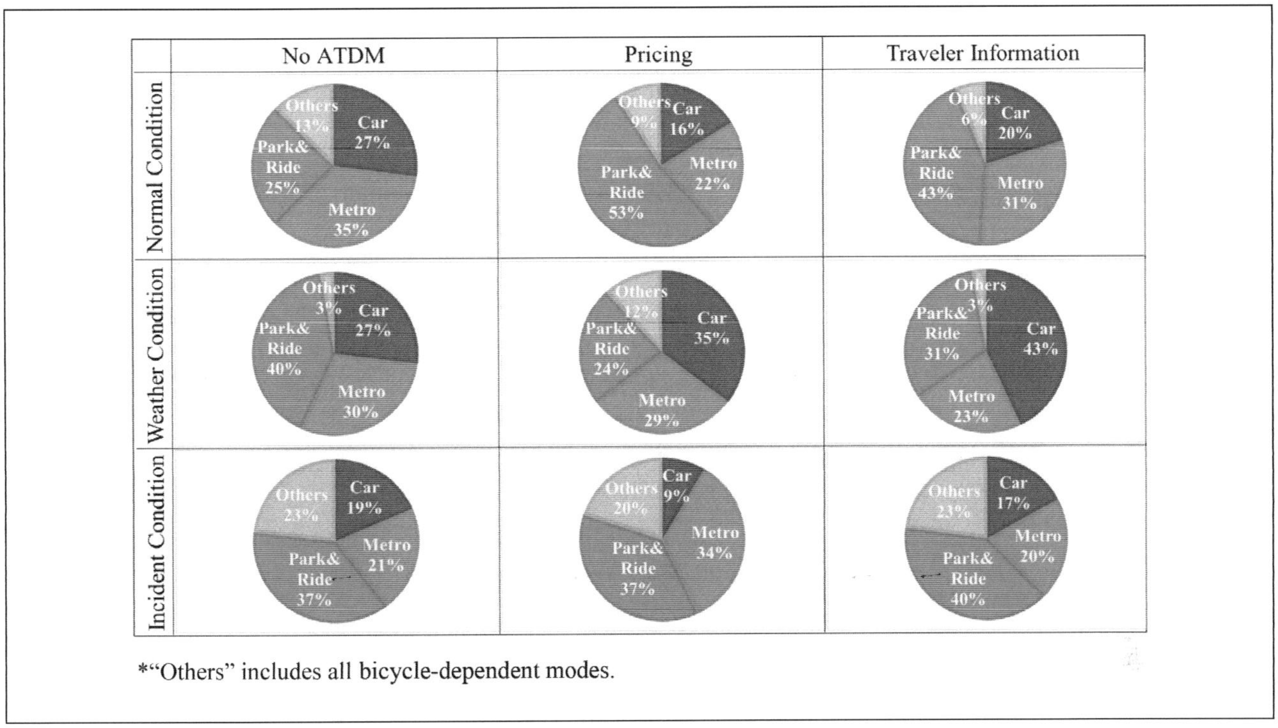

Figure 3. Pie charts illustrating the mode use distribution for each scenario.

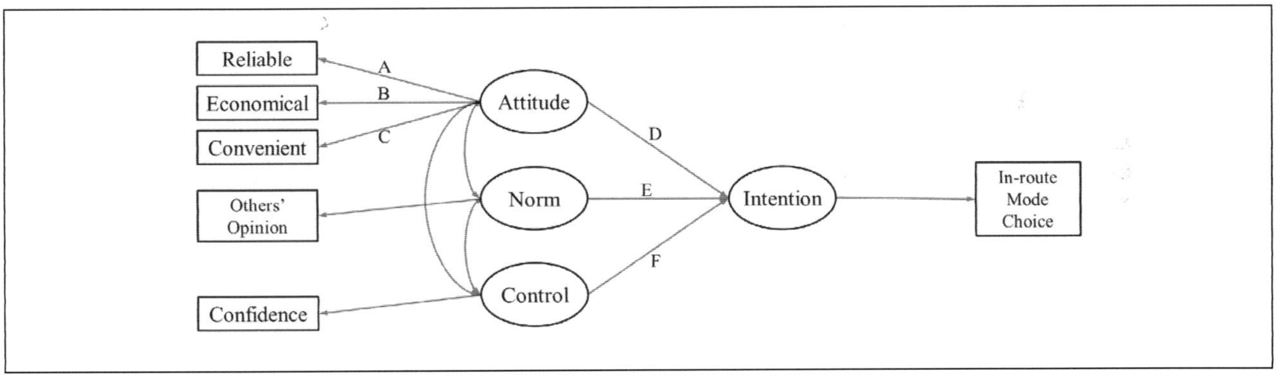

Figure 4. Path diagram for modeling virtual environment results.

modeling results are not applicable as the error variances between some manifest variables and latent ones are negative, denoting that the linearity between these variables could not be identified. In addition, the TPB modeling was not applied to bicycle use in the game as the structure of the transportation network used in this study largely suppressed the bicycle-correlated modes.

Cross-comparison of the results of the virtual environment scenarios allows the following conclusions. Participants were relatively indifferent to economy when they chose to drive a car with no pricing strategy. When the risk of income loss was noted (i.e., traveler information strategy), participants valued reliability more highly as they chose to drive; meanwhile, their attention to convenience decreased (Scenario 1 versus 3). Under inclement weather conditions, the intention of driving a car was more significantly correlated to norm and control (positive correlation), revealing that participant' decisions to drive were more consistent with their preset norms and controls; meanwhile, participants took greater account of convenience but cared less about reliability and economy as they chose to drive (Scenario 1 versus 4).

On the other hand, pricing strategy (50% discount rate) on metro led participants to take greater account of reliability but to pay less attention to convenience when

Table 5. Coefficients for Modeling Virtual Environment Results

Scenario no.	Mode	A	B	C	D	E	F
1	Car	0.80	0.096	0.53	0.62	0.08	0.01
2	Car	na	na	na	na	na	na
3	Car	0.90	0.12	0.35	0.63	–0.10	–0.18
4	Car	0.43	0.0044	0.76	0.59	0.18	0.22
5	Car	0.36	0.33	0.45	0.45	0.14	–0.09
6	Car	na	na	na	na	na	na
7	Car	na	na	na	na	na	na
8	Car	na	na	na	na	na	na
9	Car	na	na	na	na	na	na
1	Metro	0.75	0.53	0.70	0.69	–0.04	–0.03
2	Metro	0.88	0.49	0.53	0.63	0.08	–0.03
3	Metro	0.72	0.37	0.69	0.58	0.06	0.08
4	Metro	0.81	0.21	0.22	0.68	–0.12	–0.36
5	Metro	na	na	na	na	na	na
6	Metro	0.46	0.41	0.63	0.75	–0.34	–0.12
7	Metro	0.68	0.18	0.73	0.73	0.01	0.10
8	Metro	0.78	0.45	0.52	0.57	–0.05	–0.16
9	Metro	0.62	0.49	0.53	0.79	0.26	0.26

deciding whether to take metro (Scenario 1 versus 2; Scenario 7 versus 8). Under inclement weather conditions, the intention of taking metro was more significantly correlated to norm and control (negative correlation), indicating that participants might choose metro though their norms/controls were not in favor of it (in other words, they might yield their unfavorable norms/controls toward making such decisions); meanwhile, participants valued reliability more highly but paid less attention to economy and convenience as they chose to take metro (Scenario 1 versus 4). This finding can be associated with the findings of Clark et al. that transit is a more "active" mode of commuting and transit riders are more active commuters compared with car users (especially when facing disruptions) (*19*).

Conclusion

To gain insights into travelers' choices, this study applies two innovative approaches: (a) the accumulation of travelers' decisions in association with motivations and socio-economic characteristics; (b) the development of a virtual travel environment to simulate various travel conditions along with different ATDM applications. The two experiments conducted in the study—the survey and the virtual environment—enabled quantification of the motivations of travelers' strategic and tactical decisions and revealed the dynamics of motivations associated with different travel conditions and ATDM applications. Such accomplishments may offer guidelines to achieve more efficient ATDM strategies, especially when travelers are expected to play an active role in response to these strategies. For example, pricing ATDM strategies favoring transit may encourage travelers to shift travel modes; however, such strategies are temporarily effective as reliability of transit is pronounced for users under discount pricing (especially for those who "swing" between transit and other modes, i.e., choice riders). Therefore, ATDM strategies should be applied in multiple phases with a carefully designed structure. In particular, advantageous pricing for transit should be complemented with measures to improve and accentuate the improvement in reliability in order to attract more users and maintain a stable market.

One limitation of this study, however, is related to the measurement of mode choice habit/inertia (as the boundary conditions of the application of TPB). To create a more effective virtual environment that can investigate inertia, participants may need to maintain the same origin and destination (possibly their real locations of residence and work) until inertia is observed. Different travel conditions and ATDM strategies may be gradually applied afterwards, mimicking the real-life situations.

Given that participants of the virtual environment in this study were all on-campus students, faculty, or staff with relatively limited diversity in socio-economic characteristics and with preset bias toward different modes due to certain privileges/constraints (e.g., there is a transit station near the campus and some students do not own a car). For future research studies, the virtual experiment can be distributed to a more diverse set of traveler populations and the results will be more comprehensive, especially when dealing with TBP dimensions.

Finally, the findings of the study may be translated into traffic simulation tools that may be applied for both prediction and analysis purposes. Such simulation tools may be trained and may be linked to the "mobile" data collection platform. In other words, the participants in the virtual travel environment may be either real participants or virtual participants who interact in the same simulation paradigm with different interaction and collaboration levels.

Acknowledgments

The authors thank the U.S. Department of Transportation (USDOT) and the Federal Highway Administration (FHWA) for their financial support and the Battelle® personnel for their guidance and direction on the corresponding research project.

Author Contributions

The authors confirm contribution to the paper as follows: study conception and design: Samer H. Hamdar, John Campbell; data collection: Dong Pan, Amir Farrahi; analysis and interpretation of results: Dong Pan, Samer H. Hamdar; draft manuscript preparation: Dong Pan, Samer H. Hamdar, John Campbell. All authors reviewed the results and approved the final version of the manuscript.

References

1. FHWA. *Active Transportation and Demand Management Program*. https://ops.fhwa.dot.gov/atdm/. Accessed May 1, 2017.
2. Manski, C. F. Maximum Score Estimation of the Stochastic Utility Model of Choice. *Journal of Econometrics*, Vol. 3, No. 3, 1975, pp. 205–228.
3. Anderson, E., and H. Gatignon. Modes of Foreign Entry: A Transaction Cost Analysis and Propositions. *Journal of International Business Studies*, Vol. 17, No. 3, 1986, pp. 1–26.
4. Willson, R., and E. Shirazi. *Transportation Demand Management: Policy Implications of Recent Behavioral Research Symposium Summary*. University of California Transportation Center, Berkeley, CA, 1991.
5. Ajzen, I. The Theory of Planned Behavior. *Organizational Behavior and Human Decision Processes*, Vol. 50, No. 2, 1991, pp. 179–211.
6. Armitage, C. J., and M. Conner. Efficacy of the Theory of Planned Behaviour: A Meta-Analytic Review. *British Journal of Social Psychology*, Vol. 40, No. 4, 2001, pp. 471–499.
7. Sirakaya, E., and A. G. Woodside. Building and Testing Theories of Decision Making by Travellers. *Tourism Management*, Vol. 26, No. 6, 2005, pp. 815–832.
8. De Bruijn, G. J., S. P. Kremers, A. Singh, B. Van den Putte, and W. Van Mechelen. Adult Active Transportation: Adding Habit Strength to the Theory of Planned Behavior. *American Journal of Preventive Medicine*, Vol. 36, No. 3, 2009, pp. 189–194.
9. Bamberg, S., I. Ajzen, and P. Schmidt. Choice of Travel Mode in the Theory of Planned Behavior: The Roles of Past Behavior, Habit, and Reasoned Action. *Basic and Applied Social Psychology*, Vol. 25, No. 3, 2003, pp. 175–187.
10. Heath, Y., and R. Gifford. Extending the Theory of Planned Behavior: Predicting the Use of Public Transportation. *Journal of Applied Social Psychology*, Vol. 32, No. 10, 2002, pp. 2154–2189.
11. Verplanken, B., H. Aarts, and A. Van Knippenberg. Habit, Information Acquisition, and the Process of Making Travel Mode Choices. *European Journal of Social Psychology*, Vol. 27, No. 5, 1997, pp. 539–560.
12. Aarts, H., and A. P. Dijksterhuis. The Automatic Activation of Goal-Directed Behaviour: The Case of Travel Habit. *Journal of Environmental Psychology*, Vol. 20, No. 1, 2000, pp. 75–82.
13. Aarts, H., B. Verplanken, and A. Knippenberg. Predicting Behavior from Actions in the Past: Repeated Decision Making or a Matter of Habit. *Journal of Applied Social Psychology*, Vol. 28, No. 15, 1998, pp. 1355–1374.
14. Ouellette, J. A., and W. Wood. Habit and Intention in Everyday Life: The Multiple Processes by Which Past Behavior Predicts Future Behavior. *Psychological Bulletin*, Vol. 124, No. 1, 1998, p. 54.
15. Bamberg, S., I. Ajzen, and P. Schmidt. Choice of Travel Mode in the Theory of Planned Behavior: The Roles of Past Behavior, Habit, and Reasoned Action. *Basic and Applied Social Psychology*, Vol. 25, No. 3, 2003, pp. 175–187.
16. American Association of State Highway and Transportation Officials. *Commuting in America*. American Association of State Highway and Transportation Officials, Washington, D.C., 2013.
17. De Groot, J., and L. Steg. General Beliefs and the Theory of Planned Behavior: The Role of Environmental Concerns in the TPB. *Journal of Applied Social Psychology*, Vol. 37, No. 8, 2007, pp. 1817–1836.
18. Haustein, S., and M. Hunecke. Reduced use of Environmentally Friendly Modes of Transportation Caused by Perceived Mobility Necessities: An Extension of the Theory of Planned Behavior. *Journal of Applied Social Psychology*, Vol. 37, No. 8, 2007, pp. 1856–1883.
19. Clark, B., K. Chatterjee, and S. Melia. Changes in Level of Household Car Ownership: The Role of Life Events and Spatial Context. *Transportation*, Vol. 43, No. 4, 2016, pp. 565–599.

The Standing Committee on Traveler Behavior and Values (ADB10) peer-reviewed this paper (18-05179).

Article

TRR
JOURNAL OF THE TRANSPORTATION RESEARCH BOARD

Transportation Research Record
2018, Vol. 2672(47) 182–193

DOI: 10.1177/0361198118791361
journals.sagepub.com/home/trr
SAGE

Incorporating Multiple Uncertainties into Projections of Chinese Private Car Sales and Stock

I-Yun Lisa Hsieh[1], Paul Natsuo Kishimoto[2], and William H. Green[1]

Abstract

China is in a fast-growing stage of mobility development, and its increasing demand for private cars comes with growing energy consumption and pollutant emissions. Uncertainty in Chinese parameterization of car ownership models makes forecasting these trends a challenge. We develop an application of the Monte Carlo method, conditioned on historical data, to sample parameters for a model projecting aspects of private car diffusion, such as the mix of new and replacement sales. Our model includes changes in per-capita disposable income—both the mean and level of inequality—and a measure of car affordability. By incorporating multiple uncertainties, we show a distribution of possible future outcomes: a low stock of 280 million (1st decile); median of 430 million; and high of 620 million vehicles (9th decile) in 2050. This illustrates the limitations of attempts to model vehicle markets at the national level, by showing how uncertainties in fundamental descriptors of growth lead to a broad range of possible outcomes. While uncertainty in projected per-capita ownership grows continually, the share of first-time purchases in sales is most uncertain in the near term and then narrows as the market saturates. Replacement purchases increasingly capture the sales market from 2025. Our results suggest that stakeholders have a narrow window of opportunity to regulate the fuel economy, pollution and other attributes of vehicles sold to first-time buyers. These may, in turn, shape consumers' experience and expectations of car ownership, affecting their additional and replacement purchases.

Being one of the fast-growing major economies, China has seen the largest increases in passenger vehicle sales between 2005 and 2015, overtaking the U.S. and becoming the largest auto market since 2008 (*1*). However, if compared with the history of U.S. motorization, per-capita car ownership in China is just passing the level in the U.S. as of 1922 (*2*).

The strong growth in development of the automotive industry is closely associated with the economic boom in China. On the other hand, this growth has resulted in an increase in demand for fuel, and caused more pollutant and carbon dioxide emissions associated with petroleum combustion, as well as worsened traffic congestion problems in urban regions—all of which are urgent challenges for China's government.

Establishing the range of possible future outcomes allows stakeholders to decide how robust their strategies are to uncertainty, and to determine the risk of following status quo approaches. In addition to design of policies to limit the environmental impacts of transport, governments must choose to size infrastructure (roadways, parking facilities, etc.) to accommodate expected traffic volumes. Automakers must construct durable assets such as assembly plants and make supply chain investments to meet demand competitively in a future market of a certain size. The criteria for robustness differ according to an organization's tolerance for risk, and type of decision or policy under consideration; but in all cases information about distributions of future outcomes is essential.

Literature Review

The evolution of the auto market, especially private car market, is of interest to numerous agencies. Hamilton et al. (*3*) distinguishes four stages in the growth of the automobile industry: first, experiment with invention; second, luxury uses and markets; then planned price reduction and expansion of markets; and finally, price

[1]Department of Chemical Engineering, Massachusetts Institute of Technology, Cambridge, MA
[2]MIT Joint Program on the Science & Policy of Global Change, Massachusetts Institute of Technology, Cambridge, MA

Corresponding Author:
Address correspondence to William H. Green: whgreen@mit.edu

stabilization and competition in design and salesmanship. The size of China's car market is growing robustly, and the critical feature of the future growth potential of car markets has been approached and forecasted by many researchers. Different car ownership models have been applied for various purpose and different interests, as reviewed by de Jong et al. (*4*). It is suggested that the preferred model type for car stock forecasting depends on the research objectives and data availability. For those countries only having limited data, like China, some of the sophisticated methodologies may not be applied. A widely used method for application to developing countries is to express car density as a logistic function of either time (*5*), per-capita GDP (*6*) or per-capita income (*7*), because they have the lowest data requirements, and income is widely considered to be the primary motivation behind car ownership growth.

By fitting Gompertz functions to data on per-capita vehicle ownership and per-capita GDP, Wu et al. (*6*) showed that car stock level in China has followed an S-shaped curve like most other motorizing countries, and projected that the inflection point would appear around the year 2030. Instead of relying on single economic variable, Huo and Wang (*8*) built up a China-specific fleet model in more detail, simulating private car ownership on a income-level basis and taking into account explicitly the influence of car price on vehicle ownership, in order to investigate the energy demand and environmental impacts of the rising number of vehicles in China.

They separated car sales into two categories: new-growth purchases (i.e., first-time car purchases and additional purchases) motivated by increased income, and replacement purchases, which households use to replace retired or scrapped vehicles. The approach in Huo and Wang (*8*) serves as a key example for the current paper. However, we include uncertainty in the vehicle adoption trajectory, thereby addressing a key parameter and source of uncertainty in outcomes omitted in prior research.

We describe an approach that allows exploration of the probability distribution of the saturation level of car ownership at high purchasing power, and allows comparison of effects of uncertain parameters, including disposable income, Gini index, and car price. We show results by conducting Monte Carlo simulation, and discuss the sensitivity to major parameters. We conclude by stating some implications for policy makers and city planners.

Method and Data

Car Stock and Sales

With the survey of Chinese urban families with different income levels owing cars, conducted by China's Statistical Bureau, it is confirmed that it is more likely for people with higher income to own cars. Compared with income alone, per-capita income coupled with car price, called car purchasing power, was shown to have much better correlation with car ownership in China (*8*). Especially for emerging economies like China's, car purchasing power is a more suitable independent variable to be used to model car ownership than income level alone, because the automotive industry is expanding as car prices decrease in the early stage of motorization.

Let A represent an affordability index, or purchasing power, which is the ratio of disposable income to car price index. We arbitrarily choose the car price index in 2003 (p_{2003}) to be 1 and use the United Nations medium-variant population projection as our baseline (*9*). The total private car stock ($\hat{V}_i$) can be calculated by integrating across the entire population's range of incomes, x, the product of the income distribution density, I, and the propensity, g, that an individual with that affordability index owns a car, as shown in

$$\hat{V}_i = P_i \int_{\frac{x}{p_i}=0}^{\infty} I_i\left(\frac{x}{p_i}\right) g\left(\frac{x}{p_i}\right) d\left(\frac{x}{p_i}\right) = P_i \int_{A=0}^{\infty} I_i(A) g(A) dA \tag{1}$$

where
$\hat{V}_i$ = Total private car stock in year i
P_i = Population in year i [persons]
x = Income of the incremental individual [RMB_{2007}/year]
I_i = Income distribution function for year i
p_i = Car price index in year i
$A = \frac{x}{p}$ = Car affordability index
g = Deterministic relation between private car ownership (%) and affordability index

We decompose car sales, S, into two segments: new-growth purchases, S^N, associated with increases in ownership due to rising income, and replacements, S^R, for scrapped cars. New-growth purchases represent the spread of ownership to a larger portion of the population as incomes rise and are calculated as

$$S_i^N = \hat{V}_i - \hat{V}_{i-1} = P_i \int_{A=0}^{\infty} I_i(A) g(A) dA - P_{i-1} \int_{A=0}^{\infty} I_{i-1}(A) g(A) dA \tag{2}$$

Even if per-capita ownership does not rise, households must buy replacements for old vehicles that are scrapped or retired. We call these replacement purchases, and calculate them as

$$S_i^R = \sum_{j=1}^{i} S_{i-j}(\mathrm{SR}(j-1) - \mathrm{SR}(j)) \tag{3}$$

where

$$S_{i-j} = S^N_{i-j} + S^R_{i-j} \tag{4}$$
total number of new cars purchased in year $i-j$

$$\mathrm{SR}(y) = \exp\left(-\left(\frac{y}{T}\right)^b\right) \quad \text{survival ratio [dimensionless]} \tag{5}$$

y = Age of vehicle [years]
T = Average vehicle lifespan = 14.46 years
b = Scrappage intensity = 4.7

The vehicle survival rate, $SR(y)$, describes the decreasing share of vehicles of a certain age remaining in use as the vehicle age grows. Hao et al. (*10*) collected data from a field study and employed the Weibull distribution to simulate it, as shown in Equation 5. Because China has no official scrappage statistics and no compulsory scrappage standards for private cars, we adopt their values for T and b (shown above) in our analysis and assume that the pattern will remain the same in the future.

The careful reader will note certain differences in Equations 1 and 2 compared with those given in reference (*8*). In Equation 1, we use $I_i(A)g(A)$ instead of $I_i(x)g(A)$. This is necessary such that the integral properly represents the expectation of g over a population of individuals with car affordability index distributed as A. In Equation 2, the cited paper has something akin to $S_i^N = P_i \int_{M_{i-1}}^{M_i} \int_x^{\infty} I_i'(x)g(x,p_i)dxdM$, where M is the population mean income. This again mixes functions of x and A; and additionally uses a derivative of the income distribution, $I_i'(x)$. In the absence of model source code, the meaning is ambiguous; a derivative of $I(\cdot)$ with respect to x or A is negative at certain points, leading to negative projected sales in some future years, and we find that a derivative with respect to M (or μ) yields results that do not match observations. We instead use the method of finite differences described above.

The Car Ownership Function, $g(x, p_i)$

Parameter Uncertainty in Gompertz Functions. It is recognized that a primary driving force for the growth of private car ownership is income level (*11*). The statistics relating to Chinese consumer price index (*12, 13*) show that the car price in China decreased rapidly in the past decade, driven by excess capacity and competition in the rapid-growing automotive market. Based on the theory of four stages in the growth of the car industry (*3*), car price stabilizes after the stage of the planned price reduction and expansion of markets, as shown in Figure 1*a*. Compared with the early motorization stage in the U.S. (*14*), the theory suggests that both countries would have similar price trends.

A special case of the logistic function, the Gompertz curve allows for different curvatures at extreme values of the economic factors (*15*). We use a formulation (Equation 6) that represents the probability that an individual owns a car as a function of A.

$$g(x, p_i) = g(A) = \gamma \exp(\alpha \exp(\beta A)) \tag{6}$$

where
g = Probability to own a private car
A = Car affordability index
γ = Probability of owning a private car at very high income
$0 < \alpha, \beta$ = Shape parameters

In order to illustrate the limitation of point estimates, we use data from the 2008, 2009, and 2010 China urban household surveys (*16*). (Note that prior to 2010, only Shanghai limited car ownership using a license plate quota system, but in recent years a growing number of cities have adopted such policies. To avoid obscuring the underlying relationship, we omit data from 2011 onwards.) Figure 2 shows values of γ ranging from 20% to 80% or higher produce fits that are all apparently close matches to the data. A common response in the literature is to arbitrarily choose γ from another country with saturated ownership, or some plausible low or high bounding values for a few scenarios.

Economic development and car ownership in China are both behind those of motorized countries, and thus as a whole the country is still below the inflection point of g. Thus, the survey data tell us little about the eventual probability of possessing a car when people have higher income levels, even though the results show that people with higher car purchasing power are more likely to own cars, as expected. Due to the uncertainties arising from unclear future economic context and unpredictable consumer preference, the value of single-point estimates is limited.

Monte Carlo Simulation of the Distribution of $\gamma, \alpha,$ *and* β. Following (*17*), our updated algorithm to estimate the parameters of Gompertz function by incorporating uncertainty is as follows. The unknown parameters are γ, α, and β while the state variable is only $r = \gamma$, because the latter two parameters can be further determined by linear regression at a given γ. As the data are available for per-capita disposable income at national level, car price index, population, and private car stock numbers during 2007–2015, we then define the root-sum-of-square residual between back-cast private car stock ($\hat{V}_i$) by Equation 1 and actual historical data (V_i) as

$$RSS(v) = \sqrt{\sum_{i=2007}^{2015} \left(\hat{V}_i(r) - V_i\right)^2} \tag{7}$$

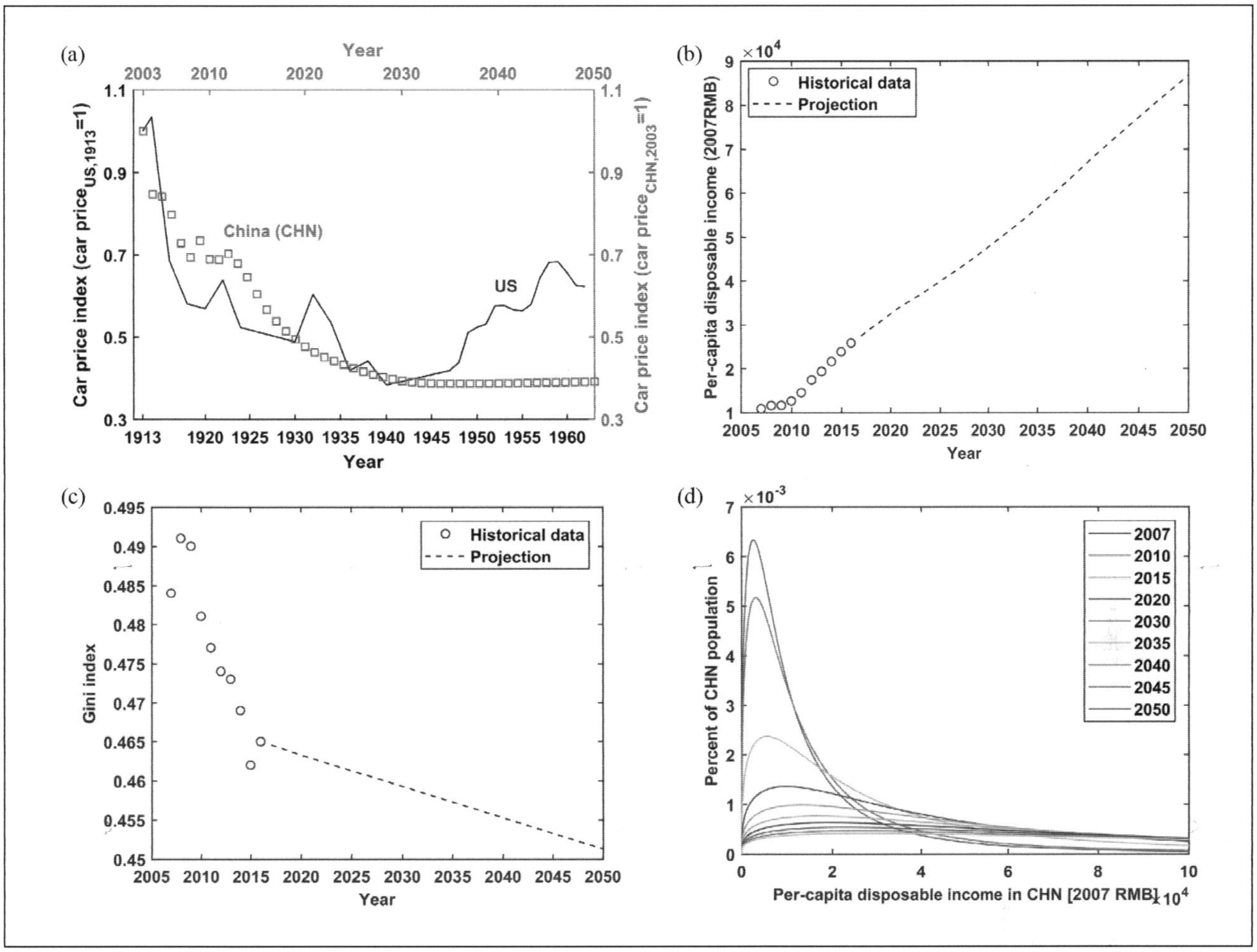

Figure 1. Major parameters and functions in car stock model: (*a*) Car price index in China (historical data 2003–2016 and future projection to 2050) and the historical data in U.S. (1913–1962); (*b*) Per-capita disposable income (2007 RMB), 2007–2050; (*c*) Gini index, 2007–2050; and (*d*) Income distribution with the log-logistic assumption, 2007–2050.

Monte Carlo simulation with 400 iterations is applied to estimate the possible outcomes of the saturation level of possibility (γ) forecasts. γ is constrained to the range between the 23% (the observed per-capita car ownership for the highest income household in 2012 (*16*)) and 100%. A discrete probability density function for $\gamma \in (23, 100)$ is constructed by randomly sampling 50 points across the range and calculating their associated $RSS(r)$, subtracting from $\max(RSS)$ and then normalizing so that the density is lowest where the residual is greatest. From the resulting distribution of r for each iteration, a single sample is drawn and accepted as the γ in that run, followed by estimating the other two parameters, α and β, from linear-regression model as Equation 8, which is converted from Equation 6 by log-linearization. As shown in Equation 8, $\ln(-\alpha)$ and β are linearly related and are regressed as a least-squares fit for time series data in each iteration.

$$\ln\left(\ln\frac{\gamma}{g(A)}\right) = \ln(-\alpha) + \beta A \quad (8)$$

The Income Distribution, I

Various probability distribution functions have been studied in research on income disparity. By comparing observations to mathematical models, researchers can determine the distribution function with the best fit. Some scholars prefer the log-normal distribution; for example, steyn (*18*) examined the inhabitants in the Orange Free State rural areas of the Union of South Africa, and showed that the logarithmic normal model is quite a good model. On the other hand, McDonald and Ransom (*19*) stated that the log-normal distribution performed the worst while considering the log-normal, gamma, beta, and Singh-Maddala functions as

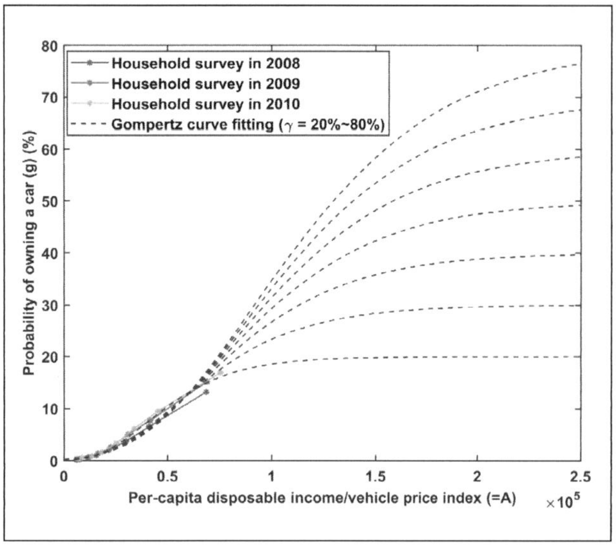

Figure 2. Private car ownership and car affordability index in China, $g(x, p_i)$, with assumed γ between 20% and 80%.

descriptive models for the distribution of family income for 1960 and 1969–1975 in the U.S.

Data availability has been a major obstacle for research in many fields in China, including research on income inequality. Chen et al. (*20*) investigated 20 sets of grouped data on family income between 2005 and 2012 in China, and demonstrated that the fitting of the log-logistic distribution is better, while the log-normal distribution function underestimates the income of the high-income group due to its thinner right tail. Following Chen et al. (*20*), we approximate a continuous income distribution for China with

$$f(x; a, b) = \frac{bx^{b-1}}{a^b\left(1 + \left(\frac{x}{a}\right)^b\right)^2} \quad (9)$$

where

x = per-capita disposable income[RMB_{2007}/year]

a,b = scale and shape parameters of the distribution

Note that if x is log-logistically distributed, then A has also a log-logistic distribution but with different scale parameter (Equation 10).

$$X \sim f(x; a, b) \quad \Rightarrow \quad A = \frac{X}{p} \sim f\left(\frac{x}{p}; \frac{a}{p}, b\right) \quad (10)$$

Given historical data and projections of the mean disposable income (M) and Gini index (G), we then solve the equations, shown as (11) and (12), analytically by computer mathematical software to simulate the income distribution in a given year.

$$G = \frac{1}{b} \quad (11)$$

$$M = \frac{a\pi}{b\sin\left(\frac{\pi}{b}\right)} \quad (12)$$

For future income, we use projections from China's 13th Five-Year Plan and the OECD, and assume that M will increase at the same rate as GDP after 2020 (*21*), shown in Figure 1*b*. Hu et al. (*22*) adopted the grey forecasting model, commonly applied in academic fields such as electrical engineering, mechanical engineering, and agriculture, to predict the Gini index in China until 2022. With the given difference between the official Gini index (0.465) in 2016 and Hu, Luo, and Yang's predicted value (0.479), we first correct those from 2017 to 2022, and then adopt the same rate of change in subsequent years to give the future Gini index projections, shown in Figure 1*c*. Figure 1*d* depicts the resulting income distributions in China for some years, *i*, between 2007 and 2050 based on our derived projected function parameters; in computing projections, we use a distinct $I_i(A)\ \forall\ i$.

Results and Analysis

Parameter Space for α, β, and γ

Figure 3 illustrates the parameter space for the Gompertz function at 400 samples of γ with the associated probability distribution from Monte Carlo simulation and regression. Note that for a randomly selected γ, our regression procedure for α, β constrains these parameters to a narrow range that matches historical data, unlike Kishimoto (*17*) which treated all three as a state vector and applied Gibbs sampling, drawing each of three parameters in sequence from a joint distribution.

Private Car Stock

Figure 4 presents the outcomes of the private car stock and ownership in China projected by our stock model for all 400 samples, with 1st decile, mean and 9th decile shown by black lines, and the probability distribution of the outcomes in 2030 and 2050. We also show the projection results, for comparison, from Huo and Wang (*8*) for stock, and from TASRI (Tsinghua Automotive Strategy Research Institute) (*23*) for ownership. Note that our model is forecasting private cars, rather than all vehicles as in TASRI. The model shows a range of possible future private car stock in China: 430 million (low of 280 million at 1st decile; high of 620 million at 9th decile) in 2050.

Table 1 gives a summary of the parameter values for 1st decile, median, 9th decile and the expectations with standard deviations, with the associated projected private car stock numbers and per-capita car ownership. As

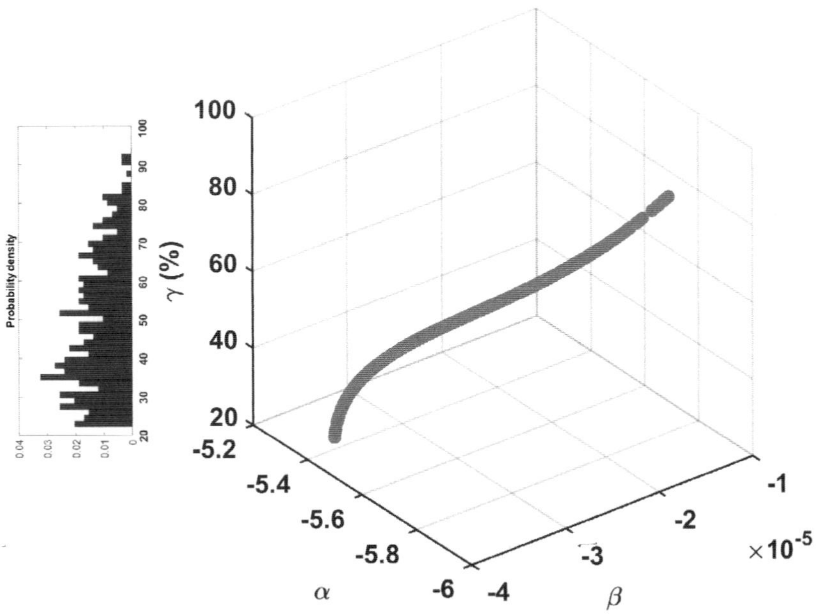

Figure 3. Parameter space for α, β, γ with the probability density distribution of γ.

shown in Table 1, uncertainty is larger in estimating the parameter γ compared with the other two, which implies that predicting the probability of private car ownership in China is challenging, as discussed above ("Parameter uncertainty in Gompertz functions").

Private Car Sales

Figure 5 draws out, for every sample, the results of private car sales in China, with 1st decile, mean and 9th decile shown in black lines, the associated shares of replacement purchase in dark red lines, and the probability distribution of the both outcomes toward 2050. It is noted that higher car sales correspond to a lower share of replacement purchases; that is, the order of the black quantile lines is reversed for the two outcomes. The predicted distribution of the private car sales in China will be centered at 30 million from 2030 to 2050 (low of 19 million at 1st decile; high of 44 million at 9th decile). While scenario analysis, which brackets uncertain outcomes of interest, is one response to improve the value of single-point estimates, our upgraded methodology provides a more detailed characterization of the uncertainty in transport projections.

Table 2 summarizes the expected value of private car sales numbers and replacement purchases share with the associated standard deviations by our model. As stated ("Method and data"), private car sales are divided into two groups due to their purchasing motivations: new-growth purchase and replacement purchase. The split between these two groups characterizes the maturity of the auto market (*8*). As shown in Figure 5, current car sales in China are mainly driven by first-time car purchases, i.e., the share of replacement purchases is still low. Our results show that replacement purchases will dominate the sales market from 2025, which is a feature of maturity. However, the replacement purchase share is still lower than 100% in 2050, implying that the car market in China will not fully mature by 2050. Unlike the uncertainty in projections of private car stock/ownership and car sales, which grow continually into the future, the distribution of the replacement purchases share narrows toward the saturated (~2050) market state. We discuss the implications below (see "Discussion of Results").

Another way to characterize the maturity of a car market is using the ratio (r) of the private car sales to the change in private car stock as an indicator. Note that "one minus the inverse of the ratio" is the proportion of the new cars sold to replace the old models which have been scrapped. Therefore, the ratio should be close to 1 in an emerging car market, as the new-growth purchases dominate sales. As a check on our model projections, we compare this indicator with data from the early stage of motorization in the U.S. (*24*), as shown in Figure 6. According to our model, the two lines (black line represents the numerator while blue dash line represents the

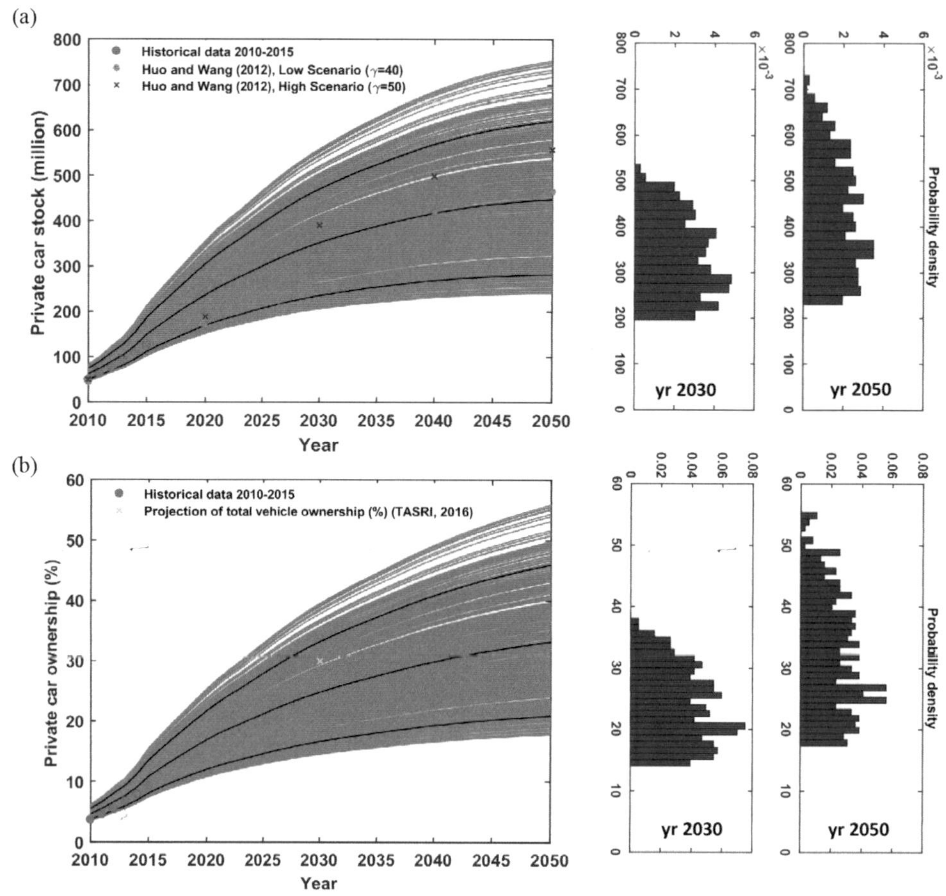

Figure 4. Possible outcomes from Monte Carlo simulation: (*a*) 2010–2050 private car ownership with the probability distribution in 2030 and 2050; and (*b*) 2010–2050 private car stock with the probability distribution in 2030 and 2050.

Table 1. Parameter Values and the Associated Projected Values for Private Car Stock

Parameter	1st decile	Median	9th decile	$E[\cdot] \pm \sigma$
γ (%)	28.206	48.090	74.538	49.94 ± 17.29
α	−5.307	−5.491	−5.796	−5.400 ± 0.184
β	3.18×10^{-5}	2.38×10^{-5}	2.00×10^{-5}	$-(3.32 \pm 0.44) \times 10^{-5}$
Stock $\hat{V}_{2030}$ (million)	235.62	342.75	470.14	423.84 ± 86.78
Stock $\hat{V}_{2050}$ (million)	282.26	434.21	620.32	504.93 ± 125.37
Vehicle per 100 $capita_{2030}$	16.634	24.213	33.213	29.942 ± 6.131
Vehicle per 100 $capita_{2050}$	20.938	32.210	46.016	37.456 ± 9.300

denominator of the ratio) are close to one another in initial years but will separate gradually as the auto market in China becomes more mature, which is the same as the observed pattern in the U.S. in the 20th century.

However, care should be taken when drawing parallels between China and the United States: while some trends may be similar, the adoption of personal vehicles in China nowadays happens in a much different context compared with the one in the United States a hundred years ago. The preferences, technology options, costs, earnings, urban situation, and policy environment faced by 21st-century Chinese households differ in important

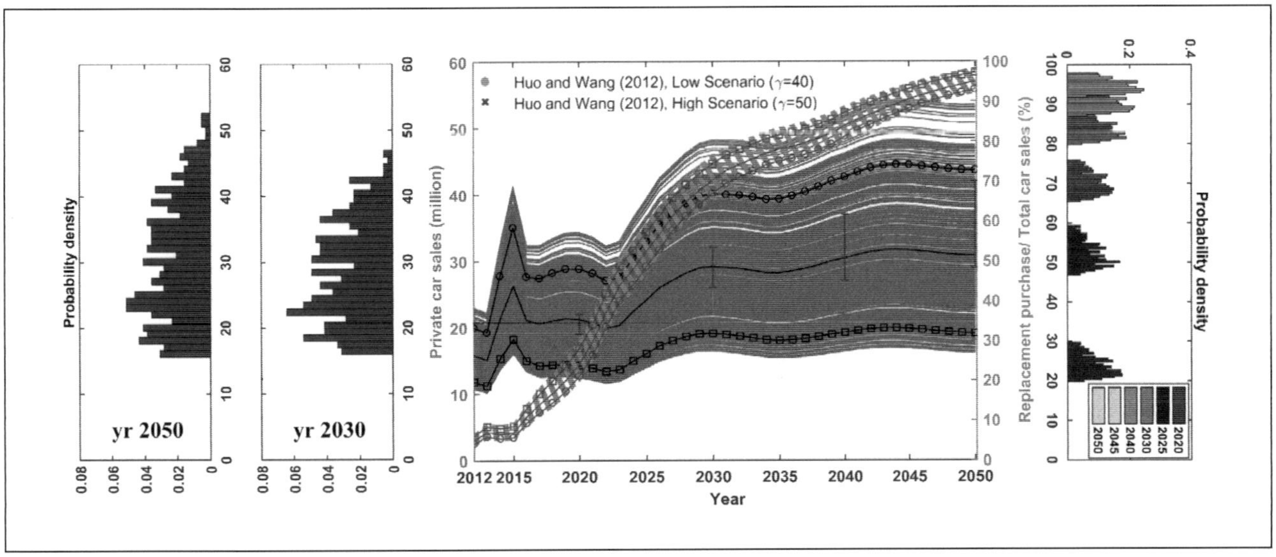

Figure 5. Possible outcomes of the private car sales and replacement purchase proportion out of total car sales in China with their probability distributions.

Table 2. Private Car Sales Projections from Car Sales Model

$E[\cdot] \pm \sigma$	Car sales (million)	Replacement purchases share (%)
2020	21.27 ± 5.39	24.16 ± 2.56
2025	24.17 ± 6.24	52.49 ± 3.17
2030	29.16 ± 7.78	70.38 ± 2.70
2040	30.40 ± 8.68	83.97 ± 2.18
2050	30.80 ± 9.13	94.95 ± 1.54

ways from those of 20th-century Westerners; our projections highlight the uncertainty these distinctions create in the trajectory of motorization.

Sensitivity Analysis

The projections presented above rely on historical data, household sample surveys, and exogenous projections of the independent variables. We conduct sensitivity analysis to help assess how much the model output values are conditioned by the latter. The major parameters here are car price (p_i), mean per-capita disposable income (M), Gini index (G_i), and the ultimate saturation level (γ). As uncertainty in γ has been investigated by Monte Carlo simulation, the influences of the other three parameters (p_i, G_i, and x) are studied, and we adopt the expectation value of γ as a base case. For brevity, we focus only on sensitivity of this central value to assumptions; however, the method we describe makes it straightforward to study sensitivity of the spread and shape of the distributions of projected outcomes, as well.

There is disagreement in the literature over the estimation of China's household income (*25*) and the precise past and present values of of the Gini index (*20*); future forecasts contain additional uncertainty. In Figure 7, we show projections from other agencies (HSBC (*26*), Standard Chartered (*27*), and Morgan Stanley (*28*)) for income level and literature (Wang et al. (*29*)) for Gini index, and our sensitivity trajectories that result in values that are $(100 + n)\%$ of the projected values in 2050, where $n \in (-40, -30, ..., 30, 40)$.

Figure 8 presents the change in car ownership as the major parameters change. As expected, car ownership increases when car purchasing power increases, due to either an increase in per-capita disposable income level or a decrease in car price index. Note that car ownership is most sensitive to car price in near-term future (2020); whereas it is nearly equally sensitive to income level along the future period.

The Gini index has a negative relationship with car ownership in near and mid-term future (2020 and 2030): as Gini index (level of income inequality) increases, the mass of income distribution shifts to the left, and thus the majority of the society becomes poorer while almost all income is held by a small fraction of very wealthy households. In this case, per-capita car ownership is lower, because a larger fraction of the population cannot afford car ownership at all. By contrast, when China's economy and car market become developed and mature, income inequality has little effect on private car ownership. While the sensitivity of car ownership to car price shown in Figure 8 is consistent with Huo and Wang (*8*), the sensitivity results with respect to per-capita

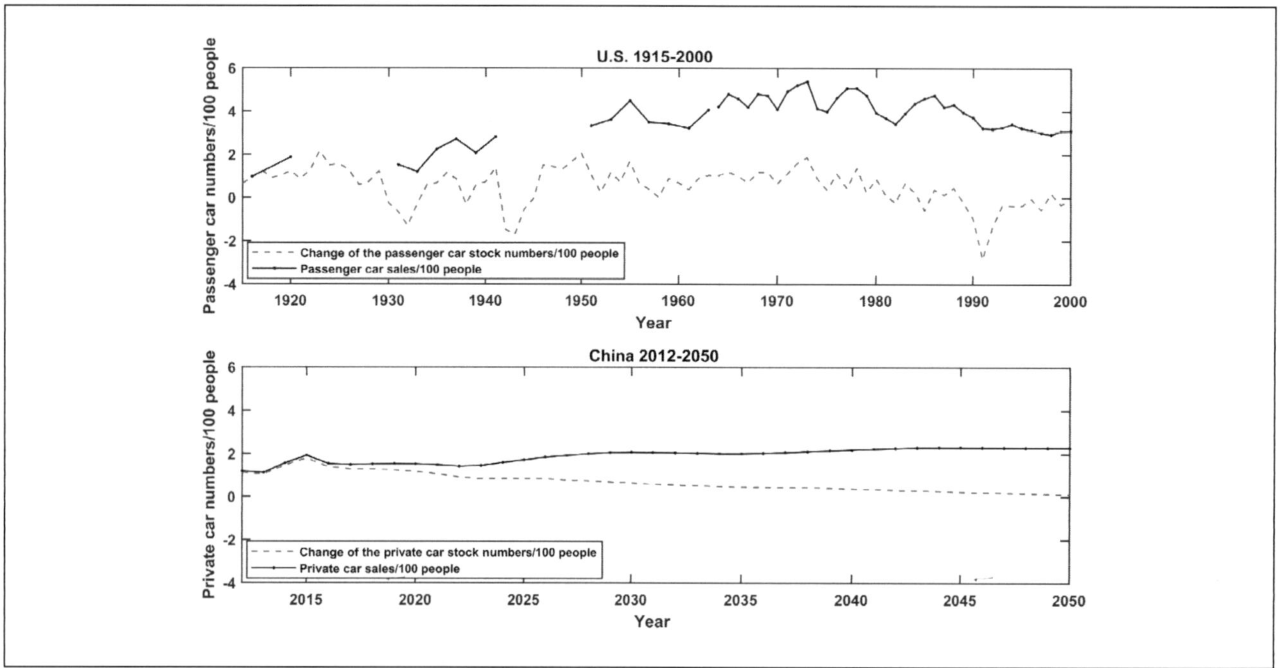

Figure 6. Comparison of our model results for China 2010–2050 with the motorization-stage in U.S. 1915–2000.

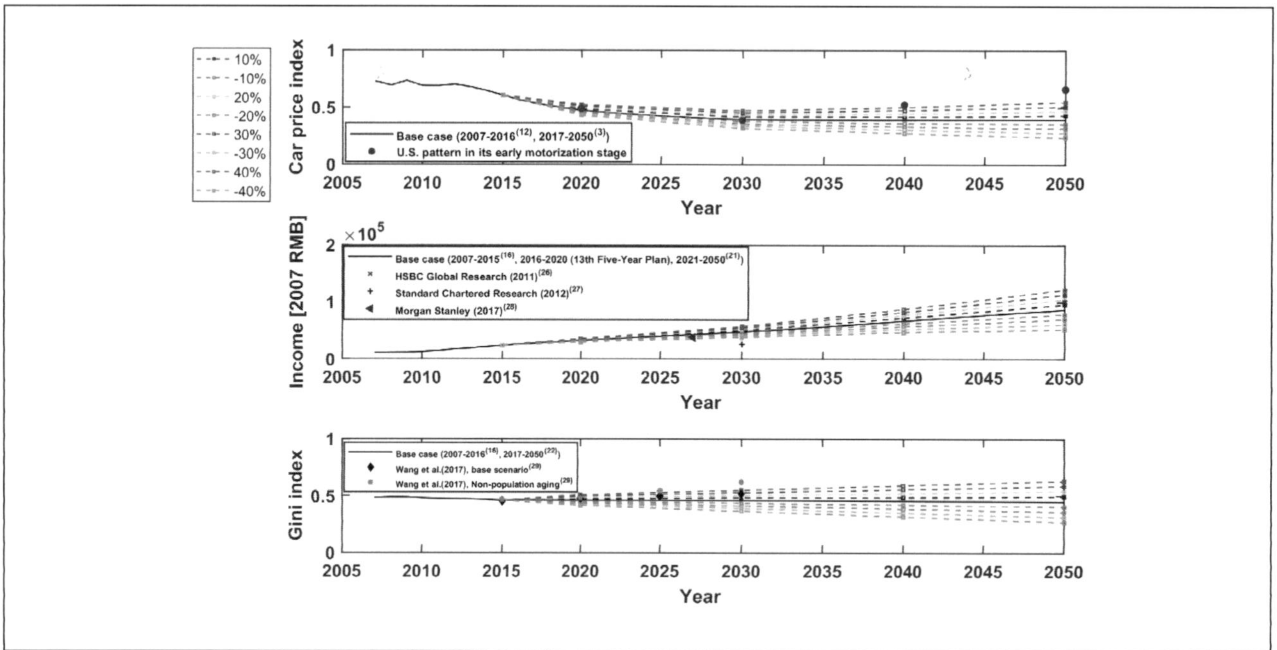

Figure 7. Trajectory of the major parameters varied in the sensitivity analysis.

disposable income and Gini index are different. Our study shows the similar sensitivity to income level toward 2050, whereas the magnitude of the changes decreases as time passes in Huo and Wang (*8*); in the current results, the sensitivity of car ownership to Gini index is not always monotonically decreasing. This result is due to the different probability distribution functions chosen to simulate the income distribution in China and the different calculation mechanisms for car stock, as described above ("Car stock and sales"). For example, the log-

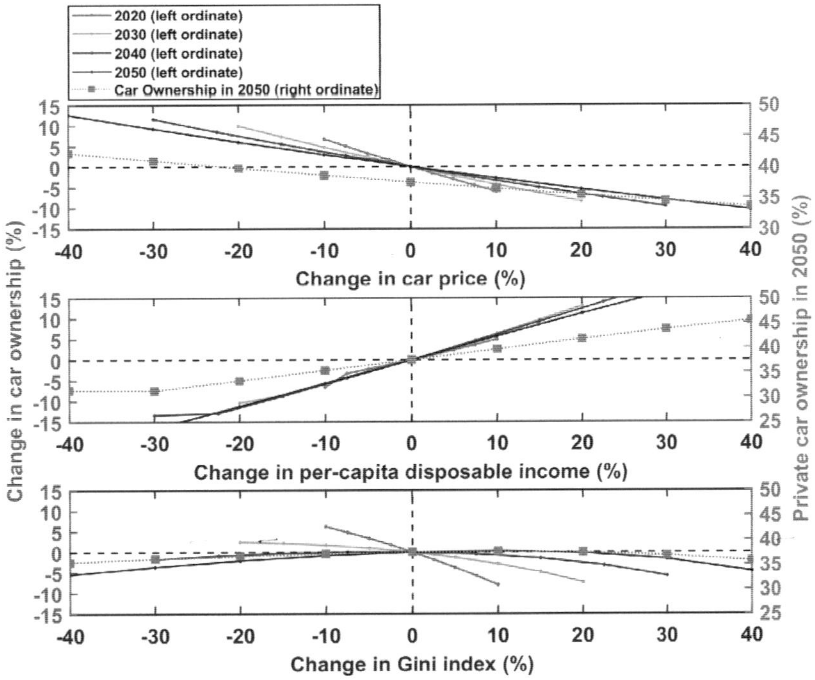

Figure 8. Sensitivity of private car ownership with respect to major parameters (solid line) and the corresponding ownership level in 2050 (dotted line).

normal distribution used in Huo and Wang (*8*) somewhat underestimates the income of the highest-income households, and thus may further underestimate the sensitivity to income level in the long-term projections as more people become wealthy.

Discussion of Results

Our analysis simulates the possible outcomes of China's growth in private car ownership and sales by characterizing the distribution of the adoption trajectory. Figure 4 shows that our model captures the pattern of growth, although the expected value slightly overestimates historical car ownership and stock. As mentioned ("Parameter uncertainty in Gompertz functions"), we are modeling the development of car market without many government interventions. However, besides household incomes, car price, and income inequality, policy also affects transport-related consumption. To reduce private vehicle travel and create more livable and sustainable cit ies, an increasing number of the largest cities in China have sharply restricted new vehicle registration by implementing quota policies (*30*). While Shanghai was a proactive adopter of license auctioning (starting in 1994), other cities did not impose their own quotas until 2011 or later, when the motorization was already in rapid progress. Such caps on vehicle purchases may increase the deviation of our projections from observations in the future; however, the likely distribution of our projections can aid in the design of robust transportation policy that relies on forecasts of demand.

Previous research has shown that the preferences of existing vehicle owners are, to some extent, anchored—they may prefer to get a vehicle that is similar to the one they are replacing, or already own (*31*); or they may even prefer to upgrade their cars in size and horsepower (*32*). Manufacturers both encourage and follow these preferences. We show (Figure 5) that the percentage of replacement sales can be expected to be large in the new car sales market around 2025. Therefore, in the near-term where new sales are a large share of the market, the government has significant opportunities to guide the type, size, fuel economy and pollution characteristics of vehicles sold. These will in turn condition households' experience and expectations of vehicle ownership, and thereby affect their future replacement and additional purchases.

According to the 13th Five-Year Plan (2016–2020), China proposed to become "a moderately prosperous society in all respects" (*33*). The government is aiming to double per capita income by 2020 from 2010 levels, and to improve the fairness of income distribution through expanded public services. If these objectives are achieved, Figure 8 reveals that the private car ownership rate in China will be higher in 2020 as the economy is more equal (i.e., the Gini index is lower). Proper transport demand management (TDM) strategies are required to reduce private vehicle travel, curb car congestion, and to make sustainable transport a reality—and our results show that they are even more urgent if the government is successful in pursuing its goal of reducing inequality.

Extensions

The regional distribution of China's current vehicle market reflects the regional heterogeneity in GDP contribution across the country and the different regional urbanization level. China's six largest civil vehicle owning provinces—Shandong, Guangdong, Jiangsu, Zhejiang, Hebei, and Henan—represented around 47% of the total vehicle fleet at the end of 2015. In particular, in terms of the contribution for incremental growth, these six provinces accounted for nearly 50% every year in the past ten years (*16*). This pattern shows that the coastal regions have dominated the national market.

In order to incorporate the significant regional differences in the level of economic development and vehicle ownership, and produce more precise projections with possibly narrower distributions, the national-level methodology shown in this study could be applied at the regional or provincial level. To do this, the car ownership function, g, would be sampled separately and survival ratio, $SR(y)$, is estimated differently for provincial-level cities and other sub-national regions, and the exogenous variables could be drawn from province-level projections.

Finally, with the car stock and car sales models at either the national or regional levels, our work allows examination of uncertainty in the future trends of life cycle energy demand and emissions from private car use in China.

Conclusion

Undoubtedly, China's car stock will continue to grow, driven by the growing economy. However, due to the short history and the country being in a fast-growing stage of transport system expansion, national observations contain little information about the eventual saturation level. Here, we improved on existing methods by incorporating adoption uncertainties and considering multiple key influencing factors, demonstrating a novel treatment of uncertainty in transport forecasting in China. A range of future forecasts in transport demands that are conditioned on observed reality was developed, producing a characterization that can help policy makers or transport infrastructure project managers create flexible designs that are robust to eventualities. On the other hand, the use of single-point estimates could lead to a "flaw of averages", by obscuring irreducible future uncertainty (*34*).

We found that the distribution of car stock/ownership is wider in 2050 than 2030; on the other hand, the distribution of the share of new-growth purchases is higher in the near term than in a saturated market, with replacement sales expected to dominate the car sales market after the next decade. Finally, our sensitivity analysis results suggest that car ownership will increase further in 2020 as the society becomes less unequal; and more so if this trend goes faster than the literature suggests. We also discussed how our method and estimates have important implications for regulators and planners, and could be a basis for regionally disaggregated projections with similar advantages.

Acknowledgments

This work was supported through the MIT Energy Initiative's Mobility of the Future study.

Author Contributions

Study conception and design: I.-Y.L.H., P.N.K., and W.H.G.; data collection: I.-Y.L.H.; analysis and interpretation of results: I.-Y.L.H., P.N.K., and W.H.G; draft manuscript preparation: I.-Y.L.H. and P.N.K. All authors reviewed the results and approved the final version of the manuscript.

References

1. OICA. *2005–2015 Sales Statistics*. International Organization of Motor Vehicle Manufacturers. http://www.oica.net/category/sales-statistics/. Accessed January 15, 2017.
2. Davis, S. C., S. W. Diegel, and R. G. Boundy. *Transportation Energy Data Book*, 33rd ed., 2014. https://info.ornl.gov/sites/publications/files/Pub50854.pdf. Accessed April 20, 2017.
3. Hamilton, W., M., Adams, A. Abrahamson, I. Till, G. Marshall, and H. E. Meiklejohn. *Price and Price Policies*. Mc Graw Hill, London, 1938, p. 27.
4. de Jong, G., J. Fox, A. Daly, M. Pieters, and R. Smit. Comparison of Car Ownership Models. *Transport Reviews*, Vol. 24, No. 4, 2004, pp. 379–408. https://doi.org/10.1080/0144164032000138733.
5. Zachariadis, T., Z. Samaras, and K. H. Zierock. Dynamic Modeling of Vehicle Populations: An Engineering Approach for Emissions Calculations. *Technological Forecasting and Social Change*, Vol. 50, No. 2, 1995, pp. 135–149. https://doi.org/10.1016/0040-1625(95)00057-H
6. Wu, T., H. Zhao, and X. Ou. Vehicle Ownership Analysis Based on GDP Per Capita in China: 1963–2050. *Sustainability*, Vol. 6, No. 8, 2014, pp. 4877–4899. https://doi.org/10.3390/su6084877

7. Dargay, J. M., D. Gately, and M. Sommer. Vehicle Ownership and Income Growth, Worldwide: 1960–2030. *The Energy Journal*, Vol. 28, No. 4, 2007, pp. 143–170. http://www.jstor. org/stable/41323125.
8. Huo, H., and M. Wang. Modeling Future Vehicle Sales and Stock in China. *Energy Policy*, Vol. 43, 2012, pp. 17–29. https://doi.org/10.1016/j.enpol.2011.09.063
9. United Nations. Department of Economic. *World Population Prospects: The 2015 Revision*, 2015. https://www.populationpyramid.net/china/2017/. Accessed April 30, 2017.
10. Hao, H., H. Wang, M. Ouyang, and F. Cheng. Vehicle Survival Patterns in China. *Science China Technological Sciences*, Vol. 54, No. 3, 2011, pp. 625–629. https://doi.org/10.1007/s11431-010-4256-1
11. Dargay, J. M. The Effect of Income on Car Ownership: Evidence of Asymmetry. *Transportation Research Part A: Policy and Practice*, Vol. 35, No. 9, 2001, pp. 807–821. https://doi.org/10.1016/S09658564(00)00018-5
12. National Development and Reform Commission. *Consumer Price Index*. http://en.ndrc.gov.cn/. Accessed May 10, 2017.
13. China Industry Information Network Website. http://www.chyxx.com. Accessed April 10, 2017.
14. Davis, S. C., S. W. Diegel, and R. G. Boundy. *Transportation Energy Data Book*, 35th edn., 2016. http://cta.ornl.gov/data/index.shtml. Accessed April 28, 2017.
15. Dargay, J., and D. Gately. Income's Effect on Car and Vehicle Ownership, Worldwide: 1960–2015. *Transportation Research Part A: Policy and Practice*, Vol. 33, No. 2, 1999, pp. 101–138. https://doi.org/10.1016/S0965-8564(98)00026-3
16. National Bureau of Statistics of China. *China Statistics Yearbook 2016*. http://data.stats.gov.cn/english/easyquery.htm?cn = C01. Accessed April 22, 2017.
17. Kishimoto, P. N. Applying Advanced Uncertainty Methods to Improve Forecasts of Vehicle Ownership and Passenger Travel in China. *Proc., 31st USAEE/IAEE North American Conference,* Austin, TX, November 4, 2012.
18. Steyn, H. S. A Model for the Distribution of Incomes. *South African Journal of Economics*, Vol. 27, No. 2, 1959, pp. 149–156. https://doi.org/10.1111/j.1813-6982.1959.tb01820.x
19. McDonald, J. B., and M. R. Ransom. Functional Forms, Estimation Techniques and the Distribution of Income. *Econometrica: Journal of the Econometric Society*, Vol. 47, No. 6, 1979, pp. 1513–1525. https://doi.org/10.2307/1914015
20. Chen, J., F. Fang, W. Hou, F. Li, M. Pu, and M. Song. Chinese Gini Coefficient from 2005 to 2012, Based on 20 Grouped Income Data Sets of Urban and Rural Residents. *Journal of Applied Mathematics*, Vol. 2015, 2015, pp. 1–16. https://doi.org/10.1155/2015/939020.
21. OECD. *Long-Term Baseline Projections, No. 95 (Edition 2014)*. OECD, 2017. https://doi.org/10.1787/data-00690-en. Accessed May 2, 2017.
22. Hu, J., Y. Luo, and X. Yang. Forecast and Policy Suggestions of Gini Coefficient in China in the Future Decade: Analysis Based on Grey Forecasting Model. *West Forum on Economy and Management*, Vol. 26, No. 1, 2015, pp. 44–49. (In Chinese)
23. Beijing Times. *Experts: 2030 China's Vehicles Per 1000 People Will Maintain the Amount of 300*. September 2016. http://epaper.jinghua.cn/html/2016-09/22/content_336471.htm. Accessed July 26, 2017. (In Chinese)
24. United States Department of Transportation. *National Transportation Statistics*. https://www.rita.dot.gov/bts/sites/rita.dot.gov.bts/files/publications/national_transportation_statistics/index.html. Accessed December 10, 2016.
25. Wang, X., and W. T. Woo. The Size and Distribution of Hidden Household Income in China. *Asian Economic Papers*, Vol. 10, No. 1, 2011, pp. 1–26. https://doi.org/10.1162/ASEP_a_00064
26. Ward, K. *The World in 2050: Quantifying the Shift in the Global Economy*. London: HSBC Global Research, 2011.
27. Daily, C. *Per Capita Income to Rise by up to 50% by 2030*. October 2011. http://www.chinadaily.com.cn/bizchina/2012-10/11/content_15810209.htm. Accessed July 5, 2017.
28. Curran, E. *China Will Avoid a Bank Crisis, Reach High Income Status: Morgan Stanley*. Febraury 2017. https://www.bloomberg.com/news/articles/2017-02-14/morganstanley-says-china-to-avoid-bank-shock-reach-high-income. Accessed July 5, 2017.
29. Wang, X., K. Z. Chen, S. Robinson, and Z. Huang. Will China's Demographic Transition Exacerbate Its Income Inequality?: CGE Modeling with Top-Down Microsimulation. *Journal of the Asia Pacific Economy*, Vol. 22, No. 2, 2017, pp. 227–252. https://doi.org/10.1080/13547860.2016.1263043
30. Xue, L. *Four Lessons from Beijing and Shanghai Show How China's Cities Can Curb Car Congestion*. WRI Ross Center for Sustainable Cities. April 10, 2015. http://www.wri.org/blog/2015/04/4-lessons-beijing-and-shanghai-show-how-china%E2%80%99s-cities-can-curb-car-congestion. Accessed December 18, 2015.
31. Sivak, M., and B. Schoettle. *What Do Current Owners of Hybrids and Non-Hybrids Think About Hybrids?* Publication UMTRI-2014-25. Transportation Research Institute, University of Michigan, Ann Arbor, 2014.
32. Knittel, C. R. Automobiles on Steroids: Product Attribute Trade-Offs and Technological Progress in the Automobile Sector. *American Economic Review*, Vol. 101, No. 7, 2011, pp. 3368–3399. https://doi.org/10.1257/aer.101.7.3368
33. Ross, J. *China's Five Year Plan to Achieve a 'Moderately Prosperous Society'*. October 2015. http://www.china.org.cn/opinion/2015-10/30/content_36935303.htm. Accessed July 25, 2017.
34. de Neufville, R., and S. Scholtes. *Flexibility in Engineering Design*. MIT Press, Cambridge, 2011.

The Standing Committee on Traveler Behavior and Values (ADB10) peer-reviewed this paper (18-05415).

Article

TRR
JOURNAL OF THE TRANSPORTATION RESEARCH BOARD

Transportation Research Record
2018, Vol. 2672(47) 194–207

DOI: 10.1177/0361198118773197
journals.sagepub.com/home/trr
SAGE

Learning Daily Activity Sequences of Population Groups using Random Forest Theory

Mohammad Hesam Hafezi[1], Lei Liu[1], and Hugh Millward[2]

Abstract

The choice of daily activity sequences differs between individuals based on their socio-demographic characteristics and their health and/or mobility status. The aim of this paper is to provide an improved methodology for learning and modeling the daily activity engagement patterns of individuals using a state-of-the-art machine learning algorithm. The dependencies between activity type, activity frequency, activity sequence, and socio-demographic characteristics of individuals are taken into account by employing a random forest model. In order to capture the heterogeneity and diversity among the predictor variables, we employed two different methods for split selection in the random forest algorithm: Classification and Regression Tree (CART) and curvature search. These two methods were examined under two different layer settings. In the first setting, the algorithm grows trees using all alternative predictor variables, whereas in the second setting the importance of the predictor variables is estimated and then the algorithm grows trees using only high-score predictor variables. The models were applied to time use data from the large Halifax Space-Time Activity Research (STAR) household travel diary survey. We evaluated the estimation accuracy of the proposed models using confusion matrix, transition matrix, and sequential alignment techniques. Results show that the random forest model with CART split selection using the first layer setting has the best accuracy in replicating activity agendas and activity sequences of individuals. The results of this paper are expected to be implemented within the activity-based travel demand model, Scheduler for Activities, Locations, and Travel (SALT).

In recent years the activity-based modeling approach has received much attention from transport modelers and policy makers. Complexities in individual travel behavior such as flexible working hours, self-employment, and online shopping have increased considerably due to rapid technological progress. Moreover, policies related to road congestion and air pollution are having greater impacts on travel behavior. Clearly, more complex and disaggregated models are needed to address the above-mentioned changes. Integrity, interdependencies, higher temporal and spatial aggregation, and behavioral basis are the four main reasons proposed by Rasouli and Timmermans (*1*) to change from the traditional four-stage travel demand models to activity-based travel demand models. Numerous modeling approaches have been developed for activity-based travel demand models. The constraints-based modeling approach was one of the earliest techniques developed, based on the theory that travel is a derived demand, originating from the need of an individual to engage in activities (*2*). With the concept that individuals desire to maximize the utility of their activity schedule, the random utility theory was then used in the development of activity-based models (*3*). More recently, the computational process modeling approach was developed using the theory of context-dependent choice preferences and emphasizing their scheduling aspect (*4*). PCATS (*5*), ALBATROSS (*4*) and ADAPTS (*6*) are some examples of activity-based travel demand models developed through the above-mentioned approaches. The broad modules for most activity-based travel demand models are: activity generator and scheduler, activity engagement, tour and trip destination, and mode choice and network assignment, among others.

The activity engagement patterns component provides explicit details on activity type, and the frequency and sequence of activities engaged in. Daily activity

[1]Department of Civil and Resource Engineering, Faculty of Engineering, Dalhousie University, Halifax, NS, Canada
[2]Department of Geography and Environmental Studies, School of the Environment, Saint Mary's University, Halifax, NS, Canada

Corresponding Author:
Address correspondence to Lei Liu: lei.liu@dal.ca

engagement patterns of individuals are crucial components in any activity-based travel demand model, as an individual's travel demand originates from their need to engage in particular activities. Many empirical and theoretical approaches have been employed for the activity engagement module, such as multinomial logit model, probabilistic grammars, decision tree, etc. (*6–10*).

Despite all of the progress made in activity engagement modules through rule-based or econometric techniques, there have been few attempts to utilize the capability of machine learning to derive daily activity engagement patterns. In this study, we developed a new model that is able to learn and replicate an individual's daily activity engagement patterns. Inspired by random forest (RF) theory, decision trees were grown using both Classification and Regression Tree (CART) and curvature search techniques. The resulting models were trained from observed activity sequences. A previous study reveals that ensemble methods performed better when they were applied to clustered data (*11*). Therefore, in this study we applied the models to 12 unique population clusters derived using a novel pattern recognition model, and compared the results with each other. We used confusion matrix, transition matrix, and sequential alignment methods to compare the estimation accuracy for replication of activity type, activity position, and activity sequences. The remainder of this study is organized as follows. First, we review relevant past research concerning daily activity engagement patterns of travelers. Second, we discuss the data used for the activity pattern recognition. The modeling approach and the planned layer settings are explained in the next section, followed by a discussion of model results. The study concludes with a summary of contributions and a brief discussion of future research directions.

Literature Review

Daily activity engagement patterns differ between individuals based on their socio-demographic characteristics and their health and/or mobility status. In a seminal early work on time geography, in 1970 Hagerstrand (*2*) proposed three types of constraints that shape individual activity sequences. The first type are capability constraints, which focus on biological needs and available resources that can require or limit an individual's participation in an activity (e.g., eating meals, drinking, and sleeping). The second type are coupling constraints, which refer to the spatial and temporal necessities for an individual who joins with other individuals to conduct a certain activity. The third type are authority constraints, which limit the individual's access to certain activity locations or times. Thus, a decision to participate in a particular activity is a combined result of several decisions and constraints (e.g., household interaction, choices of activity type, location, timing, duration, destination, etc.).

Following Hagerstrand's time-space constraints, a wide variety of modeling approaches have been employed to model various aspects of daily activity sequences, such as activity type, activity sequence, activity frequency, sequential activity location, duration, and transport mode for the next trip. A three-level structural model was proposed by Bowman and Ben-Akiva (*9*) for modeling daily activity pattern. In the first phase, the daily activity sequences choice set is established. In the next two phases, tour and trip decisions with regard to the choice of daily activity pattern are modeled. Following this concept, Bhat et al. (*12*) modeled daily activity sequences in a more disaggregate manner and considered separate models for non-workers and workers. In other studies, Bhat et al. (*13*) and Leszczyc and Timmermans (*14*) estimated activity durations using the multiple discrete continuous extreme value (MDCEMV) and hazard models, respectively. Daily activity sequences can be modeled within the tour formulation. Initially, the primary activity and its associated tour type (i.e., activity destination, sequence, and number of stops) is modeled, and in the next phase the number and purpose of secondary tours are estimated (*15, 16*).

In addition to inclusive developed econometric and rule-based models, in recent years there has been growing interest in applying machine learning practices to activity-based modeling (*7, 11, 17*). A wide range of applications has been developed using machine learning techniques, mostly in the computer science and statistics fields, but to date there have been very limited applications of the technique in activity-based modeling, and especially in modeling of activity engagement patterns. Li and Lee (*7*) modeled individuals' activity sequences by employing probabilistic context-free grammars. They adapted language concepts to the daily activity sequences and estimated activity sequences. In another study by Allahviranloo and Recker (*11*), daily traveler activity engagement patterns were modeled using support vector machines (SVM). They employed several machine learning techniques in their proposed model. Using conditional random fields, the dependencies between activity sequence, activity type, and socio-demographic data were identified, and then the sequential choice of activities was captured by employing the Markov chain model.

In this study we propose an algorithm inspired by random forest theory to model the activity engagement patterns of individuals. The existing applications of random forest theory in transportation fields are limited to transport mode recognition and traffic incident detection (*18, 19*). The proposed conceptual model in this study is able to learn and replicate individuals' daily activity engagement patterns with regard to heterogeneity characteristics. The application of this framework to activity-based modeling not only demonstrates the efficiency of machine learning to model daily traveler activity engagement patterns, but also contributes additional insights to the linkage between activity agenda formation and activity scheduling modules.

Data

The data used in this study is drawn from the Space-Time Activity Research (STAR) survey undertaken in Halifax, Canada. The STAR survey was a joint household activity survey and travel survey, and the world's first large-scale employment of global positioning system (GPS) technology for tracking and confirmation of out-of-home activities. A brief description follows, and full descriptions of the survey design and the socio-demographic features of respondents can be found in (*20–22*). The survey period was between April 2007 and May 2008, equally spaced through days of the week and months of the year. It yielded fully geo-referenced two-day (i.e., 48-h) time diary data from 1,971 randomly designated primary respondents, aged 15 years or older. Respondents carried a GPS data logger, maintained a daily "activity log" during that period, and completed a day-after computer-assisted telephone interview time diary survey. The respondents' descriptions of their out-of-home activities were prompted and confirmed by the GPS data.

In this study, we use data from 12 identified unique clusters of respondents as a result of applying a novel pattern recognition modeling framework to the STAR data. Interested readers are referred to previous research by the authors (*23, 24*) for more details on clustering methods and execution of the pattern recognition framework. In general, the pattern recognition model identified six clusters for workers (cluster#1: extended worker, cluster#3: 8–4 worker, cluster#6: shorter worker, cluster#7: 7–3 worker, cluster#10: evening worker and cluster#11: 9–5 worker), one cluster for students (cluster#12: students), four clusters for non-worker and non-student (cluster#2: non-worker midday activities, cluster#4: non-worker evening activities, cluster#8: non-worker morning shopping and cluster#9: non-worker afternoon shopping), and one cluster for individuals who mostly spend their time at home (cluster#5: stay-at-home). Individuals in each cluster differ notably from individuals in other clusters in terms of their socio-demographic characteristics. Moreover, individuals within each cluster have homogenous activity sequences whereas there is a heterogeneous diversity between clusters in terms of their distributions of start time, activity duration, activity type, and socio-demographic characteristics.

Methods

Random forest (RF) theory is based on the use of many decision trees that have been grown using the bagging method (*25, 26*). Each tree plays as a weak learner in the algorithm and the combination of these weak inputs builds up a robust ensemble learning model. Results obtained from the RF model are based on the majority votes in the ensemble models. One of the main challenges in most decision tree models is to select the best split predictor method (*25, 27*). In this study, we introduced two new techniques: CART and curvature search, each with two-layer settings for split predictor selection. Although the RF method is used in other transportation fields such as transport mode recognition and traffic incident detection, to the best of our knowledge establishing RF models using the CART and curvature search have not been explored in activity engagement pattern modeling or travel behavior studies.

The Random Forest (RF) Model

The RF structure for predicting a set of activity types in an individual's agenda is shown in Figure 1. Practically, the RF model takes the following steps. Initially, cross-validation partitions for train data (D_r) and test data (T_e) are performed. Prior to the growing of each tree, r_n observations are randomly drawn from the sampled data points $r_n \in \{1, \ldots, n\}$. Next, at each node in the decision tree a number of predictor variables are randomly selected from all the available predictor variables (Q). The default number of selected predictor variables is the square of the total number of existing predictor variables ($\sqrt{Q}$). Best split at each node (B_{lit}) is accomplished using the split predictor selection search over b_{lit} directions. As the tree grows, this process is repeated and continues until the information is saturated. In the RF model, trees are not required for cost complexity pruning. The new arrival at the testing stage $t_n \in \{1, \ldots, e\}$ is propagated down to all of the trees grown (m trees) in the RF model, and a set of prediction results from all trees is obtained $\{\hat{P}_1, \hat{P}_2, \hat{P}_3, \cdots, \hat{P}_M\}$. The final prediction result gained is based on majority votes $\hat{P}$.

Decision Splits: CART and Curvature Search

In this section, we describe how the CART and curvature search methods are operated in the RF models. We provide here only a brief overview of the algorithm, and interested readers are referred to (*25, 27*) for more details. We define a training sample $D_r = ((X_i, Y_1), \ldots, (X_i, Y_n))$ where X_i represents the response variable X = [home chores, home leisure, night sleep, workplace, shopping & services, school/college, organizational/hobbies, entertainment, sports] and Y_n represents the set of predictor variables (selected from Table 1). In this respect, we assume $E_n(C)$ as a subset of training data D_r where C is a pair of (u, w). u is the randomly selected predictor

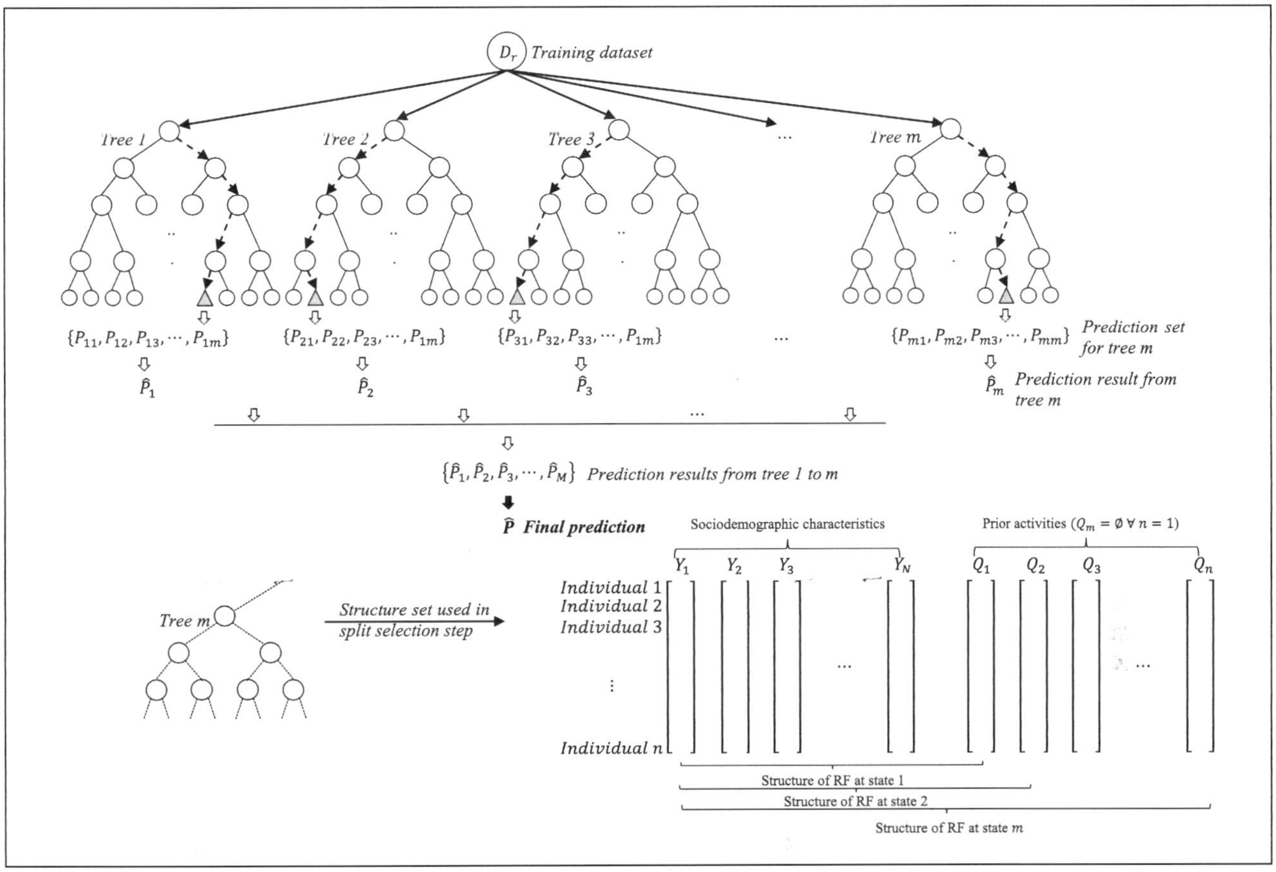

Figure 1. Random forest model structure for activity pattern sequence.

variable, and w is the place of the split along the u-th correspondent, within the limits of C. We postulate the set of all such feasible cuts in C as T_C. The split criterion $S_{clas,n}(u, w)$ is calculated as follows:

$$S_{clas,n}(u,w) = \frac{1}{E_n(C)} \sum\nolimits_{i=1}^{n} (Y_i - \bar{Y}_C)^2 \exists_{X_i \in C} - \frac{1}{E_n(C)} \sum\nolimits_{i=1}^{n} \left(Y_i - \bar{Y}_{C_L} \exists_{X_i^{(u)} < w} - \bar{Y}_{C_R} \exists_{X_i^{(u)} \geq w}\right)^2 \exists_{X_i \in C} \tag{1}$$

where

$$X_i = \left(X_i^{(1)}, \ldots, X_i^{(p)}\right) \forall (u, w) \in T_C \tag{2}$$

$$C_L = \left\{x \in C : x^{(u)} < w\right\} \tag{3}$$

$$C_R = \left\{x \in C : x^{(u)} \geq w\right\} \tag{4}$$

$\bar{Y}_C$ is the average of the Y_i such thatX_i belongs to C. In the CART, the best cut (u_n^*, w_n^*) is computed as follows:

$$u_n^*, w_n^* \in \underset{\substack{u \in B_{lit} \\ (u,w) \in T_C}}{argmax} \; S_{clas,n}(u, w) \tag{5}$$

argmax represents the maximization $S_{clas,n}(u, w)$ over B_{lit} and T_C. In the curvature search, the ideal split is computed where the density curves intersect, extracted from the roots (R_1,R_2).

Variable Importance Measures: Mean Decrease Accuracy (MDA)

RF models have the capability to measure the importance of variables and rank them in order to use in the algorithm. Mean decrease accuracy (MDA) and mean decrease impurity (MDI) are the two well-known measures of variables significance (*25, 27*). In this study, we used the MDA technique that relies on the out-of-bag (OOB) error estimate. The MDA of the variable $X^{(u)}$ is calculated by balancing the difference in OOB error (O_n) prior and subsequent to the permutation over all trees. The prior OOB error of each tree is computed by testing

Table 1. Proposed Predictor Variables for Learning Daily Activity Engagement Patterns

Predictor	Subcategories
Gender	Male, female
Age	15–19, 20–24, 25–29, 30–34, 35–39, 40–44, 45–49, 50–54, 55–59, 60–64, 65–69, 70–74, 75–79, 80–84, 85 +
Marital status	Married, living common-law, widowed, separated, divorced, single/never married
Household size	1, 2, 3, 4, 5, 6
Highest education level	Masters or earned doctorate; bachelor or undergraduate degree; diploma or certificate; some university; some community college; trade, technical or business college; high school/secondary; other
Full/part-time student	Full-time student, part-time student
Paid/self-employed	Paid worker, self-employed, unpaid family worker
Flexible work schedule	No, yes
Work at home	No, yes
Total personal income	Under \$20,000, \$20,000–\$39,999, \$40,000–\$59,999, \$60,000–\$79,999, \$80,000–\$99,999, \$100,000 or more
Total household income	Under \$20,000, \$20,000–\$39,999, \$40,000–\$59,999, \$60,000–\$79,999, \$80,000–\$99,999, \$100,000 or more
Dwelling type	Single unit residential, duplex or semi-detached, townhouse, multi-unit residential-less than 6 stories, multi-unit residential-6 or more stories, mobile dwelling, other-specify
Dwelling owned/rented	Owned, rented
Valid driver's license	No, yes
Bus pass	No, yes
Number household vehicles	0, 1, 2, 3, 4, 5, 6
Number household motorcycles	0, 1, 2, 3, 4, 5, 6, 7, 8, 9, 10
Number household bicycles	0, 1, 2, 3, 4, 5, 6, 7, 8, 9, 10
Usual mode to work	Car, truck or van—as driver; car, truck or van—as passenger; public transit; walk to work; bicycle; motorcycle; taxicab; other method
Usual mode to school	Car, truck or van—as driver; car, truck or van—as passenger; public transit; walk to work; bicycle; motorcycle; taxicab; other method
State of health	Excellent, very good, good, fair, poor
Prior activities	Home chores *(working at home, eating/meal preparation, indoor or outdoor cleaning, interior or exterior home maintenance, child care, other in-home activities)*; Home leisure *(watchingTV/listening to radio, reading books/newspapers, etc.)*; Night sleep; Workplace *(work/job, all other activities at work, work related, etc.)*; Shopping & services *(shopping for goods and services, routine shopping)*; School/college *(class participation, all other activities at school)*; Organizational/hobbies *(organizational, voluntary, religious activities; hobbies done mainly for pleasure, cards, board games; all other hobby activities)*; Entertainment *(eat meal outside of home, all other entertainment activities)*; Sports *(walking, jogging, bicycling, all sports-related activities)*.

the RF model using OOB data. The later OOB error of each tree is calculated by adding noise to the sample data of the feature randomly and retesting the OOB error. The MDA for randomly selected variable $X^{(u)}$ is calculated as follows:

$$MDA(X^{(u)}) = \frac{1}{m}\sum_{l=1}^{m}\left[O_n\left[d_n(.;\Theta_m), I^u_{m,n}\right] - O_n[d_n(.;\Theta_m), I_{m,n}]\right] \quad (6)$$

$$O_n[m_n(.;\Theta_l), I] = \frac{1}{|I|}\sum_{i:(X_i,Y_i)\in I}(Y_i - m_n(X_i;\Theta_l))^2 \quad (7)$$

where

$I_{m,n}$ is the OOB data set of the m-tree,
$I = I^u_{m,n}$ is the permuted data for variable u,
$O_n(.;\Theta_m)$ is the estimation for the m-th tree.

Model Calibration and Validation

Compared with other machine learning algorithms, such as support vector machine (SVM) and back propagation neural network (BPNN), in the RF algorithm few parameters need to be initialized and calibrated. The initial value of m (number of trees) is set to 1,000. However, it needs to be validated and adjusted if the model cannot be converged within this value. B_{lit} (best split at each node) needs to be calibrated. In this study we used CART and curvature search techniques to determine the best split at each node. T_C (cutoff) is a vector of length equal to the number of classes. T_C comprises three sub-parameters ($c1$, $c2$, $c3$) that are initially each randomly selected in the range between [0,1] with the condition of total sum equal to one. For all the above parameters, we used the OOB error rate to cross-validate and obtain the best parameter values for the RF models.

Discussion of Results

In order to analyze the impacts of decision splits on the RF model, we trained RF with two different layer settings. In the first layer, we grew trees where predictor variables were randomly selected from all nominated variables. For simplicity's sake in this study, we refer to this setting's layer as RF_CART_I and RF_CURV_I for CART and curvature search techniques, respectively. In the second layer, we estimated the variable importance and included only those variables in the RF model that had more than 70% importance compared with others. For simplicity's sake, in this study we refer to this setting's layer as RF_CART_II and RF_CURV_II for CART and curvature search techniques, respectively. The response variables are defined as nine main activities, as described in Table 1. Previous studies reveal that choice of activities in the schedule not only depends on an individual's socio-demographic characteristics but also is influenced by prior activities (*28*). Recent empirical work by Allahviranloo and Recker (*11*) also shows that SVM as a robust unsupervised machine learning technique has better performance when response variables are associated to both socio-demographic characteristics and previous activities. In this respect, we set up predictor variables in the RF models to choose from a set of socio-demographic variables and prior activities conducted by individuals in a given day. Hence, depending on an activity's position in the agenda, the number of predictor variables varies. Furthermore, additional predictor variables such as flexibility level of work schedule and health state of individuals enter into the analysis. The list of nominated response variables is shown in Table 1.

The OOB error rate cross-validation indicates that after $m > 850$ (m is the number of trees) the OOB error rate tends to be stable. Therefore, it is realistic to accept m as 1,000 at first. On that basis, we obtain the best optimal parameters for RF models as follows: $c1 = 0.25$, $c2 = 0.35$ and $c3 = 0.40$. Figure 2 illustrates the distribution of daily activity engagement for all 12 clusters. As can be seen, most individuals in the worker clusters (#1, #3, #6, #7, #10, #11), in addition to in-home activities, only participate in the out-of-home work activity. A similar result is achieved for the student cluster (#12), where most students only participate in out-of-home school

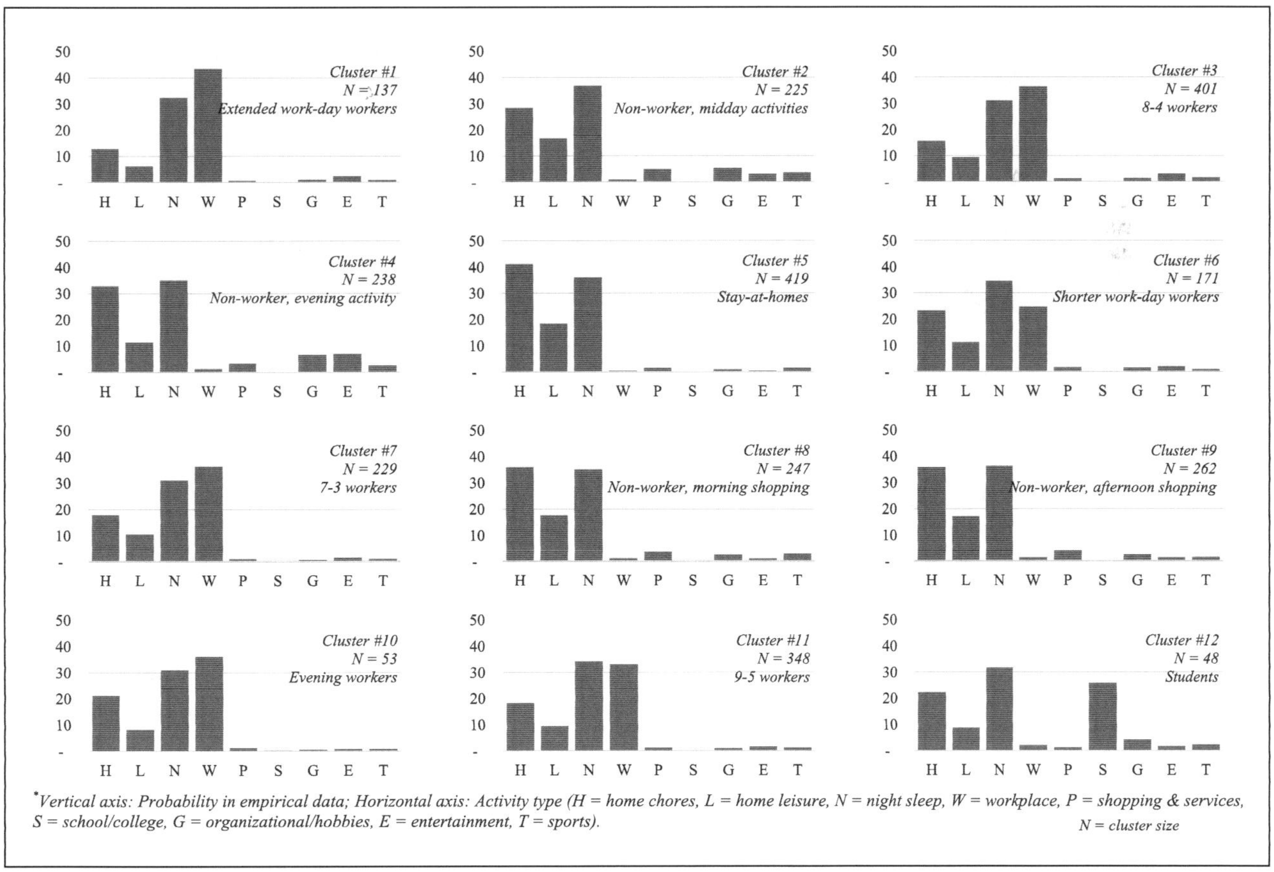

Figure 2. Distribution of daily activity engagement for all 12 clusters.

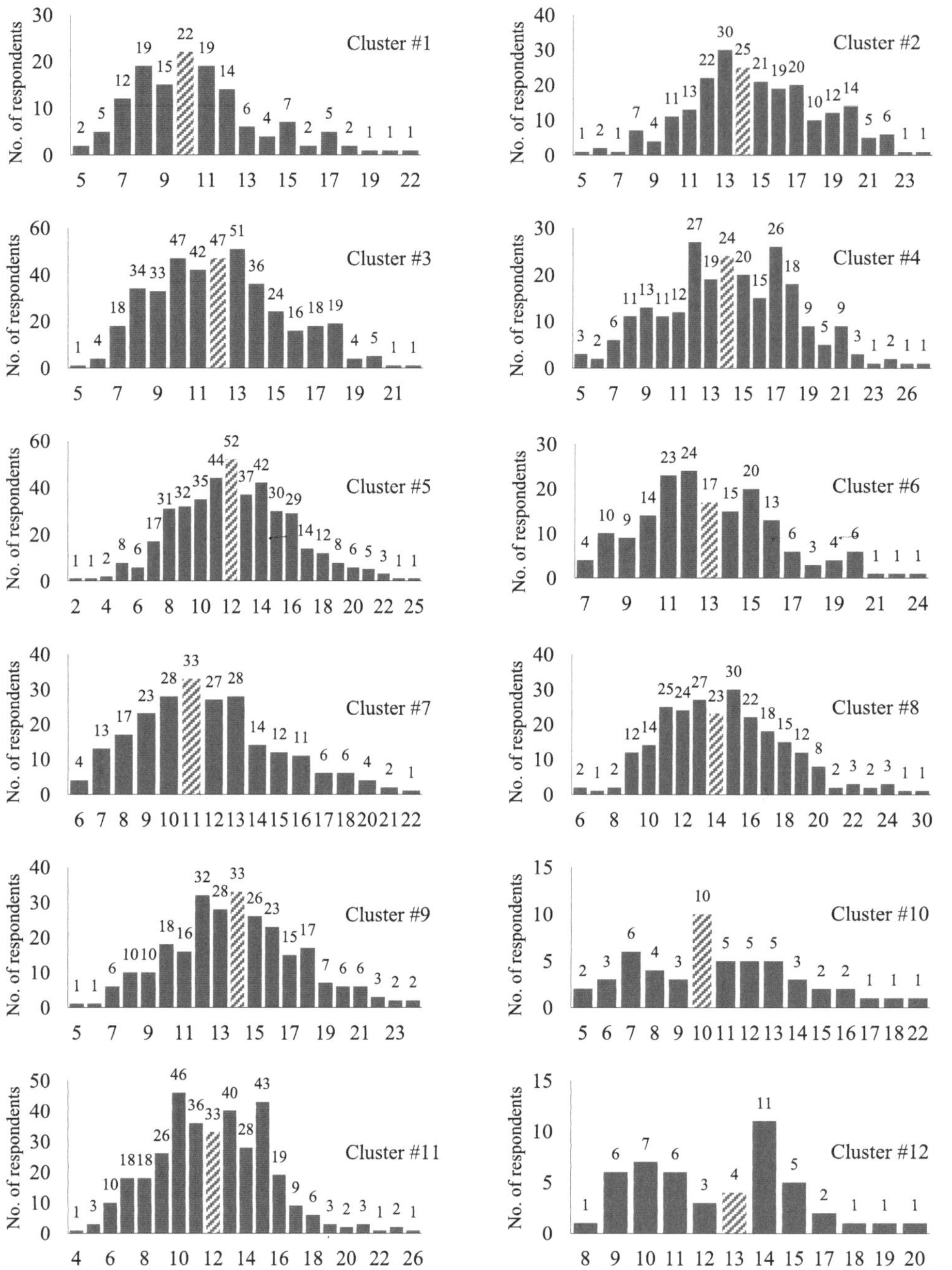

Horizontal axis: No. of separate activity episodes (of t minutes or more); Dashed column represents cluster median

Figure 3. Number of respondents against number of activity episodes for all 12 clusters.

activity in addition to in-home activities. Individuals in other clusters (non-worker and non-school goer) were found to have different out-of-home activity participation patterns (mostly engaging in entertainment or/and recreation activities).

Distribution patterns of number of daily activity episodes for individuals in all 12 clusters are shown in Figure 3. Our intent is to relate modeling to the most typical member in each cluster. In this respect, we assume the median number of episodes in each cluster as the

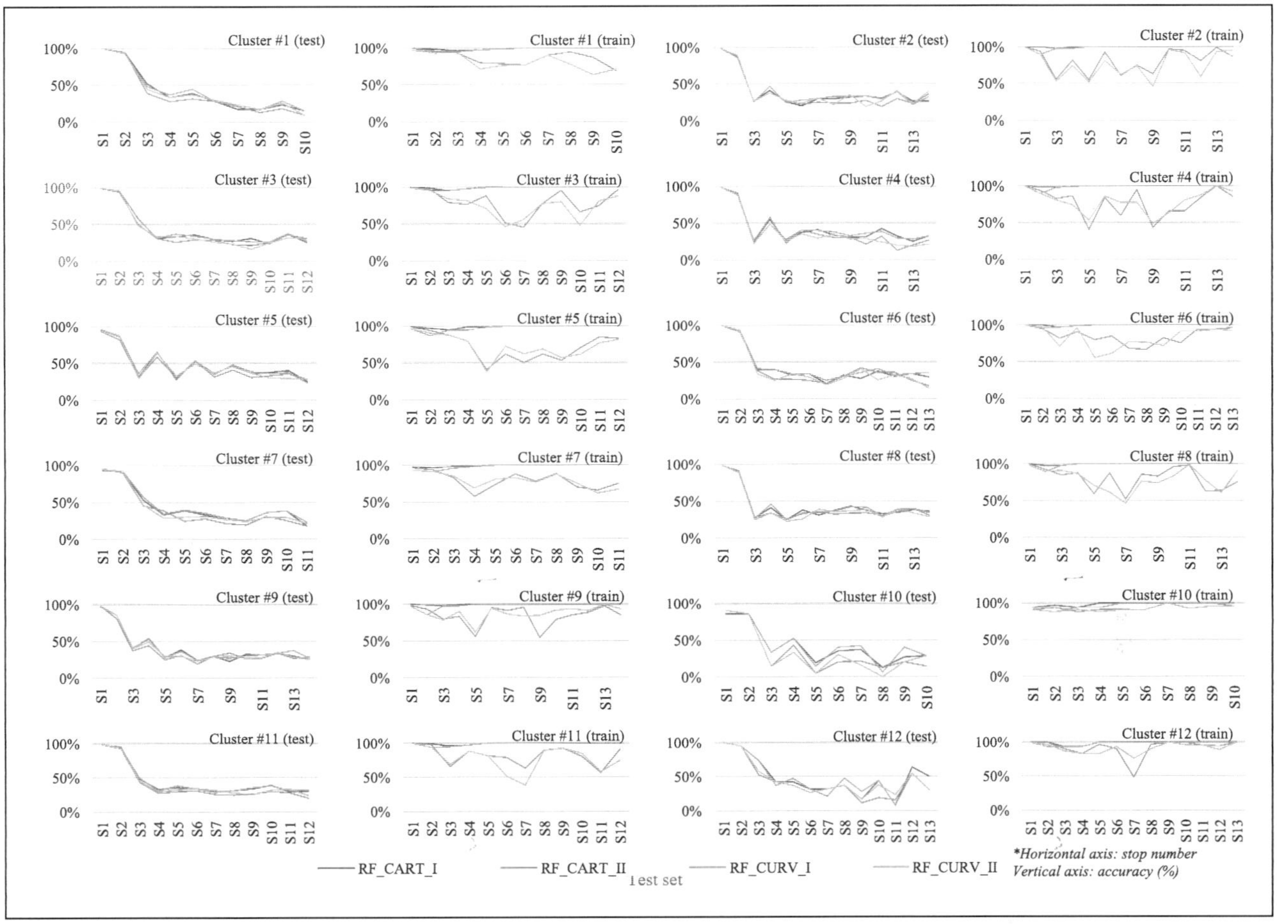

Figure 4. Comparison of the estimation precision between four RF models for test and training dataset.

alternative state number for daily activity participation in the modeling. Moreover, we assume the ultimate activity state for the day as the night sleep. Among 12 clusters, cluster #2 (non-worker midday activity) has the most complex activity pattern and cluster #12 (students) has the least complex activity pattern.

We performed a cross-validation partition to our dataset which resulted in 70% for training and 30% for testing model performance. Comparison of the estimation accuracy between four proposed RF models for test dataset and training dataset is shown in Figure 4. Based on the estimated accuracy, we can conclude that RF_CART_I provides the best results for most states, followed by RF_CURV_I.

Explicitly, estimation accuracy of RF_CART_I for the test and training sets exceeds 69% for all clusters, and is particularly high for clusters 5 (stay-at-homes) and 12 (students) (74% each). The lowest accuracy is for cluster #2 (non-worker midday activities, at 69%). The average for reproducing activity types using the RF_CART_I model is 71.4% for the entire population. Allahviranloo and Recker (*11*) modeled daily activity engagement using the SVM technique with 84% accuracy in the replicating agenda. In their study, five activity types for modeling nine states were used, and the modeling period was between 05:00 and 23:00 hours. Given that we are modeling so many more activity types (three in-home and six out-of-home activities) for a period of 24 hours, we can conclude that our model is capable of replicating agenda within a reasonable accuracy range.

A comparison of estimation accuracy between the 12 clusters in Figure 4 demonstrates that prediction percentages decline from first state to successive states for all clusters. The decline is most noticeable in more complex activity sequences, suggesting that for these patterns the algorithm needs more sophisticated structure. Further work will include improving the estimation accuracy of the RF models by adding an optimization tool such as Bayesian optimization to tune the hyperparameters in the split selection phase.

The confusion matrix for the best-performing RF model, RF_CART_I, is shown in Table 2. The confusion matrix determines the ability of the RF model to replicate the existence of each activity type in the dataset. We computed the confusion matrix for all 12 clusters. At each state, the observed activity of the individual is compared with the individual's activity in the estimation, regardless of whether the individual's prior or following activity was replicated properly. The estimation accuracy in each cluster is calculated by the sum of crosswise cells of the confusion matrixes. According to the results presented in Table 2, the highest estimation accuracy is obtained for clusters 12 and 5 (80%) and the lowest estimation accuracy obtained is for cluster 2 (76%). Overall, the RF_CART_I model estimation accuracy is 77.7% for the entire population.

For each cluster, the agenda episodes and frequency of the observed data and the results of the best-performing RF models (RF_CART_I), including both in-home and out-of-home activity episodes, were examined by constructing a transition matrix. A transition matrix between successive activity episodes illustrates the likelihood that a consecutive episode of a certain class will happen, given an episode of a current class (*29–32*). Furthermore, it presents detailed information on trip chaining patterns of different market sections in a compact way. Table 3 provides the comparison results between observed and replicated patterns through activity episode transitions matrixes for all 12 clusters. The columns in Table 3 signify the class of the subsequent activity episode (difference between observed and replicated patterns), while the rows in transition matrix signify the class of the current activity episode (difference between observed and simulated patterns). According to the results presented in Table 3, the mean absolute error (MAE) for all 12 clusters is 7.26%, signifying that RF models could successfully replicate episodes and position in agenda.

In order to test for similarity of activity sequence between observed and replicated patterns, the order of activities was compared using sequential alignment methods (*32*). In this comparison, the activities are assumed to be independent of start times or activity duration. The distance between activity sequences is computed as the number of phases needed to align two orders of activities. The smaller distance between strings designates higher similarity. Figure 5 illustrates the distribution patterns of edit distance between observed and replicated activity sequences (outcomes of the RF_CART_I) for all 12 clusters, where the dashed line indicates the mean distance. In general, RF_CART_I successfully replicated activity sequences of more than 70% of the population in each cluster, with the mean distance equal to 0.47.

Our empirical evidence from modeling 12 unique clusters shows that performance of the RF models is meaningfully correlated with the sample size used in the training model. This is most evident in cases comprising complex patterns. Further studies to overcome this limitation associated with the cluster data with small sample size are planned, in particular employing the population synthesis technique.

Conclusion

Daily activity pattern models play an essential role in many activity-based travel demand models. Individuals' travel demands originate from the need to participate in certain activities, and accordingly they plan their daily activity sequences based on these needs. The choice of daily activity sequences differs between population groups with varying socio-demographic characteristics. In this study, we developed a new model using RF theory to learn and replicate the activity engagement patterns of population groups. The RF models were trained with two different decision split techniques, CART and curvature search. Each of the proposed methods was tested under two different layer settings. In the first setting, RF_CART_I and RF_CURV_I, the algorithm randomly selects predictor variables from all nominated variables and grows trees, whereas in the second setting, RF_CART_II and RF_CURV_II, the algorithm selects only predictor variables with high importance and grows trees. Various predictor variables, such as socio-demographic characteristics (age, gender, education level, income level, number of vehicles, etc.), health and/or mobility status of the individual, and preceding activities are employed in the training of RF models.

We applied the proposed RF models to 12 unique population clusters. Individuals in each cluster differ significantly from individuals in other clusters in terms of socio-demographic characteristics and daily activity sequences. Our results demonstrate that the RF model with the CART split selection method (RF_CART_I) can replicate the activity sequences and agenda of the entire population with 70.0% and 77.7% accuracy, respectively. We considered nine activity categories (three in-home and six out-of-home) and modeled them for a 24-hour period. Comparison of estimated activity sequences for the 12 unique clusters demonstrated that RF model performance is associated with level of complexity patterns. In situations involving complex patterns, the algorithm requires more sophisticated structure in terms of predictor and split selections. Relative to other alternative machine learning algorithms, such as SVM and BPNN, the RF model is less likely to overfit. This is due to the generalization error that converges to a particular amount when the forest size gradually increases. It can be used to assess the importance of attributes, which is suitable to analyze the features of the studied system. Furthermore, fewer parameters require calibration in the

Table 2. Confusion Matrixes for Random Forest Model (RF_CART_I)

#1	H	L	N	W	P	S	G	E	T	Accuracy #1 = 77.3
H	.267	.009	.008	.022	.002	.000	.002	.002	.000	
L	.016	.065	.002	.014	.000	.000	.002	.002	.000	
N	.010	.002	.145	.006	.001	.000	.000	.000	.001	
W	.025	.005	.001	.166	.003	.000	.007	.000	.000	
P	.011	.001	.002	.007	.040	.000	.000	.002	.000	
S	.000	.000	.000	.000	.000	.000	.000	.000	.000	
G	.010	.000	.002	.012	.002	.000	.048	.000	.000	
E	.012	.000	.000	.006	.002	.000	.002	.030	.000	
T	.006	.001	.001	.006	.001	.000	.000	.001	.011	

#2	H	L	N	W	P	S	G	E	T	Accuracy #2 = 76.0
H	.309	.018	.004	.000	.016	.000	.004	.000	.002	
L	.037	.113	.001	.000	.007	.000	.003	.000	.001	
N	.010	.004	.103	.000	.001	.000	.000	.000	.000	
W	.003	.001	.000	.009	.001	.000	.000	.000	.000	
P	.033	.011	.000	.000	.112	.000	.004	.002	.001	
S	.001	.000	.000	.000	.001	.000	.000	.000	.000	
G	.019	.007	.000	.000	.008	.000	.059	.000	.000	
E	.014	.002	.000	.000	.005	.000	.001	.028	.000	
T	.010	.004	.000	.000	.002	.000	.001	.000	.025	

#3	H	L	N	W	P	S	G	E	T	Accuracy #3 = 77.8
H	.278	.014	.004	.018	.003	.000	.001	.001	.000	
L	.028	.087	.001	.008	.001	.000	.001	.001	.001	
N	.013	.004	.122	.001	.000	.000	.000	.000	.000	
W	.022	.003	.000	.150	.002	.000	.002	.003	.001	
P	.013	.004	.000	.009	.045	.000	.001	.000	.000	
S	.000	.000	.000	.001	.000	.000	.000	.000	.000	
G	.010	.003	.000	.008	.001	.000	.040	.001	.000	
E	.012	.003	.001	.005	.001	.000	.000	.034	.001	
T	.007	.003	.000	.004	.000	.000	.001	.000	.022	

#4	H	L	N	W	P	S	G	E	T	Accuracy #4 = 77.6
H	.342	.011	.002	.001	.011	.000	.003	.002	.001	
L	.032	.112	.001	.000	.006	.000	.003	.002	.001	
N	.008	.002	.108	.000	.001	.000	.001	.002	.000	
W	.005	.001	.000	.013	.001	.000	.000	.000	.000	
P	.031	.007	.001	.000	.082	.000	.002	.001	.000	
S	.002	.001	.000	.000	.000	.001	.000	.000	.000	
G	.024	.006	.000	.000	.002	.000	.059	.000	.001	
E	.021	.002	.000	.000	.004	.000	.000	.039	.000	
T	.009	.004	.000	.000	.002	.000	.000	.001	.021	

#5	H	L	N	W	P	S	G	E	T	Accuracy #5 = 79.5
H	.390	.027	.003	.000	.003	.000	.001	.000	.001	
L	.059	.167	.002	.000	.005	.000	.001	.000	.001	
N	.015	.006	.123	.000	.000	.000	.000	.000	.000	
W	.002	.001	.001	.005	.000	.000	.000	.000	.000	
P	.024	.008	.001	.000	.055	.000	.001	.000	.000	
S	.001	.000	.000	.000	.000	.000	.000	.000	.000	
G	.010	.005	.001	.000	.000	.000	.024	.000	.000	
E	.004	.002	.000	.000	.000	.000	.000	.006	.000	
T	.009	.007	.000	.000	.001	.000	.000	.000	.024	

#6	H	L	N	W	P	S	G	E	T	Accuracy #6 = 77.4
H	.311	.019	.003	.013	.004	.000	.001	.000	.001	
L	.035	.100	.001	.007	.001	.000	.001	.000	.000	
N	.013	.003	.118	.000	.000	.000	.000	.000	.000	
W	.021	.004	.000	.111	.001	.000	.002	.002	.000	
P	.021	.003	.000	.007	.057	.000	.000	.000	.000	
S	.000	.000	.000	.000	.000	.000	.000	.000	.000	
G	.012	.001	.000	.010	.001	.000	.041	.000	.000	
E	.011	.002	.000	.003	.001	.000	.001	.021	.000	
T	.008	.001	.000	.000	.000	.000	.000	.000	.013	

#7	H	L	N	W	P	S	G	E	T	Accuracy #7 = 77.9
H	.289	.023	.006	.017	.002	.000	.001	.000	.000	
L	.032	.106	.002	.010	.002	.000	.000	.000	.001	
N	.014	.003	.124	.001	.001	.000	.000	.000	.000	
W	.023	.005	.001	.143	.001	.000	.003	.003	.000	
P	.009	.006	.000	.003	.040	.000	.000	.000	.000	
S	.000	.000	.000	.001	.000	.000	.000	.000	.000	
G	.007	.002	.000	.009	.001	.000	.033	.000	.000	
E	.009	.002	.000	.006	.000	.000	.000	.028	.000	
T	.007	.001	.000	.003	.000	.000	.000	.000	.013	

#8	H	L	N	W	P	S	G	E	T	Accuracy #8 = 77.6
H	.339	.027	.001	.001	.011	.000	.003	.000	.002	
L	.049	.143	.002	.000	.007	.000	.001	.000	.001	
N	.010	.004	.105	.000	.001	.000	.000	.000	.000	
W	.005	.001	.000	.011	.001	.000	.000	.000	.000	
P	.028	.010	.000	.000	.092	.000	.002	.001	.001	
S	.001	.000	.000	.000	.000	.003	.000	.000	.000	
G	.015	.007	.000	.000	.003	.000	.047	.000	.000	
E	.007	.002	.000	.000	.001	.000	.000	.010	.000	
T	.009	.005	.000	.000	.002	.000	.001	.000	.026	

#9	H	L	N	W	P	S	G	E	T	Accuracy #9 = 76.8
H	.320	.024	.003	.001	.019	.000	.001	.000	.001	
L	.044	.144	.001	.000	.010	.000	.001	.000	.000	
N	.014	.005	.109	.000	.001	.000	.000	.000	.000	
W	.003	.002	.000	.013	.002	.000	.000	.000	.000	
P	.034	.012	.000	.000	.111	.000	.001	.000	.000	
S	.001	.000	.000	.000	.000	.001	.000	.000	.000	
G	.012	.007	.001	.000	.006	.000	.040	.000	.000	
E	.005	.002	.000	.000	.004	.000	.000	.014	.000	
T	.008	.003	.000	.000	.003	.000	.000	.000	.017	

#10	H	L	N	W	P	S	G	E	T	Accuracy #10 = 76.6
H	.307	.017	.002	.021	.002	.000	.000	.000	.000	
L	.023	.117	.002	.002	.006	.000	.000	.000	.000	
N	.004	.010	.134	.006	.000	.000	.000	.000	.002	
W	.025	.004	.006	.096	.004	.000	.000	.002	.000	
P	.013	.006	.002	.006	.056	.000	.000	.002	.000	
S	.000	.002	.000	.002	.000	.000	.000	.000	.000	
G	.008	.006	.000	.010	.000	.000	.029	.000	.000	
E	.006	.002	.000	.008	.002	.000	.000	.015	.000	
T	.008	.002	.000	.004	.002	.000	.000	.000	.013	

#11	H	L	N	W	P	S	G	E	T	Accuracy #11 =78.1
H	.290	.012	.002	.019	.003	.000	.001	.000	.001	
L	.029	.086	.000	.008	.002	.000	.001	.000	.001	
N	.013	.004	.125	.002	.000	.000	.000	.000	.000	
W	.023	.006	.001	.146	.002	.000	.002	.001	.000	
P	.016	.003	.001	.006	.046	.000	.001	.001	.000	
S	.001	.000	.000	.000	.000	.000	.000	.000	.000	
G	.012	.002	.001	.008	.000	.000	.039	.000	.001	
E	.009	.002	.000	.006	.001	.000	.001	.027	.000	
T	.006	.002	.000	.004	.001	.000	.001	.000	.022	

#12	H	L	N	W	P	S	G	E	T	Accuracy #12 = 79.4
H	.318	.014	.002	.000	.004	.016	.002	.005	.004	
L	.016	.095	.002	.000	.000	.004	.002	.002	.000	
N	.009	.009	.113	.000	.000	.000	.000	.000	.000	
W	.004	.002	.000	.011	.000	.004	.004	.000	.000	
P	.011	.005	.000	.000	.034	.004	.002	.000	.000	
S	.018	.000	.000	.000	.000	.122	.002	.000	.005	
G	.016	.004	.000	.000	.000	.007	.050	.000	.004	
E	.002	.002	.000	.000	.000	.005	.004	.023	.000	
T	.005	.000	.000	.000	.000	.005	.002	.002	.029	

Note: H = home chores; L = home leisure; N = night sleep; W = workplace; P = shopping & services; S = school/college; G = organizational/hobbies; E = entertainment; T = sports.
Shaded areas represent mean proportional error; #1, #2, #3, #4, #5, #6, #7, #8, #9, #10, #11, #12: cluster number.

Table 3. Activity Episode Transitions Matrix: Comparison between Observed and Replicated Patterns (in %)

#1	H	L	N	W	P	S	G	E	T
H	0.0	4.0	7.8	35.2	4.6	0.0	4.2	2.1	4.6
L	26.6	0.0	7.8	11.7	9.0	0.0	7.0	5.3	1.3
N	11.5	0.5	0.0	5.8	1.6	0.0	2.1	1.6	0.1
W	22.1	4.6	13.7	0.0	2.7	0.0	7.2	23.2	7.4
P	15.7	12.3	16.2	35.5	0.0	0.0	1.3	7.4	0.0
S	0.0	0.0	0.0	0.0	0.0	0.0	0.0	0.0	0.0
G	11.2	0.6	18.2	22.3	7.0	0.0	0.0	0.4	0.2
E	24.4	11.1	17.4	52.7	1.4	0.0	1.0	0.0	2.6
T	7.4	7.4	13.3	18.5	0.0	0.0	9.6	13.3	0.0

MAE cluster 1 = 7.8

#2	H	L	N	W	P	S	G	E	T
H	0.0	5.3	2.8	0.2	23.4	0.0	5.1	3.8	0.0
L	25.3	0.0	4.1	1.0	16.0	0.3	8.4	3.6	0.1
N	0.1	3.1	0.0	0.0	2.4	0.0	0.4	0.0	0.4
W	33.3	20.8	20.8	0.0	16.7	0.0	13.9	11.1	0.0
P	8.0	5.3	8.3	16.7	0.0	2.9	11.1	21.3	0.4
S	0.0	0.0	0.0	0.0	0.0	0.0	0.0	0.0	0.0
G	8.7	2.4	3.2	0.4	14.0	0.0	0.0	1.6	8.0
E	30.1	11.3	3.8	4.9	9.5	0.0	0.2	0.0	30.8
T	8.4	10.8	0.0	0.0	18.3	0.0	6.7	7.4	0.0

MAE cluster 2 = 6.6

#3	H	L	N	W	P	S	G	E	T
H	0.0	23.4	4.8	1.8	21.2	0.1	5.1	4.5	0.9
L	18.4	0.0	9.8	1.0	10.6	0.0	9.9	5.2	1.4
N	1.2	2.3	0.0	0.0	1.5	0.0	1.4	0.4	0.3
W	8.1	7.6	17.5	0.0	1.6	0.0	14.6	10.0	2.1
P	18.1	3.1	11.8	16.8	0.0	0.0	10.6	3.7	1.9
S	5.0	0.0	0.0	0.0	0.0	0.0	0.0	5.0	0.0
G	20.8	5.6	20.2	1.1	5.0	1.4	0.0	3.6	4.0
E	20.9	12.4	17.7	7.0	4.7	0.0	8.9	0.0	30.3
T	18.1	1.8	7.6	1.5	16.5	0.0	4.8	10.6	0.0

MAE cluster 3 = 6.8

#4	H	L	N	W	P	S	G	E	T
H	0.0	8.9	7.6	32.6	6.5	0.0	5.7	3.5	1.9
L	25.4	0.0	1.1	8.9	7.6	0.0	3.5	4.5	0.2
N	7.6	3.9	0.0	0.6	1.8	0.0	0.0	0.8	0.5
W	29.3	10.9	5.4	0.0	2.6	0.0	7.8	25.1	10.0
P	30.1	8.3	16.7	28.0	0.0	4.7	8.4	12.1	1.9
S	0.0	0.0	0.0	0.0	0.0	0.0	0.0	0.0	0.0
G	9.4	2.7	9.9	34.3	5.2	0.0	0.0	4.4	2.7
E	29.9	14.7	6.1	56.1	5.3	0.6	4.6	0.0	5.2
T	21.5	6.1	13.0	11.7	4.0	0.0	8.0	2.1	0.0

MAE cluster 4 = 8.2

#5	H	L	N	W	P	S	G	E	T
H	0.0	18.4	4.6	0.6	11.4	0.0	3.3	1.3	2.0
L	14.9	0.0	4.2	0.3	7.9	0.0	1.7	0.6	0.2
N	3.0	1.7	0.0	0.9	1.3	0.0	0.2	0.0	0.7
W	17.6	7.3	19.2	0.0	2.0	0.0	5.4	2.7	3.8
P	9.3	20.0	12.1	17.2	0.0	0.0	3.7	14.9	5.6
S	0.0	0.0	0.0	0.0	0.0	0.0	0.0	0.0	0.0
G	5.8	1.5	3.6	2.8	5.2	0.0	0.0	0.8	0.5
E	25.9	20.0	10.0	15.4	31.8	0.0	6.7	0.0	15.4
T	22.9	3.7	12.8	0.8	5.4	0.0	0.7	0.5	0.0

MAE cluster 5 = 5.7

#6	H	L	N	W	P	S	G	E	T
H	0.0	5.2	1.2	24.4	7.4	0.0	5.6	2.3	3.3
L	21.9	0.0	1.0	6.4	6.3	0.0	5.8	3.4	1.0
N	7.5	3.3	0.0	1.0	1.5	0.0	2.0	0.0	0.3
W	11.3	7.1	2.4	0.0	1.3	0.0	2.3	18.4	3.4
P	24.4	5.3	12.4	19.3	0.0	0.0	16.6	0.6	6.8
S	0.0	0.0	0.0	0.0	0.0	0.0	0.0	0.0	0.0
G	14.1	5.1	11.3	4.8	12.4	2.9	0.0	6.3	0.7
E	26.2	12.6	0.0	53.9	5.1	0.0	8.0	0.0	2.0
T	9.9	0.3	22.2	5.7	4.8	0.0	3.7	3.7	0.0

MAE cluster 6 = 6.4

#7	H	L	N	W	P	S	G	E	T
H	0.0	4.9	1.6	26.7	5.3	0.0	4.2	3.1	0.6
L	21.8	0.0	0.2	10.0	5.4	0.0	4.0	2.7	0.0
N	9.2	5.0	0.0	1.9	0.7	0.0	0.9	0.4	0.3
W	21.0	7.0	6.6	0.0	2.8	0.7	2.0	26.0	8.7
P	6.4	27.2	9.7	21.0	0.0	1.0	2.8	0.9	6.7
S	0.0	0.0	0.0	100.0	0.0	0.0	0.0	0.0	0.0
G	8.3	7.5	7.9	24.6	1.6	0.0	0.0	2.5	3.1
E	26.4	22.0	12.5	9.9	6.9	0.0	0.9	0.0	1.2
T	6.7	4.2	9.1	13.5	4.6	0.0	0.8	9.9	0.0

MAE cluster 7 = 7.6

#8	H	L	N	W	P	S	G	E	T
H	0.0	32.1	3.4	1.6	20.8	0.4	4.6	2.0	0.6
L	21.3	0.0	2.7	0.1	12.4	0.2	5.6	1.8	1.3
N	3.1	0.0	0.0	0.7	0.7	0.0	0.7	0.4	0.6
W	8.5	0.9	2.8	0.0	14.8	0.0	7.5	5.3	2.8
P	29.6	19.7	6.3	20.4	0.0	0.0	11.7	20.1	3.4
S	20.8	12.5	0.0	0.0	41.7	0.0	0.0	33.3	0.0
G	17.6	3.9	6.3	7.4	11.8	1.2	0.0	1.2	5.3
E	23.9	7.6	0.0	18.5	10.9	0.0	5.5	0.0	18.5
T	5.0	7.1	9.7	3.2	17.3	1.1	10.8	0.0	0.0

MAE cluster 8 = 8.0

#9	H	L	N	W	P	S	G	E	T
H	0.0	31.4	1.4	0.2	29.4	0.2	4.8	1.6	3.0
L	20.3	0.0	0.8	1.0	15.1	0.0	3.2	1.3	0.4
N	0.3	3.0	0.0	0.4	2.7	0.0	0.4	0.7	0.0
W	9.7	4.2	4.4	0.0	6.5	0.0	3.8	0.8	0.8
P	34.9	15.3	10.6	17.2	0.0	2.1	21.6	11.3	8.7
S	0.0	0.0	0.0	0.0	0.0	0.0	0.0	0.0	0.0
G	19.8	3.6	13.8	0.7	9.2	0.0	0.0	2.3	3.8
E	34.4	10.0	13.3	3.3	3.3	0.0	28.9	0.0	22.2
T	29.0	7.9	7.5	2.2	29.4	0.0	0.2	2.4	0.0

MAE cluster 9 = 7.2

#10	H	L	N	W	P	S	G	E	T
H	0.0	3.9	8.9	32.3	5.5	0.0	3.3	3.1	3.6
L	22.3	0.0	5.3	19.7	8.1	0.0	4.3	1.3	3.1
N	8.0	3.0	0.0	7.6	1.3	0.0	0.6	2.5	1.1
W	31.3	2.3	7.2	0.0	5.5	0.0	5.0	19.7	0.5
P	4.8	0.9	21.7	20.1	0.0	0.0	6.7	19.2	9.6
S	0.0	0.0	0.0	0.0	0.0	0.0	0.0	0.0	0.0
G	17.9	13.0	23.9	42.9	6.0	12.5	0.0	6.5	0.0
E	19.1	0.2	19.4	67.3	22.6	0.0	6.5	0.0	0.0
T	26.3	20.0	12.5	15.0	0.0	0.0	18.8	0.0	0.0

MAE cluster 10 = 9.1

#11	H	L	N	W	P	S	G	E	T
H	0.0	6.4	4.8	31.2	5.5	0.1	4.9	3.2	3.5
L	20.1	0.0	1.3	11.4	5.9	0.0	2.2	2.5	0.5
N	10.0	3.8	0.0	1.6	2.1	0.0	1.1	0.7	0.7
W	24.8	7.1	7.1	0.0	0.2	0.0	3.1	26.9	9.1
P	6.5	13.3	17.6	31.2	0.0	0.0	5.0	3.5	2.3
S	0.0	0.0	0.0	0.0	0.0	0.0	0.0	0.0	0.0
G	1.6	10.2	11.2	27.7	2.8	0.0	0.0	3.9	1.9
E	24.6	14.2	5.6	57.7	3.5	0.0	8.2	0.0	1.5
T	4.2	5.7	10.1	8.2	9.5	0.0	1.8	1.1	0.0

MAE cluster 11 = 6.9

#12	H	L	N	W	P	S	G	E	T
H	0.0	3.5	8.7	2.1	3.1	26.7	6.9	5.5	2.0
L	18.6	0.0	5.7	1.9	3.8	0.8	11.3	1.9	4.6
N	4.9	1.3	0.0	0.0	0.0	0.0	3.6	0.0	0.0
W	16.7	0.0	0.0	0.0	0.0	16.7	0.0	0.0	0.0
P	5.0	10.0	10.0	25.0	0.0	10.0	0.0	5.0	5.0
S	1.5	6.6	1.6	4.3	9.7	0.0	1.8	25.1	5.0
G	5.6	7.5	1.4	4.7	9.1	23.3	0.0	1.6	6.1
E	3.3	11.1	11.1	5.6	5.6	2.2	20.0	0.0	14.4
T	7.3	13.6	18.2	0.0	9.1	13.9	9.1	6.7	0.0

MAE cluster 12 = 6.8

Note: H = home chores; L = home leisure; N = night sleep; W = workplace; P = shopping & services; S = school/college; G = organizational/hobbies; E = entertainment; T = sports. Shaded areas represent mean absolute error; italic values: Main diameter (cross of two same activities); #1, #2, #3, #4, #5, #6, #7, #8, #9, #10, #11, #12: cluster number.

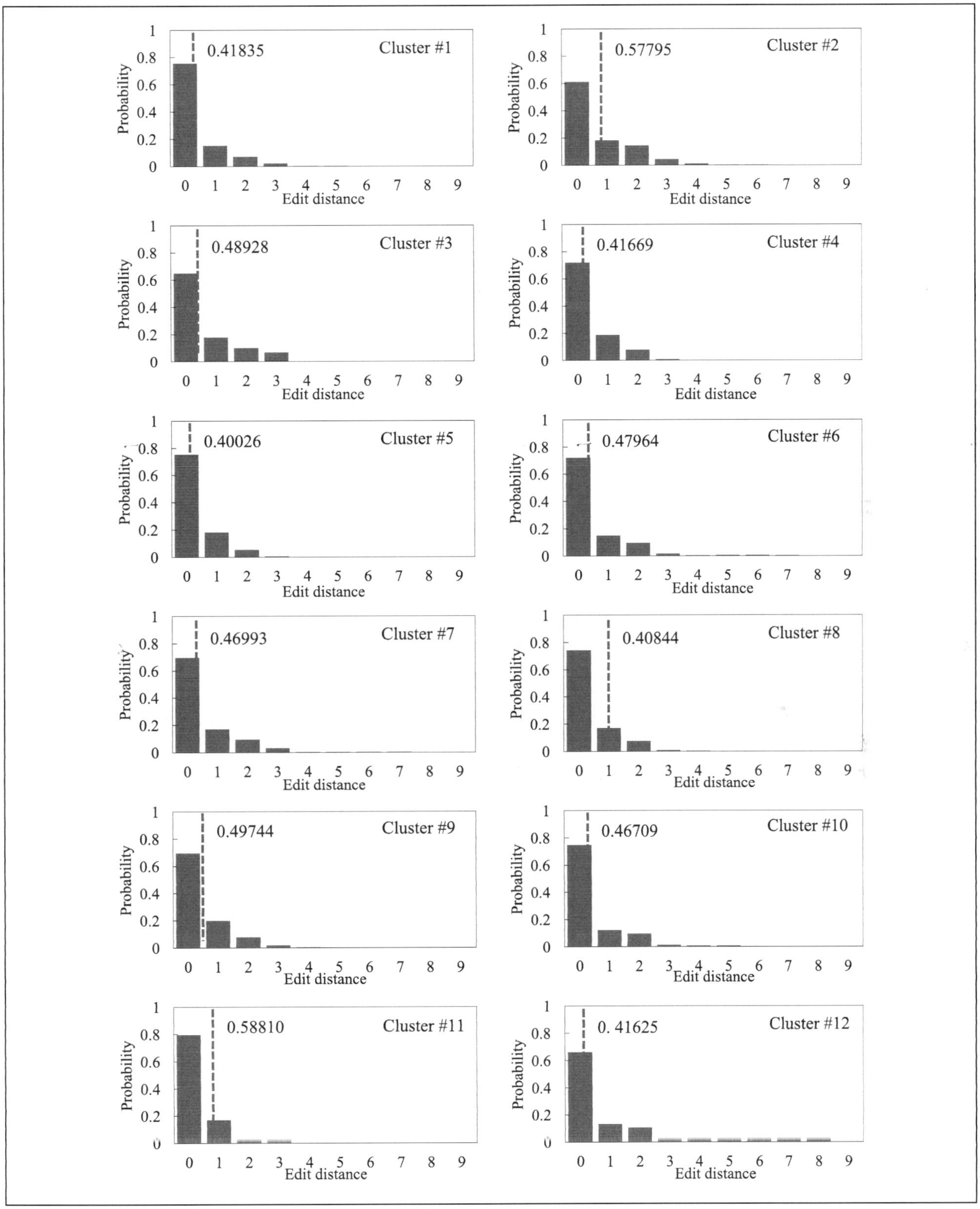

Figure 5. Comparison of edit distance between observed and replicated activity sequences in 12 clusters.

model, and the algorithm may be applied to datasets containing a large number of hidden attributes (*25*).

This study illustrates the utility of the RF machine learning technique to modeling of sequential activity selection, an application not previously employed in travel behavior analysis. The developed RF models can yield the type and frequency of activities in the schedule, and their sequential order, for use in activity-based travel demand modeling. Compared with multinomial logit regression modeling, which yields interpretable coefficients, the RF model is designed as a black box. It is trained to evaluate the labels of the data and its result is a set of support data points and their respective weights. In addition, compared with other artificial intelligence methods such as the SVM, the proposed method in this study is able to handle missing values in the algorithm automatically. Furthermore, variables do not need to be transformed, very few parameters need adjustment, and the algorithm does not overfit easily. The proposed RF model is also very efficient in computational time.

The empirical results from modeling 12 population clusters show that although RF_CART_I provides superior results to other proposed models, its performance is correlated with the sample size used in the training model, and also to heterogeneity and diversity among the predictor variables. Moreover, the model with random selection of predictor variables performed better than including only high importance predictor variables. Further behavioral investigation can be made for interpretation of such a result. Further improvement involves adding an optimization tool to tune the hyperparameters in the split selection step of RF models. In this study we only addressed the activity type and activity sequencing. Various other aspects of daily activity sequences, such as sequential activity location, duration, and mode, can be synthesized using the RF models. Moreover, incorporating a probabilistic model such as a hidden Markov model into the RF model structure can be explored in future extensions of this study. Several studies have illustrated the importance of intra-household interaction, in which decisions of activity participation by one household member are often constrained or correlated to the decisions of other household members. Extensions to our modeling procedure are required to account for such correlations. Finally, the results of this study are expected to be implemented within the activity-based travel demand model, Scheduler for Activities, Locations, and Travel (SALT) for Halifax, Nova Scotia.

Acknowledgments

The authors thank the Natural Sciences and Engineering Research Council (NSERC) for its contribution in supporting the research. The authors also thank the members of the Dalhousie Transportation and Environmental Simulation Studies (TESS) group for their valuable suggestions. The authors also thank the anonymous reviewers for their very useful comments and suggestions which helped to improve the quality of this paper. Data for this research was provided by the Halifax STAR Project, supported through the Atlantic Innovation Fund from the Atlantic Canada Opportunities Agency, Project No. 181930.

Author Contributions

The authors confirm contributions to the paper as follows: study conception and design: Mohammad Hesam Hafezi, Hugh Millward, Lei Liu; data collection: Hugh Millward; analysis and interpretation of results: Mohammad Hesam Hafezi; draft manuscript preparation: Mohammad Hesam Hafezi. All authors reviewed the results and approved the final version of the manuscript.

References

1. Rasouli, S., and H. Timmermans. Activity-Based Models of Travel Demand: Promises, Progress and Prospects. *International Journal of Urban Sciences*, Vol. 18, No. 1, 2014, pp. 31–60. https://doi.org/10.1080/12265934.2013.835118.
2. Hagerstrand, T. What about People in Regional Science? *Papers in Regional Science*, Vol. 24, No. 1, 1970, pp. 7–24. https://doi.org/10.1111/j.1435-5597.1970.tb01464.x.
3. McFadden, D. Econometric Models for Probabilistic Choice among Products. *The Journal of Business*, Vol. 53, No. 3, 1980, pp. 13–29. http://www.jstor.org/stable/2352205.
4. Arentze, T., and H. Timmermans. Conceptual Framework. In *ALBATROSS: A Learning Based Transportation Oriented Simulation System*, European Institute of Retailing and Services Studies (EIRASS), Technische Universiteit Eindhoven, Netherlands, 2000.
5. Kitamura, R., C. Chen, R. M. Pendyala, and R. Narayanan. Micro-Simulation of Daily Activity-Travel Patterns for Travel Demand Forecasting. *Transportation*, Vol. 27, No. 1, 2000, pp. 25–51. https://doi.org/10.1023/A:1005259324588.
6. Auld, J., and A. Mohammadian. Framework for the Development of the Agent-Based Dynamic Activity Planning and Travel Scheduling (ADAPTS) Model. *Transportation Letters*, Vol. 1, No. 3, 2009, pp. 245–255. https://doi.org/10.3328/TL.2009.01.03.245-255.
7. Li, S., and D. H. Lee. Learning Daily Activity Patterns with Probabilistic Grammars. *Transportation*, Vol. 44, No. 1, 2017, pp. 49–68. https://doi.org/10.1007/s11116-015-9622-1.
8. Arentze, T., and H. Timmermans. Parametric Action Decision Trees: Incorporating Continuous Attribute Variables into Rule-Based Models of Discrete Choice. *Transportation Research Part B: Methodological*, Vol. 41, No. 7, 2007, pp. 772–783. https://doi.org/10.1016/j.trb.2007.01.001.
9. Bowman, J. L., and M. E. Ben-Akiva. Activity-Based Disaggregate Travel Demand Model System with Activity Schedules. *Transportation Research Part A: Policy and*

Practice, Vol. 35, No. 1, 2001, pp. 1–28. https://doi.org/10.1016/S0965-8564(99)00043-9.
10. Daisy, N. S., M. H. Hafezi, L. Liu, and H. Millward. Understanding and Modeling the Activity-Travel Behavior of University Commuters at a Large Canadian University. *Journal of Urban Planning and Development*, Vol. 4, No. 2, 2018, p. 04018006. https://doi.org/10.1061/(ASCE)UP.1943-5444.0000442.
11. Allahviranloo, M., and W. Recker. Daily Activity Pattern Recognition by Using Support Vector Machines with Multiple Classes. *Transportation Research Part B: Methodological*, Vol. 58, 2013, pp. 16–43. https://doi.org/10.1016/j.trb.2013.09.008.
12. Bhat, C., J. Guo, S. Srinivasan, and A. Sivakumar. Comprehensive Econometric Microsimulator for Daily Activity-Travel Patterns. *Transportation Research Record: Journal of the Transportation Research Board*, 2004. 1894: 57–66. https://doi.org/10.3141/1894-07.
13. Bhat, C., R. K. G. Goulias, R. M. Pendyala, R. Paleti, R. Sidharthan, L. Schmitt, and H. H. Hu. A Household-Level Activity Pattern Generation Model with an Application for Southern California. *Transportation*, Vol. 40, No. 5, 2013, pp. 1063–1086. https://doi.org/10.1007/s11116-013-9452-y.
14. Leszczyc, P. T. L. P., and H. Timmermans. Unconditional and Conditional Competing Risk Models of Activity Duration and Activity Sequencing Decisions: An Empirical Comparison. *Journal of Geographical Systems*, Vol. 4, No. 2, 2002, pp. 157–170. https://doi.org/10.1007/s1010902 00083.
15. Ho, C., and C. Mulley. Tour-Based Mode Choice of Joint Household Travel Patterns on Weekend and Weekday. *Transportation*, Vol. 40, No. 4, 2013, pp. 789–811. https://doi.org/10.1007/s11116-013-9479-0.
16. Ben-Akiva, M. E., and J. L. Bowman. Activity Based Travel Demand Model Systems. In *Equilibrium and Advanced Transportation Modeling*, Springer, Boston, Mass., 1998, pp. 27–46. https://doi.org/10.1007/978-1-4615-5757-9_2.
17. Hafezi, M. H., L. Liu, and H. Millward. Modeling Activity Scheduling Behavior of Travelers for Activity-Based Travel Demand Model. Presented at 97th Annual Meeting of the Transportation Research Board, Washington, D.C., 2018.
18. Shafique, M. A., and E. Hato. Use of Acceleration Data for Transportation Mode Prediction. *Transportation*, Vol. 42, No. 1, 2015, pp. 163–188. https://doi.org/10.1007/s11116-014-9541-6.
19. You, J., J. Wang, and J. Guo. Real-Time Crash Prediction on Freeways using Data Mining and Emerging Techniques. *Journal of Modern Transportation*, Vol. 25, No. 2, 2017, pp. 116–123. https://doi.org/10.1007/s40534-017-0129-7.
20. Spinney, J. E., and H. Millward. Weather Impacts on Leisure Activities in Halifax, Nova Scotia. *International Journal of Biometeorology*, Vol. 55, No. 2, 2011, pp. 133–145. https://doi.org/10.1007/s00484-010-0319-z.
21. Millward, H., and J. Spinney. Time Use, Travel Behavior, and the Rural-Urban Continuum: Results from the Halifax STAR Project. *Journal of Transport Geography*, Vol. 19, No. 1, 2011, pp. 51–58. https://doi.org/10.1016/j.jtrangeo.2009.12.005.
22. TURP. *Halifax Regional Space Time Activity Research (STAR) Survey: A GPS-Assisted Household Time-Use Survey, Survey Methods*. Saint Mary's University, Halifax, 2008.
23. Hafezi, M. H., L. Liu, and H. Millward. Identification of Representative Patterns of Time Use Activity through Fuzzy C-Means Clustering. *Transportation Research Record: Journal of the Transportation Research Board*, 2017. 2668: 38–50.
24. Hafezi, M. H., L. Liu, and H. Millward. A Time-Use Activity-Pattern Recognition Model for Activity-Based Travel Demand Modeling. *Transportation*, New York, US, 2017, pp. 1–26. https://doi.org/10.1007/s11116-017-9840-9
25. Breiman, L. Random Forests. *Machine Learning*, Vol. 45, No. 1, 2001, pp. 5–32. https://doi.org/10.1023/A:1010933404324.
26. Suthaharan, S. *Machine Learning Models and Algorithms for Big Data Classification: Thinking with Examples for Effective Learning*. Springer. New York, US, 2015. https://doi.org/10.1007/978-1-4899-7641-3
27. Biau, G., and E. Scornet. A Random Forest Guided Tour. *TEST*, Vol. 25, No. 2, 2016, pp. 197–227.
28. Kitamura, R., C. Chen, and R. Pendyala. Generation of Synthetic Daily Activity-Travel Patterns. *Transportation Research Record: Journal of the Transportation Research Board*, 1997. 1607: 154–163.
29. Lockwood, A., S. Srinivasan, and C. Bhat. Exploratory Analysis of Weekend Activity Patterns in the San Francisco Bay Area, California. *Transportation Research Record: Journal of the Transportation Research Board*, 2005. 1926: 70–78.
30. Hafezi, M. H., N. S. Daisy, L. Liu, and H. Millward. Daily Activity and Travel Sequences of Students, Faculty, and Staff at a Large Canadian University. *Transportation Planning and Technology*, 2018. https://doi.org/10.1080/03081060.2018.1469286.
31. Daisy, N. S., H. Millward, M. H. Hafezi, and L. Liu. Trip-Chaining and Tour Complexity: Contrasts between Worker and Non-Worker Groups in Halifax, Nova Scotia. *Presented at the 28th Atlantic Division of the Canadian Association of Geographers (ACAG/ACGA) Conference*, Halifax, Canada. 2017.
32. Daisy, N. S., H. Millward, and L. Liu. Trip Chaining and Tour Mode Choice of Non-Workers Grouped by Daily Activity Patterns. *Journal of Transport Geography*, Vol. 69, 2018, pp. 150–162. https://doi.org/10.1016/j.jtrangeo.2018.04.016.
33. Needleman, S. B., and C. D. Wunsch. A General Method Applicable to the Search for Similarities in the Amino Acid Sequence of Two Proteins. *Journal of Molecular Biology*, Vol. 48, No. 3, 1970, pp. 443–453. https://doi.org/10.1016/0022-2836(70)90057-4.

The Standing Committee on Traveler Behavior and Values (ADB10) peer-reviewed this paper (18-05702).

Article

TRR
JOURNAL OF THE TRANSPORTATION RESEARCH BOARD

Transportation Research Record
2018, Vol. 2672(47) 208–218

DOI: 10.1177/0361198118798478
journals.sagepub.com/home/trr
SAGE

Exploring the Self-Reported Long-Distance Travel Frequency of Millennials and Generation X in California

Rosaria M. Berliner[1], Lisa Aultman-Hall[2], and Giovanni Circella[1,3]

Abstract

Long-distance travel research is limited because of the lack of robust data and the complexity of defining a long-distance trip. The patterns of infrequent long-distance trips are poorly understood especially compared with the better studied (and understood) local daily travel patterns. This study contributes to filling that gap by investigating the factors that affect the frequency of long-distance trips of Californian millennials (18–34 years old, in 2015) and members of the preceding Generation X (35–50 years old, in 2015). Data collected with an online survey administered in fall 2015 are used to study the mobility of these age groups. The survey collected information on several travel-related variables, including the number of long-distance trips (defined as trips longer than 100 miles, one way) made by various modes during the previous 12 months. The authors estimate six negative binomial regression models of long-distance travel separated by purpose (*business* or *leisure*) and mode (*overall travel* versus *air*). The study explores the relationship of long-distance trip formation with several sociodemographic, land use, and attitudinal variables. Consistent with expectations, individual income positively affects the number of long-distance trips made by each individual. Among the attitudinal variables, the individuals who are adventurers, have higher "variety seeking" attitudes and are more interested in adopting new technologies are found to make a larger number of long-distance trips. However, those who prefer to shop in brick-and-mortar stores rather than online are found to have lower levels of long-distance travel.

The data and models used for travel demand forecasting in the U.S. have typically focused on the daily, more routine, trips or tours made by individuals in their home regions. This focus is consistent with metropolitan planning organizations' (MPOs') planning interests, as they conduct planning in large urban areas where congestion management has dominated the infrastructure investment priorities since the 1950s. Although many states have statewide travel demand models that simulate all components of travel demand (including long-distance travel), limited data are available on long-distance travel to calibrate these models. For example, the 2009 and 2017 National Household Travel Surveys (NHTS) did not collect information on long-distance or intercity travel within the main surveys. In 2017, six "add-on" agencies collected limited long-distance data using their extra questions, and asked respondents to report the number of trips over 50 or 75 miles that they had taken in the previous 8 weeks (*1*). To date, only a limited number of states—Utah, Ohio, Michigan, and California—have conducted dedicated long-distance travel data collection (*2*).

Many recent reports, such as Circella et al, have addressed current changes in travel behavior, including the leveling of vehicle miles of travel (VMT) per person through the late 2000s (*3*). However, few of these evaluations explore the relationship between daily travel and the infrequent, but much longer, trips between regions, many of which include overnight stays. Within this landscape of limited data, indications are that long-distance travel is increasing. In 2016, more than 3.2 trillion miles were driven on U.S. roadways. In December 2016, drivers put more than 263.6 billion miles on roadways, which represents an increase of 0.5% from the previous year (*4*). In April 2017, air travel load factors crept up to 84%, and enplanements hit 70 million (*5, 6*). Highway

[1]Institute of Transportation Studies, University of California, Davis, CA
[2]School of Engineering, University of Vermont, Burlington, VT
[3]School of Civil and Environmental Engineering, Georgia Institute of Technology, Atlanta, GA

Corresponding Author:
Address correspondence to Rosaria M. Berliner: aberliner@ucdavis.edu

and interstate congestion and the need for more targeted public planning and investments in airports, intercity rail, and other modes necessitate that researchers improve their understanding of the factors influencing long-distance travel.

This paper exploits an existing data set collected with an online survey of Californians in which questions were added about self-reported annual long-distance travel to answer the following research questions:

1. What are the characteristics of long-distance travel of Californian millennials and Generation Xers?
2. What is the association of sociodemographic, geographic, and attitudinal factors with the frequency of long-distance trips?
3. What is the relationship between the attributes of one's local daily travel and one's long-distance travel?

The remainder of the paper is organized as follows. After a summary of the literature describing the factors that are associated with one's level of long-distance travel, the authors describe the survey data source and the attitudinal variables used in the study. A set of six negative binomial regression models is then presented, and the authors discuss the factors that are found to affect the self-reported long-distance trip rates. Finally, the conclusions of the paper present some summary remarks and implications for future research, focusing on the need for additional data to better understand long-distance travel.

Literature Review

Long-distance travel patterns have received limited attention by researchers and modelers. Challenges plague researchers, from defining what constitutes a long-distance trip to the availability of robust data. Thus, the gaps in knowledge are numerous, especially compared with the better-understood local daily travel patterns. There is no standard long-distance trip length definition. Further, long-distance trips are sometimes defined temporally and sometimes spatially. The American Community Survey defines a "long commute" as 60 min or longer in one direction (*7*). Other studies define long-distance travel spatially where one-way distances are longer than 50 or 100 miles (*8–11*). Additionally, some studies take a hybrid temporal-spatial approach that focuses on trip distance and trip duration to define long-distance trips that include an "overnight stay" (*8*, *12*). A larger number of long-distance travel studies have been conducted in Europe where long-distance data sets are more common and the smaller physical size of countries allows the definition to be based on whether a destination is out of country. In this study, a long-distance trip is defined as a one-way trip of 100 miles or longer; therefore, focus is given to previous studies that define long-distance travel in a similar context. Attention is paid to studies that focus on the factors related to long-distance travel (*11–14*). Furthermore, interest is given to studies that separate leisure from business travel as our survey did (*8*, *11*, *12*, *14–17*).

Of great interest to this research are the factors that affect long-distance travel. Many studies rely on nationally available data and trip diaries—to date, no study uses attitudinal factors to estimate long-distance travel. In a study of long-distance trips made by residents of the UK and the Netherlands in 1998, Limtanakool et al used binary logit models, showing that men and individuals with higher income make more long-distance trips by private car and train (*13*). They also observed that women are less likely to engage in medium- and long-distance commuting (*13*). Similarly, Aultman-Hall et al used data from a longitudinal panel of 628 individuals from the U.S. and Canada. These individuals reported information on their long-distance overnight trips monthly with a series of online surveys for approximately 1 year between February 2013 and February 2014 (*8*). Aultman-Hall et al found that men make more long-distance work trips and air trips annually than women. However, a different study involving the analysis of 6.4 million flight bookings in 2014 found that women make more long-distance airplane business trips than men (*18*). Using data from the 1995 American Travel Survey, Georggi and Pendyala estimated linear regression models of long-distance trip generation, segmented by household income per person, which showed that individuals with higher income made more business trips and overall a higher number of trips per year (*11*). Using data from random-day trip diaries and long-distance trips questionnaires from the 2008 NHTS "Mobility in Germany" (MID), Holz-Rau et al estimated Heckman models and ordinary least squares regression models that found that residents in low-density neighborhoods make fewer long-distance trips than those living in high-density neighborhoods (*12*). Using an ordered probit to model the frequencies of long-distance business and long-distance leisure trips on data from self-reported retrospective surveys collected from about 1,200 respondents from the U.S. in 2013, LaMondia et al observed that education and income increased most types of long-distance travel but age generally decreased long-distance trip frequency (*14*).

Trip type is also an important factor in long-distance travel: how do business travel and personal travel differ? In some cases, respondents have difficulty separating the two trip types. Aultman-Hall et al observed that in their data set 14% of tours had mixed purposes (*8*). Moreover,

16% of total tours were spatially complex "hub and spoke," and an additional 4.5% were "circular chains"; meaning that trips are complex with reference to purpose and space. Using negative binomial regressions, Aultman-Hall et al estimated six models for long-distance travel: all long-distance travel; work travel for full-time workers; work travel for those who traveled overnight once during the survey year; personal travel; air travel; and personal air travel (*8*). For the air, work, and leisure travel, age and household income over $100,000 positively affected the number of tours (*8*). Additionally, longer daily commutes indicate a higher number of personal long-distance tours but this is insignificant in determining work-related tours (*8*). In another study, Georggi and Pendyala observed that the frequency of long-distance trip purposes were different across income and age groups—high income households made more business trips but fewer recreational trips than low income households, and younger age groups made more recreation and business trips than those in older age groups (*11*).

In summary, prior literature has generally found that higher income and traveling for work increase overall levels of long-distance tours. Some studies find gender and age are important. Long-distance travel is complex, requiring more data and studies to better understand the phenomena.

Data and Methods

Data Description

This paper stems from a detailed study of young adult mobility in California (*19*). As part of this project, a detailed cross-sectional online survey was designed and administered to a sample of more than 1,900 California residents between the ages of 18 and 50 years, recruited through an online opinion panel. The survey was administered between September and December 2015. The survey collected information on personal attitudes and preferences toward travel, technology, social media, the environment, life satisfaction, land use, and so forth, as well as information on lifestyles, residential location and travel patterns, adoption of shared mobility, current travel behavior, past travel, driver's licensing, level of vehicle ownership, sociodemographic traits, and several additional factors that may affect respondents' mobility and vehicle ownership. The final data set available for the project includes 1,975 valid cases after removing cases with severely incomplete or inconsistent information. For additional information about the project, the survey content, and the data collection process, see Circella et al (*20*).

The members of the research team geocoded the data using the U.S. Census Master Address File/TIGER version 2015 (*21*), the ESRI database (*22*), and information available from Google (*23*). Using the self-reported information (addresses) on the respondents' residential and work locations, the research team was able to geocode almost all observations in the sample with a good level of precision (block group or even individual block for most respondents). Once the respondents' home and work locations were geocoded, neighborhood types were assigned to each respondent's home and work or school (if applicable) location, relying on the classifications from Salon (*24*). Moreover, additional land use data from external sources, such as the U.S. Environmental Protections Agency's Smart Location data (*25*), were imported, based on the geocoded residential location. The additional data allowed the authors to control for land use characteristics and population densities in the place of residence.

Table 1 summarizes selected descriptive statistics of the final sample used in this paper, which was reduced to 1,512 for this modeling effort based on disqualification metrics described in detail in the following section. The average person in the final data set is female, employed, almost 34 years old, has a bachelor's degree, and lives in a household with an annual income of $76,879. However, as evident in Table 1, the sample contains a broad range of income and education levels for adults aged 18–50 years old. Most respondents were workers although just over a quarter of the sample did not work.

Data Cleaning/Disqualification Metrics

This study tries to gain a deeper understanding of long-distance travel among California millennials. As part of this, the authors looked at self-reported round-, long-distance trips by mode, described as one-way trips of over 100 mi, in the last 12 months. Although many of the long-distance trips reported fell within a reasonable number of trips, based on prior studies, it was found that several respondents indicated an excessively large number of long-distance trips (*8*). Moreover, a significant number of people indicated no trips compared with existing databases, suggesting that respondents skipped these questions that were near the end of the survey because of burden.

To address these issues, the team developed disqualification metrics that were applied to clean the sample. In general, observations with more than 100 long-distance car trips were removed from this analysis if the commute distances and self-reported weekly VMT could not explain such large amounts of car travel. The authors recognize that self-reported VMT is typically underreported for low mileage groups and overreported for high mileage groups (*23, 26*): to account for such good-faith underreporting of a respondent's VMT, it was decided that a case would be disqualified if the self-reported VMT did not cover at least 50% of the minimum

Table 1. Descriptive Statistics of the Final Working Data Set (*N* = 1,512)

Characteristic	*N* (%)
Gender	
Male	611 (*40.4*)
Age	
18–24	260 (*17.2*)
25–34	584 (*38.6*)
35–44	449 (*29.7*)
45–50	219 (*14.5*)
Education level	
Some grade/high school	23 (*1.52*)
High-school diploma	158 (*10.5*)
Some college/technical school	405 (*26.8*)
Associate's degree	174 (*11.5*)
Bachelor's degree	530 (*35.1*)
Graduate degree (e.g., MS, PhD, etc.)	165 (*10.9*)
Professional degree (e.g., JD, MD, etc.)	52 (*3.44*)
Decline to state	5 (*0.33*)
Neighborhood type	
Urban	290 (*19.2*)
Suburban	744 (*49.2*)
Rural	478 (*31.6*)
Employment	
Employed full-time	771 (*51.0*)
Employed part-time	261 (*17.3*)
Two or more jobs	34 (*2.25*)
Unpaid work	28 (*1.85*)
Homemaker/unpaid caregiver	258 (*17.1*)
Does not work	160 (*10.6*)

Characteristic (sample/population size)		*N* (%)
Annual household income		
Less than $15,000		86 (*5.69*)
$15,000 to $24,999		111 (*7.34*)
$25,000 to $34,999		129 (*8.53*)
$35,000 to $49,999		202 (*13.4*)
$50,000 to $74,999		364 (*24.1*)
$75,000 to $99,999		247 (*16.3*)
$100,000 to $149,999		239 (*15.8*)
More than $150,000		134 (*8.86*)
Continuous variables	Mean	SD
Age (years)	33.9	8.7
Household income	$76,879	$51,192
Household size	3.08	1.56
Number of household vehicles	1.80	0.97
Long-distance business trips	3.77	13.37
Long-distance leisure trips	5.74	8.45
Total long-distance trips	9.49	18.03
Long-distance airplane business trips	0.69	2.47
Long-distance airplane leisure trips	1.01	1.77
Total long-distance airplane trips	1.70	3.49

Note: SD = standard deviation.

mileage needed for the reported number of long-distance car trips. For cases in which the total long-distance round trips with count was over 100 per year, the authors reviewed the case to check for validity and consistency throughout the survey (e.g., variables measuring employment and work trips, to check whether the individual is employed and makes work trips, has a driver's license, has access to a vehicle, etc.); as a result of this process, in most instances the cases were excluded from the analysis. Additionally, the authors looked at observations in which the total count of long-distance round trips with "non-auto" modes, for example, train, airplane, and so forth, was over 30 per year. The authors also focused on cases in which the reported annual household income was below $20,000 but the respondent made more than five commercial airplane trips (for either leisure or business). Similarly, households with an annual income below $20,000 who reported any business trips made by plane were also identified and investigated. The last "overreported" disqualification metric was aimed at individuals who indicated that they "do not work" and they are not in school, but who reported that they made any business trips in the previous 12 months. Individuals who are homemakers, unpaid caregivers, and volunteers are included in the analysis as any of these professions could incur "work" (or non-leisure or non-personal) travel. The "overreported" long-distance trip disqualification metric flagged 100 cases for review; 76 of which resulted in exclusion. The remaining 24 cases were mostly individuals who fell into the low-income household category—although these cases were examined for unusual travel, there was nothing in their responses that further disqualified them from being included in our analysis.

As a result of the coding system used in the online platform, the authors were unable to distinguish between missing values and inputted zeros for the long-distance questions. Consequently, developing metrics for underreported long-distance travel proved challenging. Although it is reasonable to check for consistency for overreported trips, data from a recent 1-year panel showed that whereas three people in the 628-person panel reported no long-distance travel in their responses, their actual monthly travel reported over a year differed—the three people who reported no typical long-distance travel in a year made at least one overnight trip in the panel year (8). In looking at underreported long-distance trips, the team concluded that all respondents who reported both no long-distance business trips and no long-distance leisure trips would be excluded from the analysis; meaning that an additional 367 individuals were excluded from the analysis. Therefore, the models presented here should be considered as an analysis of travel frequency for those reporting at least one long-distance trip of more than 100 miles, one way, in a year.

Model Specification

For this analysis, the authors estimated negative binomial regression models to explore the relationships between the number of self-reported, long-distance, round trips, by purpose and mode, taken within the last 12 months, and sociodemographic and economic characteristics, residential location, and personal attitudes.

The authors estimated six negative binomial regression models: (i) total number of long-distance business trips; (ii) total number of long-distance leisure trips; (iii) total number of long-distance business airplane trips; (iv) total number of long-distance leisure airplane trips; (v) total number of long-distance plane trips; and (vi) total number of long-distance trips.

Dependent Variables

Survey respondents were asked to report the number of long-distance round trips made in the past 12 months by either car, airplane, train, bus, or another mode. Long-distance trips were defined as a trip longer than 100 miles, one way. Those who used multiple modes for the long-distance trip were asked to only report the mode that was used for the longest part of the trip to avoid double counting long-distance trips. For instance, if a respondent drove 110 miles to get to their closest airport to fly 800 miles, and did the same for the trip back, that individual would have recorded this round trip in the airplane category. Furthermore, respondents were asked to categorize trips as either business/work related or leisure/personal.

As one might expect, most individuals made very few long-distance trips. Generally, respondents made more leisure trips than business trips. In fact, on average, respondents made approximately 3.7 business trips and 5.7 leisure trips. Unsurprisingly, the number of airplane trips was considerably lower than the total number of long-distance trips. An average of 1.0 long-distance leisure airplane trips and 0.7 long-distance business airplane trips were made by respondents. This information is summarized in Table 1.

Explanatory Variables

The authors tested the inclusion of several groups of explanatory variables. The final version of the models included a total of 21 variables, distributed throughout the six models. These variables were selected on their expected relationship with long-distance travel and were found to be statistically significant in the estimated models. In addition, the correlation between predictors was checked and found to be generally negligible—only rather low correlations (lower than 0.25) were observed among the predictors.

The 21 variables can be grouped into four categories: (1) individual characteristics; (2) residential location neighborhood type; (3) factor scores extracted from attitudinal variables from the data set; and (4) level of travel including modes used for daily travel and total VMT in the average week (note this VMT excludes air travel and therefore reduces the risk of endogeneity).

The authors controlled for sociodemographic characteristics in the models by using variables for age, gender, and income. A continuous variable for income was used to evaluate if individuals with higher incomes had a larger number of long-distance trips. Residential location was explicitly controlled using a home neighborhood type variable that was geocoded based on a respondent's home address. It was hypothesized that individuals living in urban neighborhood types made more long-distance trips based on background literature.

Attitudinal Variables. There were 65 separate statements in the survey included to measure individuals' attitudes about several dimensions related to the environment, travel, technology adoption, multitasking, life satisfaction, land use, the role of government regulations, and so forth. The research team conducted a multiround factor analysis using principal axis factoring with an oblique rotation to extract the final 14 factors that combine the information available in the attitudinal sections of the study. More specifically, when the researchers attempted to initially extract factors, they included all 65 attitudinal variables. From this first attempt, 19 factors were extracted. However, there were nine attitudinal variables that did not load heavily (at least 0.25) onto any one factor. The researchers wanted to have a factor analysis in which all variables used to extract factors would load onto at least one factor with a 0.25 or greater loading. Therefore, those nine variables were removed and the factor analysis was rerun with the same factoring method and rotation. This procedure was continued until every attitudinal variable being analyzed loaded onto at least one factor with a 0.25 loading. The factors presented in this paper are the result of a five-round factor analysis. The final set of 14 factors explains 58.6% of the

Table 2. Factor Analysis Loadings Table

Factors and associated statements	Factor loadings
Traditional shopper	
I prefer to shop in a store rather than online.	1.029
I enjoy shopping online.	–0.368
I enjoy the social aspects of shopping in stores.	0.332
Pro-environmental policies	
We should raise the price of gasoline to reduce the negative impacts on the environment.	0.918
We should raise the price of gasoline to provide funding for better public transportation.	0.854
To improve air quality, I am willing to pay a little more to use a hybrid or other clean-fuel vehicle.	0.254
The government should put restrictions on car travel in order to reduce congestion.	0.339
Tech-savvy	
Having Wi-Fi and/or 3G/4G connectivity everywhere I go is essential to me.	0.719
Getting around is easier than ever with my smartphone.	0.594
I like to be among the first people to have the latest technology.	0.38
Social media (e.g., Facebook) makes my life more interesting.	0.359
Traditional thinking	
Greenhouse gases from human activities are creating major problems.	–0.71
Any climate change that may be occurring is part of a natural cycle.	0.682
It is pointless for me to try too hard to be more environmentally friendly because I am just one person.	0.419
It is more important for men than for women to have a high-paying career.	0.384
At work, it is perfectly fine for women to have authority over men.	–0.371
Adventurer	
I like trying things that are new and different.	0.562
I like to juggle two or more activities at the same time.	0.323
Commute-loving	
My commute is stressful.	–0.793
My commute is generally pleasant.	0.66
Traffic congestion is a major problem for me personally.	–0.576
The time I spend commuting is generally wasted time.	–0.538
Getting stuck in traffic does not bother me that much.	0.328
Pro-suburban	
I prefer to live in a spacious home, even if it is farther from public transportation and many places I go to.	0.771
I prefer to live close to transit even if it means I will have a smaller home and live in a more crowded area.	–0.676
I like the idea of living somewhere with large yards and lots of space between homes.	0.443
I like the idea of having different types of businesses (such as stores, offices, restaurants, banks, library) mixed in with the homes in my neighborhood.	–0.342
Established in life	
I am already well-established in my field of work.	0.733
I am still trying to figure out my career (e.g., what I want to do, where I will end up).	–0.623
I am generally satisfied with my life.	0.409
Long-term suburbanite	
I picture myself living long term in a suburban setting.	0.829
A house in the suburbs is the best place for kids to grow up.	0.556
I picture myself living long term in an urban setting.	–0.323
Practical (anti-materialistic)	
I prefer to minimize the material goods I possess.	0.439
To me, a car is just a way to get from place to place.	0.42
I would/do enjoy having a lot of luxury things.	–0.413
The functionality of a car is more important to me than its brand.	0.355
For me, a lot of the fun of having something nice is showing it off.	–0.325
I like to be among the first people to have the latest technology.	–0.289

variance. The final models included 10 factor scores as listed in Table 2.

Levels of Local Routine Travel. In addition to the sociodemographic, residential choice, and attitudinal variables, the authors also controlled for a person's overall level of travel in our models. For both leisure and business long-distance trips, the self-reported weekly vehicle miles traveled were used as an explanatory variable. It was recognized that there was a potential risk of endogeneity with the inclusion of self-reported weekly VMT variables in the non-airplane models of long-distance travel: in fact, many travelers might report a higher level of self-reported VMT because they make more long-distance

Table 3. Parameter Estimates for Leisure, Business, and Total Long-Distance Trip Models

	Number of total long-distance trips		Number of long-distance leisure trips		Number of long-distance business trips	
	Exp(β)	*p*	Exp(β)	*p*	Exp(β)	*p*
Intercept	10.147	<0.0001	12.469	<0.0001	0.549	0.0377
Age of respondent	0.986	<0.0001	0.990	0.0003		
Sex (male)			0.823	0.0002		
Individual Income (base: $10k)	1.033	0.0005			1.088	<0.0001
Neighborhood type						
Urban	1.517	<0.0001			2.656	<0.0001
Suburban	1.053	0.4119			1.326	0.0573
Commute distance					0.016	<0.0001
Weekly VMT	1.001	<.0001	1.001	<.0001		
Primary commute mode						
Active mode (e.g., walk, bike, skateboard, etc.)					2.494	0.0055
Private vehicle (e.g., drive alone, carpool, etc.)					2.009	0.0039
Never used car-sharing			0.762	0.0145		
Made no long-distance business trips			0.784	<.0001		
Doesn't bike for short-distance personal trips	0.8070	0.005	0.851	0.0236		
Factor: Traditional thinking	1.057	0.0234	0.936	0.003	1.352	<.0001
Factor: Established in life	1.068	0.0106				
Factor: Long-term suburbanite	0.894	<.0001	0.926	0.0003	0.871	0.0091
Factor: Adventurer	1.137	<.0001	1.090	<.0001	1.173	0.0015
Factor: Commute-loving	0.912	0.0002			0.782	<.0001
Factor: Pro-suburban	1.054	0.0406			1.232	0.0007
Factor: Tech-savvy	1.073	0.0034			1.253	0.0001
Factor: Traditional shopper	0.938	0.0163			0.845	0.006
Dispersion	0.860	<.0001	0.706	<.0001	3.260	<.0001
Observations	1,406		1,498		1,034	
Generalized *R*-squared	0.20		0.10		0.17	

trips by car (and not the opposite). Accordingly, model specifications were also tested in which this variable was dropped from the model estimation. Removing the weekly VMT variable from the model had a limited impact on the magnitude of the other coefficients, which remain rather stable, and this reassured the authors about the limited effects that the inclusion of this variable has in affecting the final model solution.

When modeling business long-distance trips, the authors specifically controlled for primary commute mode and commute distance. For long-distance leisure trips, we controlled for the patterns of short-distance trips by mode and the adoption of shared mobility. Further, to uncover a relationship between leisure and business travel, a dummy variable was used to indicate if a business trip had been made when estimating the number of long-distance leisure trips. Based on Aultman-Hall et al, the authors hypothesize that long-distance business and leisure travel patterns are related entities (*8*).

In future extensions of the research, the authors plan to expand on this topic and better capture the relationships between long-distance business and leisure trips through the estimation of joint models of long-distance business and leisure trips.

Model Results

Based on initial exploratory analysis and descriptive statistics on the dependent variables, the authors found that the variables have a negative binomial distribution and thus negative binomial regression was used. Six models were estimated (Tables 3 and 4) to understand which factors affect long-distance travel.

Table 3 contains the models for the number of long-distance leisure trips, the number of long-distance business trips, and the number of total long-distance trips. The columns are labeled by dependent variable. The columns labeled exp(β) represent the odds-ratio of the parameter. Interpretation is similar to that of a Poisson model: for dummy variables, if exp(β) is 1.2, this can be interpreted as that group (who has that characteristic) making 20% more long-distance trips. For continuous variables, such as income or age, an exp(β) value of 1.10 means that each unit increase is interpreted as a 10% increase in the number of long-distance trips.

The sociodemographic characteristics used in these models provided interesting insights into long-distance travel. As expected, age is negatively correlated with the total number of long-distance trips and the number of

Table 4. Parameter Estimates for Long-Distance Airplane Trips

	Number of total long-distance air trips		Number of long-distance air leisure trips		Number of long-distance air business trips	
	Exp(β)	*p*	Exp(β)	*p*	Exp(β)	*p*
Intercept	2.220	0.0003	0.865	0.4334	0.120	<0.0001
Age of respondent	0.979	<0.0001	0.987	0.0079		
Household size	0.875	<0.0001				
Individual income (base: $10k)	1.140	<0.0001	1.069	<.0001	1.235	<0.0001
Neighborhood type						
Urban	2.122	<0.0001	1.973	<.0001	3.040	<0.0001
Suburban	1.087	0.3808	1.104	0.2942	1.259	0.1799
Weekly VMT			1.001	0.0146		
Primary commute mode						
Active mode (e.g., walk, bike, skateboard, etc.)					2.901	0.0073
Private vehicle (e.g., drive alone, carpool, etc.)					3.144	0.0002
Studies/works exclusively from home					3.252	0.0032
Number of long-distance business trips by air			1.105	<.0001		
Doesn't bike for short-distance personal trips	0.698	0.0007	0.718	0.001		
Factor: Established in life					1.322	<.0001
Factor: Practical (anti-materialistic)	0.921	0.0066			0.893	0.0257
Factor: Long-term suburbanite					1.174	0.0102
Factor: Pro-environmental policies	1.125	0.0015	1.155	<.0001		
Factor: Pro-suburban	0.922	0.0278	0.908	0.0057	0.777	0.0002
Factor: Tech-savvy	1.109	0.0045				
Factor: Traditional shopper	0.898	0.0062	0.888	0.0019	0.855	0.0181
Dispersion	1.283	<.0001	0.838	<.0001	2.826	<.0001
Observations	1,406		1,498		1,034	
Generalized *R*-squared	0.20		0.10		0.17	

long-distance leisure trips—older individuals in this sample are not traveling as much as young individuals. Gender plays a role in the number of long-distance leisure trips: men make fewer long-distance leisure trips than women and not by a trivial amount— approximately, 17.1% fewer trips. Both Aultman-Hall et al and Limtanakool et al find that men on average make more long-distance trips than women (*8*, *13*); however, this analysis yields the opposite result. Individual income is positively correlated with the number of long-distance business trips and the total number of long-distance trips. Those living in an urban neighborhood, when compared with a rural neighborhood, make more long-distance business trips and have a higher number of overall trips. Urban workers may work for companies that require more business travel and may make more long-distance business trips than those in rural environments.

An individual's overall level of vehicle travel as measured by commuting and VMT also affected the number of long-distance trips made in the prior 12 months. Longer commutes were indicative of more long-distance business travel. Since long-distance travel did not explicitly exclude long commutes, the number of long-distance trips reported may include commute trips. Similarly, the reported weekly VMT had a small but positive impact on the total number of long-distance leisure and overall trips. For the long-distance business trip model, the authors also considered the primary commute mode. Those who commuted by an active mode (e.g., bike, walk, etc.) made more long-distance business trips than those who commuted by transit/work-provided shuttle (transit). Similarly, those who commuted by private vehicle made more long-distance business trips than those who commuted by transit. Those who had never used car-sharing services (e.g., Zipcar) made fewer leisure trips than those who had used them. All these effects might signal the impact of some latent construct associated with more traditional lifestyles and perhaps belonging to less affluent and less dynamic sociodemographic groups.

In this paper, in place of estimating a joint model for long-distance leisure and long-distance business trips (which remains a direction for future research for this study), the authors used an indicator variable for business trips in the leisure model. It was found that those who made no long-distance business trips tended to make 21% fewer long-distance leisure trips. Like the active-mode commuters, those who bike for short-distance trips made more long-distance leisure trips than those who do not bike.

To the best of the authors' knowledge, this is the first study that incorporates attitudinal factors into an analysis of long-distance travel. The inclusion of factors extracted from the attitudinal variables provided more insights into further understanding long-distance travel. Individuals who scored highly on the "traditional thinking" factor made fewer long-distance leisure trips, whereas those who scored highly on the "established in life" factor made more long-distance trips. The "traditional thinking" factor has the largest effect on long-distance business travel. In fact, its effect on long-distance business travel is almost twice as much as the most influential total long-distance travel factor and almost three times more influential than the strongest long-distance leisure factor—both are highly influenced by the "adventurer" factor.

The second most influential factor on long-distance travel is the "tech-savvy" factor—those who score highly on the "tech-savvy" factor make more long-distance trips than those who do not score highly on that factor. Further, individuals who feel as though they do not have a grasp on their career, life, or are not generally satisfied with the state of their life may not be interested in or able to take long-distance trips. Those who scored highly on the "adventurer" factor made more long-distance trips—adventurers may not be satisfied with business as usual and perhaps look at travel (for whatever reason) for variety. However, those who scored highly on the "traditional shopper" factor made fewer trips. As in the overall long-distance travel models, those who live in predominantly urban neighborhoods make more long-distance airplane trips; this could be because of the proximity to airports or the emergence of lifestyles that are more common among urban dwellers.

Table 4 presents the final models for the business, leisure, and total long-distance airplane trips. These second models are very similar to those in Table 3 with regards to explanatory variables. Sociodemographic characteristics, land use, travel choices, and attitudinal factors all play a role in the number of long-distance trips by airplane for leisure and business purposes. The predictor that has the most influence on long-distance travel by airplane, excluding categorical variables, is the "established in life" factor, followed by individual income. Those who feel as though they are established in their career and satisfied with their life make more long-distance business trips by air than those who are pro-suburban. Additionally, higher individual income is associated with a higher number of long-distance airplane business trips.

Age and individual income negatively affect the number of long-distance airplane trips. Age has a negative effect for total, leisure, and business long-distance air trips, whereas household size has a negative impact on total long-distance plane trips. Each additional household member is associated with a reduction in the number of long-distance airplane trips of 12%. This is consistent with prior research showing that the presence of children reduces the amount of travel because of the increased household obligations. The decrease for age is not extreme: an increase in age of 1 year is associated in a 2% reduction in the number of long-distance airplane trips. This finding is unexpected because the age range was capped at 50 years old, so this requires more investigation.

An individual's general level of travel also has an impact on the number of long-distance airplane trips. As it was for overall long-distance travel, those who make short-distance leisure trips by bike have a higher number of long-distance trips by air. Similarly, the number of long-distance air trips for business is positively associated with the number of long-distance air trips for leisure. For example, as many business flyers earn frequent flyer miles for their business trips, they can buy tickets for their vacation travel for little to no cash, spending air miles instead. Higher weekly VMT positively affects long-distance leisure airplane trips. An individual's primary commute mode has an impact on the number of long-distance business trips taken by airplane. Those who primarily telecommute make more long-distance business airplane trips annually than those who do not primarily telecommute: a telecommuter's corporate office or main office may be located in a different city or even state, which might necessitate frequent in-person meetings and long-distance trips. Individuals who commute by transit tend to make fewer long-distance airplane trips than those who commute by other modes: those who commute by private vehicle and active modes make higher numbers of long-distance business plane trips. These findings point to the impact of the knowledge economy and the system of a global mobile elite that includes a level of mobility not accessible to everyone.

Attitudinal factors affect the number of long-distance airplane trips reported in the 12-month period. Identifying as "established in life" has the greatest influence on the number of long-distance business trips by air and total long-distance trips by air compared with any other factor. Individuals who scored highly on the "pro-environmental policies" factor made more long-distance trips by air—as the authors do not explicitly control for education this could be capturing the influence of educated individuals who travel by air. "Traditional shoppers" and those that are "tech-savvy" have opposite trends in long-distance airplane travel. "Traditional shoppers" made fewer long-distance airplane trips, whereas those who are more "tech-savvy" made more long-distance airplane trips. "Tech-savvy" individuals may have more access to purchasing of air travel online, whereas "traditional shoppers" may dislike travel by airplane or purchasing air travel online—online shopping is not something they enjoy doing.

Conclusions

Long-distance travel is difficult to study. Before this study, only two year-long measures of long-distance travel had been collected in the U.S.: the 1995 American Travel Survey and the relatively small Longitudinal Survey of Overnight Travel (*8*). Presently, there are only a handful of states that have collected long-distance travel data, usually over 8–12 week time periods. This work aims to help fill the long-distance data gap in the U.S. by contributing models of annual long-distance travel by California millennials and Generation X members. This study focuses on the effects of sociodemographic traits, residential location, and individual attitudes on long-distance travel. Using data collected in fall 2015 from the California Millennials Dataset, the authors model the number of long-distance trips of 1,512 respondents aged 18–50 years who made at least one long-distance trip in the last 12 months.

Californian millennials and Generation X members have a wide range of levels of long-distance travel. The oldest individuals in this sample made, on average, 8.4 total long-distance trips compared with the youngest persons in the sample who made, on average, 13.5 long-distance trips per year. Furthermore, this study suggests that women may be starting to make more long-distance trips than men. It is uncertain whether future long-distance travel of the younger millennials will remain higher than those in Generation X or if it will decrease to a similar level. It could be that the youngest cohort will always make more long-distance trips or that as individuals age their sense of adventure or desire to travel is reduced. In the meantime, local and regional planners need to balance systems and programs for residents and visitors while preparing to increase consideration of long-distance travel as a growing component of volume and factor in assessing mobility.

The authors' negative binomial regressions estimated the effect of the explanatory variables on the number of long-distance trips made for business, leisure, and their sum. The results indicate that age, being male, and household size have negative effects on long-distance trip rates. Individual income positively affects the number of long-distance trips made. An individual's residential location also affects the number of long-distance trips made; urbanites made more business trips than rural dwellers. The factors that represent the largest impact on the number of long-distance trips per year are those developed from the attitudinal questions. This is an important core finding of this study that informs approaches for future data collection and studies. Attitudes are important to long-distance travel. Being a "traditional thinker," "tech-savvy," and "established in life" were significant predictors. Those who scored highly on the "traditional thinking" factor made fewer leisure but more business trips, whereas those who scored highly on the "established in life" factor made more total trips. Weekly VMT and the use of certain modes for short-distance and commute travel also affect long-distance travel, with individuals who usually commute by transit making fewer long-distance airplane trips than those commuting by other modes.

Another new type of relationship that was explored, but not a large explanatory factor in the authors' models, is the one between long-distance and short-distance travel. In the development of the analyses for the study, short-distance trip patterns (in particular, for car trips) were analyzed and compared with the number of long-distance annual (car) trips; a statistically significant relationship between the frequency of leisure short-distance car trips and the frequency of leisure long-distance car trips could not be found.

This study demonstrates that measuring and modeling long-distance travel is viable and that future data collection efforts and models should incorporate this travel to achieve more accurate overall models of the transportation system and its impact on sustainability and quality of life. Challenges related to the burden of the survey suggest that more streamline measures of long-distance travel may be needed. However, although many are advocating for passive or semi-passive mobile device data collection for long-distance travel measurement, the attitudinal variables and resultant factors, collected and modeled here, demonstrate the sustained importance of collecting personal variables that require survey collection to understand travel and forecast it.

Acknowledgments

Funding for this project was provided by the National Center for Sustainable Transportation, the California Department of Transportation (CalTrans) and the UC Davis Sustainable Transportation Energy Pathways (STEPS) Program. We would like to thank Eric Gudz, Kate Tiedeman, Yongsung Lee, and Farzad Alemi for their support and contribution to this project. We would also like to thank Freshta Pirzada who assisted with data cleaning. Lew Fulton, Susan Handy, and Patricia Mokhtarian heavily contributed to the development of the survey content and data collection process.

Author Contributions

The authors confirm contribution to the paper as follows: study conception and design: RB, LAH, GC; data collection: RB, GC; analysis and interpretation of results: RB, LAH, GC; draft manuscript preparation: RB. All authors reviewed the results and approved the final version of the manuscript.

References

1. Westat. *Task C: Sample Design, Final: December 31*, 2015.
2. California Department of Transportation. *2010-2012 California Household Travel Survey Final Report Appendix*. California Department of Transportation, Sacramento, CA, 2013.
3. Circella, G., K. Tiedeman, S. Handy, F. Alemi, and P. L. Mokhtarian. *What Affects U.S. Passenger Travel? Current Trends and Future Perspectives*. National Center for Sustainable Transportation, University of California, Davis, CA, 2016.
4. Federal Highway Administration. *3.2 Trillion Miles Driven on U.S. Roads in 2016*. https://www.fhwa.dot.gov/pressroom/fhwa1704.cfm. Accessed September 30, 2018.
5. Bureau of Transportation Statistics. *Load Factor for U.S. Air Carrier Domestic and International, Scheduled Passenger Flights*. https://www.transtats.bts.gov/osea/seasonaladjustment/?PageVar=ASM. Accessed September 30, 2018.
6. Bureau of Transportation Statistics. *Enplanements for U.S. Air Carrier Domestic and International, Scheduled Passenger Flights*. https://www.transtats.bts.gov/osea/seasonaladjustment/?PageVar=ASM. Accessed September 20, 2018.
7. McKenzie, B. Out-of-State and Long Commutes: 2011. *American Community Survey Reports*, No. ACS-20, February 2013.
8. Aultman-Hall, L., C. Harvey, J. Sullivan, and J. J. LaMondia. The Implications of Long-Distance Tour Attributes for National Travel Data Collection in the United States. *Transportation*, Vol. 45, No. 3, 2016, pp. 875–903.
9. Outwater, M., M. Bradley, N. Ferdous, R. Pendyala, V. Garikapati, C. Bhat, S. Dubey, J. LaMondia, S. Hess, and A. Daly. *Foundational Knowledge to Support a Long-Distance Passenger Travel Demand Modeling Framework Implementation Report*. Federal Highway Administration, Washington, D.C., 2015.
10. Bricka, S., and E. Sabina. Evaluation of Key Design Elements of Long-Distance Survey of Front Range Travel Counts. *Transportation Research Record: Journal of the Transportation Research Board*, 2012. 2285: 110–118.
11. Georggi, N. L., and R. M. Pendyala. *Analysis of Long-Distance Travel Behavior of the Elderly and Low Income*. Master's thesis. University of South Florida, Tampa, FL, 2001.
12. Holz-Rau, C., J. Scheiner, and K. Sicks. Travel Distances in Daily Travel and Long-Distance Travel: What Role Is Played by Urban Form? *Environment and Planning A*, Vol. 46, No. 2, 2014, pp. 488–507.
13. Limtanakool, N., M. Dijst, and T. Schwanen. On the Participation in Medium- and Long-Distance Travel: A Decomposition Analysis for the UK and the Netherlands. *Tijdschrift voor Economische en Sociale Geografie*, Vol. 97, No. 4, 2006, pp. 389–404.
14. LaMondia, J., L. Aultman-Hall, and E. Greene. Long-Distance Work and Leisure Travel Frequencies. *Transportation Research Record: Journal of the Transportation Research Board*, 2014. 2413: 1–12.
15. Van Nostrand, C., V. Sivaraman, and A. R. Pinjari. Analysis of Long-Distance Vacation Travel Demand in the United States: A Multiple Discrete–Continuous Choice Framework. *Transportation*, Vol. 40, No. 1, 2013, pp. 151–171.
16. Van Nostrand, C., A. R. Pinjari, and V. Siveraman. A Household-Level Annual, Long-Distance, Vacation Travel Demand Model: A Multiple Discrete-Continuous Choice Framework. Presented at 91st Annual Meeting of the Transportation Research Board, Washington, D.C., 2012.
17. Steiner, R., and H. Cho. *Florida Long-Distance Travel Characteristics and Their Potential Impacts on the Transportation System*. Gainsville, FL, 2013.
18. Ciobanu, C., J. Donna, G. Lungu, and G. Veramendi. Women Book Business Travel Earlier, Saving Companies Millions. *Harvard Business Review*, Apr 13, 2016.
19. Circella, G., F. Alemi, K. Tiedeman, R. M. Berliner, Y. Lee, L. Fulton, P. L. Mokhtarian, and S. Handy. *What Affects Millennials' Mobility? Part II: The Impact of Residnetial Location, Individual Preferences and Lifestyles on Young Adults' Travel Behavior in California*. National Center for Sustainable Transportation, University of California, Davis, CA, 2017.
20. Circella, G., F. Alemi, R. M. Berliner, and K. Tiedeman. *What Affects Millennials ' Mobility? PART I: Investigating the Environmental Concerns, Lifestyles, Mobility-Related Attitudes and Adoption of Technology of Young Adults in California*. National Center for Sustainable Transportation, University of California, Davis, CA, 2016.
21. U.S. Census, B. TIGER/Line® Shapefiles and TIGER/Line® Files. https://www.census.gov/geo/maps-data/data/tiger-line.html. Accessed September 30, 2018.
22. ESRI. ArcGIS. https://www.esri.com/en-us/home. Accessed September 30, 2018.
23. Google. Google Maps API. https://developers.google.com/maps/documentation/geocoding/start. Accessed September 30, 2018.
24. Salon, D. Heterogeneity in the Relationship between the Built Environment and Driving: Focus on Neighborhood Type and Travel Purpose. *Research in Transportation Economics*, Vol. 52, 2015, pp. 34–45.
25. Ramsey, K., and A. Bell. *Smart Location Database*. https://www.epa.gov/sites/production/files/2014-03/documents/sld_userguide.pdf. Accessed September 30, 2018.
26. Staplin, L., K. W. Gish, and J. Joyce. "Low Mileage Bias" and Related Policy implications—A Cautionary Note. *Accident Analysis & Prevention*, Vol. 40, No. 3, 2008, pp. 1249–1252.

The Standing Committee on Traveler Behavior and Values (ADB10) peer-reviewed this paper (18-05960).

Article

TRR
JOURNAL OF THE TRANSPORTATION RESEARCH BOARD

Transportation Research Record
2018, Vol. 2672(47) 219–230

DOI: 10.1177/0361198118796402
journals.sagepub.com/home/trr
SAGE

Time-Dependent Intermodal A* Algorithm: Methodology and Implementation on a Large-Scale Network

Ömer Verbas[1], Joshua Auld[1], Hubert Ley[1], Randy Weimer[1], and Shon Driscoll[1]

Abstract

This paper proposes a time-dependent intermodal A* (TDIMA*) algorithm. The algorithm works on a multimodal network with transit, walking, and vehicular network links, and finds paths for the three major modes (transit, walking, driving) and any feasible combination thereof (e.g., park-and-ride). Turn penalties on the vehicular network and progressive transfer penalties on the transit network are considered for improved realism. Moreover, upper bounds to prevent excessive waiting and walking are introduced, as well as an upper bound on driving for the park-and-ride (PNR) mode. The algorithm is validated on the large-scale Chicago Regional network using real-world trips against the Google Directions API and the Regional Transit Authority router.

Transportation network companies (TNC) such as Uber and Lyft, car and bike sharing companies, on-demand transit services, the forthcoming connected and autonomous vehicle (CAV) technologies, as well as the increasing availability of real-time traffic and transit information, give travelers the opportunity to evaluate their multiple routing options and make better-informed decisions. The advent of real-time control and management technologies, and vehicle-to-vehicle (V2V) and vehicle-to-infrastructure (V2I) communication technologies provide opportunities to increase mobility, accessibility, throughput, and safety in the entire transportation network. These advancements call for a comprehensive modeling of the transportation system as an integrated multimodal network.

Most of the existing transportation network modeling literature focuses either on the vehicular traffic network, or the transit network. The full integration of the two major modes is usually limited to small hypothetical networks, which is not practical for large cities. At the large scale, the integration is performed in an ad-hoc fashion by which separate models communicate with each other at designated outer iterations. The drawbacks of this approach can be listed as follows:

- The interaction between the transit traffic and vehicular traffic cannot be modeled properly. For instance, transit buses share the street network with passenger cars and affect the performance characteristics of each other.
- The modeling of intermodal routing such as park-and-ride (PNR), kiss-and-ride (KNR), taxi/TNC/CAV to transit, and taxi/TNC/CAV after transit is limited.
- The modeling of en-route mode switching is limited.
- As a result of all of the above, the integration of the transportation supply model with the activity-based demand models (ABM) is limited.

This paper proposes a flexible intermodal routing algorithm that can provide time-dependent shortest paths for conventional modes such as passenger car and walk-to-transit, as well as any feasible intermodal combination such as PNR, KNR, taxi/TNC/CAV before/after transit, and so on.

The rest of the paper is structured as follows: The following section provides a brief literature review on the shortest path problem and its multiple branches,

[1]Argonne National Laboratory, Lemont, IL

Corresponding Author:
Address correspondence to Ömer Verbas: omer@anl.gov

followed by the multimodal network representation. Afterwards, the proposed time-dependent intermodal A* (TDIMA*) algorithm is introduced. This section is followed by the large-scale numerical examples for validation on the Chicago Regional network. The final section concludes the paper.

Literature Review

Shortest path algorithms with label setting (*1*) and label correcting (*2*, *3*) algorithms date back to 1950s, but in these the link costs are static and deterministic. The first time-dependent algorithm was introduced by Cooke and Halsey in 1966 (*4*), whereas the first hybrid label setting/correcting algorithm was introduced by Glover et al. in 1985 (*5*).

A series of large-scale implementations were introduced in (*6–9*). A time-dependent large-scale implementation was presented in (*6*), whereas a static implementation with intersection movement penalties and prohibitions was introduced in (*8*). A multimodal, time-dependent algorithm with movement and transfer penalties was introduced in (*9*). In this paper, every link carries multiple modes and services. At every node, time-dependent transfer costs are defined between every adjacent link/service pair (*7*, *9*). Another approach to a multimodal shortest path algorithm is the "divide-and-conquer" technique (*10–12*). In this approach, the network is divided into subnetworks for every mode, and transfer between the subnetworks is allowed only at certain nodes. First, nondominated subpaths are found within every mode's subnetwork (divide). Then, these are combined to form nondominated multimodal paths (conquer) (*10–12*).

In conventional shortest path algorithms, the output is a link-path incidence matrix with binary values: a link either belongs to a path (1) or not (0). In a transit network, a node can serve multiple routes which can take a traveler to their destination. In this case, the binary condition is relaxed to be a probability between 0 and 1 to represent the likelihood of a certain link-route (service) pair being boarded. This phenomenon was first introduced by Chriqui and Robillard in (*13*). A decade later, Nguyen and Pallottino (*14*, *15*), and Spiess and Florian (*16*) introduced solution algorithms to solve the common bus lines problem. The former introduced the term "hyperpath" (*14*, *15*), whereas the latter used the term "optimal strategy" (*16*). Later on, capacity constraints were introduced into the transit shortest path algorithms (*17–20*). Verbas and Mahmassani introduced a transit hyperpath algorithm that is time-dependent on a frequency-based network with seat and standing capacities (*21*, *22*).

The initial transit hyperpath algorithms were frequency-based. In more recent years, a schedule-based literature has emerged (*23*). Some exceptions aside (*21*, *22*), frequency-based algorithms are static, whereas schedule-based algorithms are time-dependent (*24*). Schedule-based algorithms incorporated capacity penalties, congestion pricing, and the differentiation between seated and standing passengers (*25–27*). Most of the schedule-based algorithms require the temporal expansion of the network. However, Noh, Hickman, and Khani developed a schedule-based algorithm that does not require the said expansion (*28*).

Another branch of the shortest path algorithms deals with user heterogeneity (*29–32*). In their seminal work, Lu and Mahmassani propose a bi-criterion time-dependent shortest path algorithm in which the different value of time (VOT) classes emerge naturally from a continuous distribution.

All of the algorithms introduced up to this point are tree-based algorithms, in which the shortest path from (to) a single origin (destination) is found to (from) all destination (origin) nodes. In 1968, Hart et al. introduced the first A* algorithm that makes it possible to find an exact shortest path algorithm between any two nodes in a static network (*33*). In A* algorithms, the node label has two components: (i) the actual cost from the origin to the node, and (ii) the estimated (heuristic) remaining cost from the node to the destination. Usually, the estimated component is the Euclidean distance divided by the maximum allowed speed, which is a lower bound on the actual remaining cost to the destination (*33*). The authors prove admissibility of the algorithm as long as the estimated code is never larger than the actual cost. A very low estimated cost guarantees an exact solution at the expense of many node scans. At the extreme case, the estimated code can be set to zero, which makes the algorithm identical to Dijkstra's (*1*) algorithm that terminates when the destination node is reached. Conversely, the larger the estimated cost is, the faster the algorithm terminates. However, the code may return nonexact solutions if the admissibility condition is violated (*33*).

Chabini and Lan extended the A* algorithm to the time-dependent case (*34*). Zhao et al. showed that in the time-dependent case, a small enough estimated code is a necessary but not a sufficient condition for admissibility (*35*). First-in-first-out (FIFO) conditions must also be satisfied, which is not guaranteed in transit networks. In 1991, Bander and White introduced a transit A* algorithm called interruptible A* (IA*), for which a set of island nodes are preselected, which are favored by the algorithm for faster termination (*36*). In 2015, Khani et al. introduced a trip-based A* (TBA*) algorithm for transit networks (*37*). The algorithm has two major contributions: (i) once a node (stop) belonging to a transit is scanned, all the downstream nodes are scanned and potentially updated (however, unlike in the conventional shortest path algorithms, the predecessor node is set as the boarding node, not as the previous node); and (ii) for the estimated component of the shortest path label, instead of using a Euclidean-based estimator, the authors use the resulting labels of a static, all-to-all label setting algorithm. In this estimation algorithm, no waiting or transfer times are considered. For the link costs, the minimum observed travel time throughout the time horizon from different trips on a given link is used. The authors suggest that these costs would serve as a lower bound on the actual costs, as any in-vehicle or walking time on a link would be greater than or equal to the chosen minimum. Moreover, there would be non-negative waiting times.

This paper introduces a time-dependent intermodal A* algorithm (TDIMA*), which extends Khani et al.'s TBA* algorithm (*37*) and makes the following contributions.

- The network representation is multimodal with three types of links: (i) transit, (ii) vehicular traffic, and (iii) walking. Walking links are derived from the street network, and walking is not limited to designated transfer areas (no predefined hierarchy) but allowed along all local and arterial roads.
- The TDIMA* algorithm is multimodal and finds paths for passenger cars, walking, and walk-to-transit modes.
- The TDIMA* algorithm is intermodal and finds paths for PNR, KNR, and any other feasible combinations such as PNR, KNR, taxi/TNC/CAV before/after transit, and so on.
- Like the TBA* algorithm, the TDIMA* algorithm uses the results of a label setting algorithm as estimated costs. However, instead of performing the label setting algorithm from every link to every link, the algorithm is performed from every traffic analysis zone (TAZ) to every link to reduce computational time.
- Unlike the TBA* algorithm, this algorithm does not follow a trip's further downstream stops.
- Unlike the TBA* algorithm, the TDIMA* updates the link labels as opposed to node labels and thereby more easily accounts for node switching penalties.

Network Representation

As mentioned in the previous section, the multimodal network used in the TDIMA* algorithm has three layers:

1. Transit
2. Walking
3. Vehicular traffic

The transit network is constructed using General Transit Feed Specification (GTFS) (*38*). First, a "typical" weekday is chosen to detect the used service IDs from the "calendar" file. The same service IDs are also seen in the "trips" file. Using these, the transit trips that are in service on that day are detected. Although every trip $q \in Q$ is associated with a route, in most major cities the routes have variations called "patterns" $p \in P$ (*39–41*). Depending on the transit agency, patterns are properly and uniquely linked with a shape ID in an optional file. However, to guarantee consistency, the following method is used to identify patterns: any unique sequence of stops/stations under a route is defined as a pattern. Thus, every transit trip exclusively belongs to one pattern $q \in Q_p$, and every pattern exclusively belongs to one route. The set Q_p is ordered by departure time. This is obtained from the "stop_times" file.

Generation of the nodes N is trivial: every used transit stop/station is a transit node. Similarly, any two consecutive node pairs of a pattern form a transit link $i \in A$. After all the transit links are generated, an ordered set of links is generated for every pattern $i \in A_p$. Conversely, on every link, a set of patterns $p \in P_i$ is generated. Afterwards, the process is to store the number s_{ip}, which is the sequence number of link $i \in A_p$ along pattern $p \in P_i$. Finally, using the schedule information in the "stop_times" file, the arrival $a^q_{s_{ip}}$ and departure times $d^q_{s_{ip}}$ of a trip $q \in Q_p$ are associated with the sequence number s_{ip}.

The vehicular network can be obtained using any existing network data-source available to the modeler. This study uses the existing Chicago Regional network data designed for POLARIS, an integrated activity-based modeling and traffic simulation software (*42*). The walking links are derived from the vehicular traffic network and are segmented into smaller links whenever a transit stop/station is nearby. In other words, the walking layer initially is partly a duplicate layer of the vehicular network, where the link types are arterial or local (no ramps, highway, or freeway links are allowed). Afterwards, transit nodes are projected onto these links to segment the walking links and provide connectivity between the transit and vehicular traffic layer. See Figure 1 for an illustration of the network representation.

As seen in Figure 1, the walking links run parallel with the vehicular network and provide connectivity between the vehicular and transit network. Walking through a transit node without boarding a vehicle is allowed. Because the sum of the lengths of the two segmented walking links equals the length of the corresponding vehicular link, this does not cause any detouring. In the case of rail stations far away from any vehicular link, additional walking links are added.

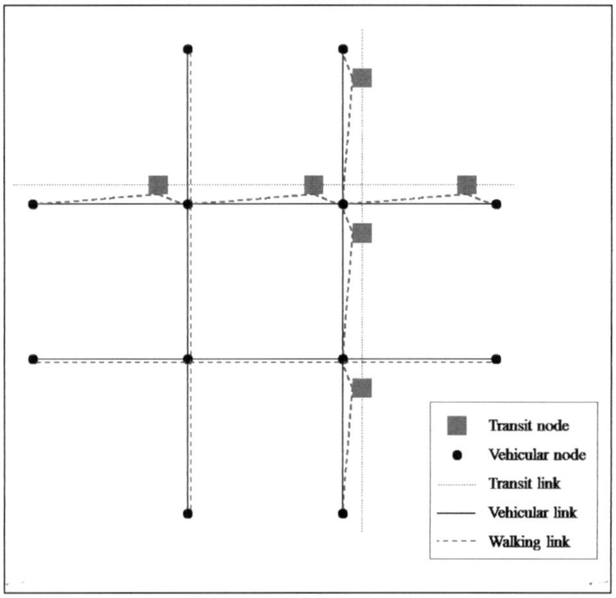

Figure 1. Multimodal network representation.

Time-Dependent Intermodal A* Algorithm

Notation

The following notation is used throughout the rest of the paper:

Sets. N = set of nodes,
A = set of links,
O = set of origin links; $O \subseteq A$,
D = set of destination links; $D \subseteq A$,
A_i^+ = set of successor links of link $i \in A$,
P = set of patterns,
P_i = set of patterns that pass through link $i \in A$; $P_i \subseteq P$,
A_p = ordered set of links that belong to pattern $p \in P$; links are ordered by their sequence traversal by the pattern,
Q = set of trips,
Q_p = ordered set of trips that belong to pattern $p \in P$; trips are ordered by departure time,
Ω = set of scan-eligible links; $\Omega \subseteq A$.

Parameters. δ = traveler's departure time from origin,
x = base transfer penalty,
w_ω = weight of time spent waiting,
w_κ = weight of time spent walking,
w_λ = weight of time spent in a transit vehicle,
w_μ = weight of time spent in a traffic vehicle,
U^ω = upper bound on individual waiting time,
U^κ = upper bound on total walking time,
U^μ = upper bound on total time in a traffic vehicle,
m_i = type of link $i \in A$; $m_i \in \{\text{Transit}, \textit{Vehicular}, \textit{Walk}\}$
l_i = length of link $i \in A$,
v_κ = walking speed,
$\tau_i(t)$ = time-dependent travel time on a vehicular link $i \in A$ at time t,
τ_i = travel time on a walking link; $\tau_i = l_i/v_\kappa$,
$\psi_{ij}(t)$ = time-dependent switching delay from link $i \in A$ at time t to link $j \in A$,
s_{ip} = sequence number of link $i \in A_p$ along pattern $p \in P_i$,
s = a generic sequence number,
$a^q_{s_{ip}}$ = arrival time of trip $q \in Q_p$ at the upstream node of link $i \in A_p$; $p \in P_i$ and s_{ip} is the sequence number of link $i \in A_p$ along pattern $p \in P_i$,
$d^q_{s_{ip}}$ = departure time of trip $q \in Q_p$ from the upstream node of link $i \in A_p$; $p \in P_i$ and s_{ip} is the sequence number of link $i \in A_p$ along pattern $p \in P_i$,
H_i = estimated cost label from link $i \in A$ to destination.

Decision Variables. π_i = selected predecessor link of link $i \in A^+_{\pi_i}$,
φ_i = selected trip on link $i \in A$; $\varphi_i \in Q_p$ and $p \in P_i$.

Endogenous Variables. G_i: Generalized cost label from origin to link $i \in A$,
F_i = total cost label from origin to destination through link $i \in A$,
ρ_i = total number of waiting occurrences from origin to link $i \in A$,
ω_i = total waiting time from origin to link $i \in A$,
χ_i = total transfer penalty from origin to link $i \in A$,
κ_i = total walking time from origin to link $i \in A$,
λ_i = total in-transit-vehicle time from origin to link $i \in A$,
μ_i = total in-traffic-vehicle time from origin to link $i \in A$,
α_i = arrival time at the downstream node of link $i \in A$.

Algorithm

It is important to emphasize once again that the transit portion of the TDIMA*algorithm is similar to Khani et al.'s TBA* algorithm (*37*) except that it does not follow the downstream stops (links) of a trip. The presented algorithm is link-based. Any label such as the generalized cost G_i is associated with the downstream node of the link. In other words, G_i is the generalized cost of the path from the upstream node of the origin link to the downstream node of link $i \in A$.

This algorithm is developed under the framework of POLARIS (*38*) although it serves as a stand-alone tool, as well. In POLARIS, a trip starts and terminates at designated activity locations as opposed to TAZs. Every activity location is projected on a walking link and a vehicular link. To prevent unnecessary U-turns or unrealistic walking back-and-forth, the reverse links are also associated with the given activity location. As a result, the algorithm query is from a set of origin links $i \in O$ (one or two elements) to a set of destination links $i \in D$ (one or two elements). If the selected mode is PNR/KNR, the mode of the origin links m_i = Vehicular, and the mode of the destination links m_i = Walking. If the selected mode is transit, the mode of the origin and destination links m_i = Walking.

This algorithm uses several upper bounds and conditionals for improved computational performance, as well as realism. All cost components such as waiting, walking, and so forth are weighted. These weights w, and the walk-speed v_κ are traveler-dependent, which enables modeling of traveler heterogeneity.

The version presented in the following pseudo-code is for walk-to-transit, and drive-to-transit modes, such as PNR and KNR. Figure 2 presents the main algorithm. The labels G_i and F_i of every link $i \in A \backslash O$ are set to infinity. The label G_i of every origin link $i \in O$ is set based on the mode (walking or driving). It is the cost of traversing the link based on the weighted travel time of walking $w_\kappa \tau_i$ or weighted time-dependent travel time of driving $w_\mu \tau_i(\delta)$ at the departure time δ. Similarly, total walking time κ_i, total in-vehicle-traffic time μ_i, and the arrival time at the downstream node α_i are updated.

H_i is the precalculated heuristic cost from the downstream node of a given link $i \in A$ to the destination links $i' \in D$. The algorithm selects the link with the least total cost $F_i = G_i + H_i$. If $H_i = 0$, the algorithm becomes a conventional label setting algorithm (*1*), where the link with the minimum cost from the origin is selected. Although this guarantees an exact solution, the algorithm will not be "guided" towards the destination, so it will require a longer computational time. Therefore, a heuristic cost H_i is used, and the algorithm picks the link with the minimum total cost F_i. F_i is the cost of a path from origin to destination that traverses link $i \in A$ (*33*). The cost G_i from origin to link $i \in A$ is exact, whereas the cost H_i from link $i \in A$ to destination is an estimate. The higher the estimate, the faster the code reaches the destination, with a caveat: if it is inadmissibly high, the solution is not the least cost path. Still, the final cost at the destination link is $F_i = G_i$, where $H_i = 0$. This means that the final cost is the "exact" cost of a suboptimal path. Traditionally, H_i is the Euclidean distance divided by the maximum possible speed, which is an admissible lower bound (*33*). Similar to Khani et al. (*37*), this study uses the results of a static label setting algorithm for the estimated cost H_i. The travel times on the static network are the minimum observed travel time on a link. Waiting times and transfer or turn penalties are not included. Therefore, the cost on a given link in this static network can never be higher than the cost of that link in the actual network, which makes the algorithm admissible (*37*). Instead of performing the label setting algorithm from every link to every link, the algorithm is performed from every TAZ to every link to reduce computational time. Therefore, H_i is the heuristic cost from link i's zone to the destination link. However, the zonal aggregation is for the heuristic cost H_i calculation only. The TDIMA*algorithm calculates an exact cost G_i from an origin location to the link $i \in A$ in question until eventually that link is one of the (two) destination links $i \in D$. Back-tracking recursively from that link terminates in one of the origin links. Therefore, the algorithm finds the least cost path between an origin–destination pair at a given departure time, where both the optimal origin and the optimal destination links are implicitly selected in the process.

Once the link $i = \underset{i \in \Omega}{\operatorname{argmin}}\{F_j\}$ with the minimum total label is selected, the algorithm starts scanning the downstream links $j \in A_i^+$. If the downstream mode m_j = Transit, then *Evaluate_Transit_Neighbor(i,j)* is called (see Figure 3). If the downstream mode m_j = Walk, then *Evaluate_Non_Transit_Neighbor(i,j)* is called (see Figure 4). Finally, if both the current mode m_i = Vehicular and m_j = Vehicular, then *Evaluate_Non_Transit_Neighbor(i,j)* is called again. The last two conditions are proper for PNR and KNR trips. No vehicle can be generated in the middle of the journey. However, this constraint is relaxed if the intermodal trip allows for taking a taxi or other services. The algorithm terminates ideally once a destination link $i \in D$ is reached, or unideally when Ω becomes empty, that is, no path could be found between the two location pairs (O, D) at departure time δ.

In the *Evaluate_Transit_Neighbor(i,j)* subroutine, the algorithm loops over all the patterns $p \in P_j$ that serve on the successor link $j \in A_i^+$. For a given pattern $p \in P_j$, it loops over all the trips $q \in Q_p$ of that pattern. First, the sequence number s_{jp} of link $j \in A_p$ along pattern $p \in P_j$ is obtained. For simplicity, it is called s. By using s, the algorithm obtains the departure time d_s^q and calculates the waiting time $\omega' = d_s^q - \alpha_i$. If $\omega' < 0$, the trip is skipped. If ω' is larger than the traveler-specific upper bound U^ω, then all the trips departing later than q will exceed the upper bound. Therefore, no further scanning of this pattern is required. In the case of the trip q in question being the same as the arrival trip φ_i, then there is no waiting. Therefore the candidate waiting count ρ' is the same as the waiting count ρ_i. The in-transit-vehicle travel time of the link is set to the difference between the arrival time a_{s+1}^q at the downstream node of link j and

TDIMA* (O, D, δ)

Initaliation

```
Ω = {}                          //the set of scan eligible links is set as empty
For every link i ∈ A\O:
    G_i = F_i = ∞               //the non-origin link labels are set as infinity

For every origin link i ∈ O:

    If m_i = Walk:              //if origin link is a walking link
        G_i = w_κ τ_i           //generalized cost label is the weighted walking time
        κ_i = τ_i               //total walking time is the link's walking time
        μ_i = 0                 //total in-traffic-vehicle time is zero
        α_i = δ + τ_i           //arrival time is departure time plus link travel time

    Else:                       //if origin link is a traffic vehicle link
        G_i = w_μ τ_i(δ)        //generalized cost label is the weighted time-dependent in-traffic-vehicle time
        κ_i = 0                 //total walking time is zero
        μ_i = τ_i(δ)            //total in-traffic-vehicle time is the time-dependent in-traffic-vehicle time
        α_i = δ + τ_i(δ)        //arrival time is departure time plus link travel time

    π_i = {}                    //there is no predecessor to origin link
    φ_i = {}                    //an origin link is never a transit link, hence there is no transit trip
    F_i = G_i + H_i             //total cost label is generalized cost label plus heuristic cost
    ρ_i = ω_i = χ_i = λ_i = 0   //all other labels are zero
    Ω = Ω ∪ {i}                 //the set of scan eligible links now includes i
```

Main

```
While Ω ≠ {}:
    Select i = argmin_{j∈Ω}{F_j}   //select the link from the set of scan eligible links with the minimum total cost

    Ω = Ω\{i}                      //remove the current link i from the set of scan eligible links

    If i ∈ D:                      //if we reached the destination terminate the algorithm
        Stop and report results!

    Else:
        For every j ∈ A_i^+ where m_j = Transit:        //scan and evaluate every successor transit link
            Call Evaluate_Transit_Neighbor(i,j)

        For every j ∈ A_i^+ where m_j = Walk:           //scan and evaluate every successor walking link
            Call Evaluate_Non_Transit_Neighbor(i,j)

        If m_i = Vehicular:
            For every j ∈ A_i^+ where m_j = Vehicular:  //scan and evaluate every successor vehicular link
                Call Evaluate_ Non_Transit_Neighbor(i,j)
```

Figure 2. Pseudo-code for time-dependent intermodal A* (TDIMA*) algorithm.

the current time α_i. Otherwise, the traveler is waiting to board, the waiting count $\rho' = \rho_i + 1$, and the in-transit-vehicle travel time μ' is the difference between the arrival time a^q_{s+1} at the downstream node of link j and the departure time d^q_s from the upstream node of link j.

The algorithm keeps track of the total number of transfers $\max\{0, \rho' - 1\}$ and progressively penalizes transfers using $\chi' = \max\{0, \rho' - 1\}x$. That is the first transfer is penalized by x, the second by $2x$, and so on. Therefore, a trip with two transfers has a total transfer penalty of $3x$,

Evaluate_Transit_Neighbor(i,j)

For every pattern $p \in P_j$:
  For every trip $q \in Q_p$:

$s = s_{jp}$ — *//get the sequence number of link j along pattern p*
$\omega' = d_s^q - \alpha_i$ — *//waiting time is departure time minus current time*

If $\omega' < 0$, go to next trip — *//if waiting time is negative skip that trip*

If $\omega' > U^{\omega}$, go to next pattern — *//if waiting time is higher than the upper bound skip that pattern*

If m_i = Transit AND $q = \varphi_i$: — *//if the candidate trip q equals to the arrival trip φ_i*
  $\omega' = 0$ — *//then there is no waiting*
  $\rho' = \rho_i$ — *//candidate waiting count equals to the value of current link i*
  $\mu' = a_{s+1}^q - \alpha_i$ — *//in-transit-vehicle time is arrival at the next stop minus current time*

Else:
  $\rho' = \rho_i + 1$ — *//then there is waiting and candidate waiting count is increased by one*
  $\mu' = a_{s+1}^q - d_s^q$ — *//in-transit-vehicle time is arrival at the next stop minus departure time*

$\chi' = \max\{0, \rho' - 1\} x$ — *//transfer penalty is base penalty times (waiting count minus one) or zero*
$G' = G_i + w_{\omega}\omega' + w_{\mu}\mu' + \chi'$ — *//candidate generalized cost is a weighted sum of previous link's cost, //waiting time, in-transit-vehicle time, and transfer penalty*

If $G' < G_j$: — *//if the candidate cost is less than the prevailing generalized cost label*

  $G_j = G'$ — *//generalized cost label is set as the candidate cost label*
  $F_j = G_j + H_j$ — *//total cost label is generalized cost label plus heuristic cost*

  $\pi_j = i$ — *//predecessor link of j is set as i*
  $\varphi_j = q$ — *//selected trip on link j is set as q*
  $\alpha_j = a_{s+1}^q$ — *//arrival time at the downstream node is set as the arrival time of the trip q*

  $\rho_j = \rho'$ — *//total number of waiting occurrences from origin is set as the candidate value*
  $\omega_j = \omega_i + \omega'$ — *//total waiting time from origin is increased by ω'*
  $\chi_j = \chi'$ — *//total transfer penalty from origin is set as the candidate value*
  $\kappa_j = \kappa_i$ — *//total walking time from origin is set as the value of current link i*
  $\lambda_j = \lambda_i + \lambda'$ — *//total in-transit-vehicle time from origin is increased by λ'*
  $\mu_j = \mu_i$ — *//total in-traffic-vehicle time from origin is set as the value of current link i*

  If $j \notin \Omega$, then $\Omega = \Omega \cup \{j\}$ — *//add link j to the set of scan eligible links if it is not already there*

Figure 3. Subroutine for evaluating a transit neighbor link.

whereas a trip with three transfers has a total transfer penalty of $6x$. Interested readers are referred to (*21*, *22*) for details. The candidate label G' is the weighted sum of the waiting time ω', in-transit-vehicle time μ', the transfer penalty χ', and the label G_i of the current link i. If $G' < G_j$ then link j should be traversed on trip $\varphi_j = q$ and be preceded by link $\pi_j = i$. The arrival time is set to $\alpha_j = a_{s+1}^q$. All the labels are updated as seen in Figure 3.

As seen in Figure 4, if the successor link j's mode m_j = Walk, then the algorithm first checks whether the potential cumulative walking time $\kappa_i + \tau_j$ is larger than the traveler-specific upper bound U^{κ}. If it is true, the algorithm skips that link. If not, the algorithm calculates the candidate label G', the candidate arrival time α', the candidate total walking time κ', and the candidate in-traffic-vehicle time μ'.

If the successor link j's mode m_j = Vehicular, the time-dependent travel time has two components: the time-dependent switching delay $\psi_{ij}(\alpha_i)$ at time α_i and the time-dependent link travel time $\tau_i(\alpha_i + \psi_{ij}(\alpha_i))$ at time $\alpha_i + \psi_{ij}(\alpha_i)$, that is, after the turn movement is completed. As for walking, the algorithm checks whether the cumulative driving time exceeds the upper bound U^{μ} on driving. This prevents the algorithm from excessive driving almost all the way up to the destination and then taking transit for just a few blocks. If the upper bound is not exceeded, the algorithm calculates the candidate label G', the candidate arrival time α', the candidate total walking time κ', and the candidate in-traffic-vehicle time μ'.

If $G' < G_j$ then link j should be preceded by link $\pi_j = i$. All the labels are updated as seen in Figure 4.

Evaluate_Non_Transit_Neighbor(i,j)

If m_j = Walk:

If $\kappa_i + \tau_j > U^{\kappa}$:
Return //*if the total walking time is larger than the upper bound return to main*

$G' = G_i + w_{\kappa}\tau_j$ //*candidate generalized cost is weighted walk time plus current link's cost*
$\alpha' = \alpha_i + \tau_j$ //*candidate arrival time is increased by walking time*
$\kappa' = \kappa_i + \tau_j$ //*candidate walking time is increased by walking time*
$\mu' = \mu_i$ //*candidate in-traffic-vehicle time is set as the value of current link i*

If m_i = Vehicular:

If $\mu_i + \psi_{ij}(\alpha_i) + \tau_j\left(\alpha_i + \psi_{ij}(\alpha_i)\right) > U^{\mu}$:
Return //*if the weighted total in-traffic-vehicle time is larger than the upper bound return to main*

$G' = G_i + w_{\mu}\left(\psi_{ij}(\alpha_i) + \tau_j\left(\alpha_i + \psi_{ij}(\alpha_i)\right)\right)$
//*candidate generalized cost is a weighted sum of previous link's cost,*
//*time-dependent switching delay, and time-dependent in-traffic-vehicle time*
$\alpha' = \alpha_i + \psi_{ij}(\alpha_i) + \tau_j\left(\alpha_i + \psi_{ij}(\alpha_i)\right)$
//*candidate arrival time is increased by the time-dependent switching delay,*
//*and the time-dependent in-traffic-vehicle time*
$\kappa' = \kappa_i$ //*candidate walking time is set as the value of current link i*
$\mu' = \mu_i + \psi_{ij}(\alpha_i) + \tau_j\left(\alpha_i + \psi_{ij}(\alpha_i)\right)$
//*candidate walking time is increased by the time-dependent switching delay,*
//*and the time-dependent in-traffic-vehicle time*

If $G' < G_j$ //*if the candidate cost is less than the prevailing generalized cost label*

$G_j = G'$ //*generalized cost label is set as the candidate cost label*
$F_j = G_j + H_j$ //*total cost label is generalized cost label plus heuristic cost*

$\pi_j = i$ //*predecessor link of j is set as i*
$\varphi_j = \{\}$ //*there is no transit trip*
$\alpha_j = \alpha'$ //*arrival time at the downstream node is set as the candidate arrival time*

$\rho_j = \rho_i$ //*total number of waiting occurrences is set as the value of current link i*
$\omega_j = \omega_i$ //*total waiting time from origin is set as the value of current link i*
$\chi_j = \chi_i$ //*total transfer penalty from origin is set as the value of current link i*
$\kappa_j = \kappa'$ //*total walking time from origin is set as candidate walking time*
$\lambda_j = \lambda_i$ //*total in-transit-vehicle time from origin is set as the value of current link i*
$\mu_j = \mu'$ //*total in-traffic-vehicle time from origin is set as candidate in-traffic-vehicle time*

If $j \notin \Omega$, then $\Omega = \Omega \cup \{j\}$ //*add link j to the set of scan eligible links if it is not already in there*

Figure 4. Subroutine for evaluating a nontransit neighbor link.

Validation Results

The time-dependent intermodal A* algorithm implementation was verified against observed trips drawn from the Chicago Metropolitan Agency for Planning (CMAP) 2008 Travel Tracker survey. A multimodal Chicago Regional network was generated using the existing POLARIS (*42*) vehicular network and the GTFS (*38*) data for the three transit agencies; Chicago Transit Authority (CTA) (urban bus and rail), PACE (suburban bus), and METRA (commuter rail). Below are the network properties:

- 54,028 nodes
  - 35,077 transit
  - 18,951 vehicular
- 217,119 links
  - 37,642 transit
  - 123,000 walking
  - 56,477 vehicular
- 173,236 activity locations
- 1,961 TAZs
- 344 transit routes
- 2,098 transit patterns
- 28,138 transit trips

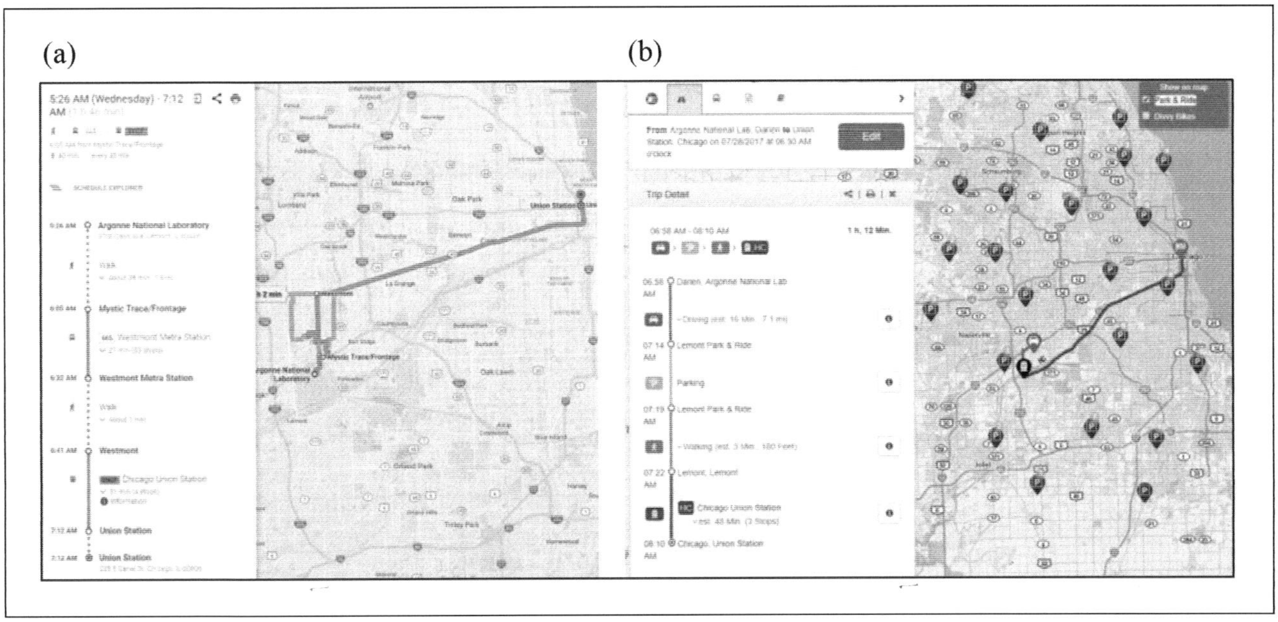

Figure 5. Online routing results from (*a*) Google API; and (*b*) RTA trip planner.

The validation trips were drawn from the set of multimodal trips observed in the survey, that is, a trip which involved transit at any point and involved at least one nonwalking leg. This subset was further limited to multimodal trips for which the access mode was PNR/KNR trips. Out of a total set of 6,500 valid transit trips, 556 trips met these criteria and passed a set of data validity checks for use in the validation analysis. A subset of 180 randomly selected trips was then routed from origin coordinate to destination coordinate at the specified departure time in three different routing engines:

- TDIMA* algorithm using PNR mode,
- Google Directions API using transit mode (*43*), as it does not have PNR/KNR option,
- Regional Transit Authority (RTA) trip planner (*44*) using PNR mode, as it provides PNR directions.

The reason for selecting 180 trips is the need to input trips manually into the RTA router. Results from the two online direction finders for a hypothetical trip are shown in Figure 5, where significant differences in overall trip time can be observed. The main reason behind this difference is that the Google router can only have walking as the access mode to transit, whereas the RTA trip planner is able to find a PNR route which selects from a set of prespecified parking locations. On the other hand, the TDIMA* algorithm is much more flexible in the sense that it does not limit the parking locations.

The TDIMA* algorithm used the following weights and upper bounds:

- Base transfer penalty; $x = 5\text{min}$,
- Weight of time spent waiting; $W_{\omega} = 2$,
- Weight of time spent walking; $W_K = 2$,
- Weight of time spent in a transit vehicle; $w_{\lambda} = 1$,
- Weight of time spent in a traffic vehicle, $w_{\mu} = 3$,
- Upper bound on individual waiting time; $U^{\omega} = 60\,\text{min}$,
- Upper bound on total walking time; $U^{\kappa} = 60\,\text{min}$; with walking speed $v_{\kappa} = 1.39\,\text{m/s}$; $U^{\kappa} = 5\,\text{km}$ in distance,
- Upper bound on total time in a traffic vehicle; $U^{\mu} = 60\,\text{min}$.

The results for each trip were compared for total travel time, total walking/driving (nontransit) time, total waiting time (including waiting during transfers), and total in-transit-vehicle time (see Figure 6). It is not known which upper bounds, weights, or transfer penalties are used by the online algorithms, or, if they are used, the values thereof. However, the Google Directions API does not return results for about 20% of the trips. Therefore, it can be inferred that there are limits on walking. Overall, the new algorithm significantly outperforms the online routing engines. Although the details of the online routers are not known in terms of upper bounds and weights, the results indicate that the TDIMA* algorithm outperforms them for total travel, waiting, transit, and

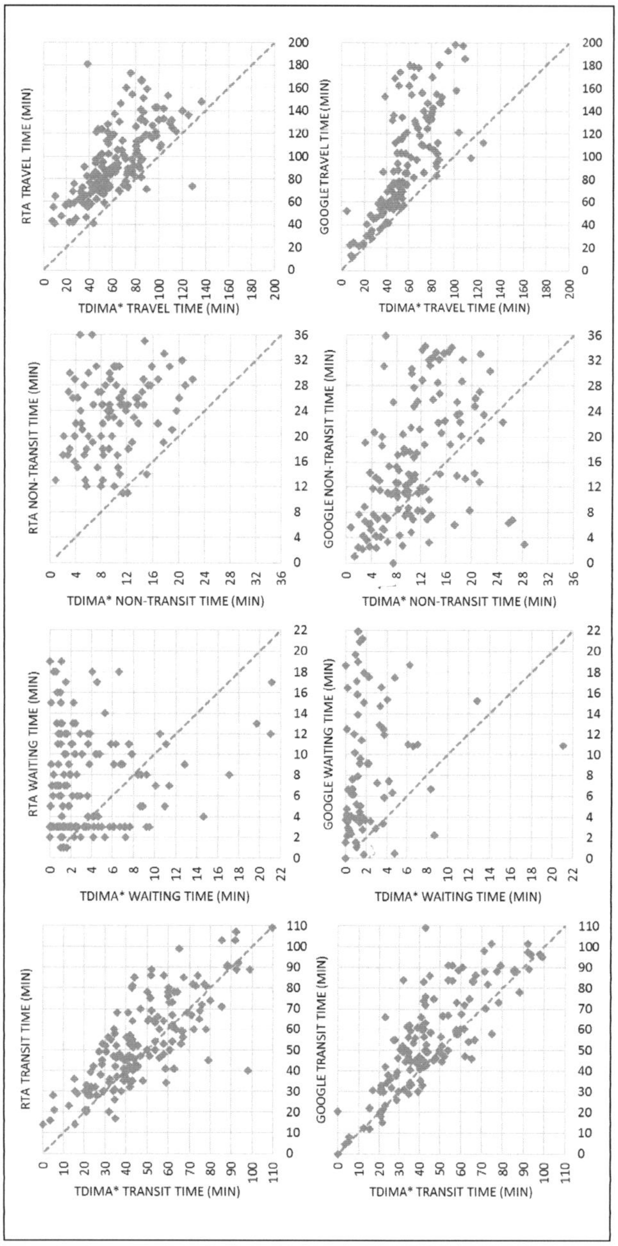

Figure 6. Comparison of TDIMA* results to online routers.

nontransit times. Therefore, there is no trade-off between the cost components across different algorithms. The TDIMA* algorithm is mostly nondominated.

Conclusion

A time-dependent intermodal A* (TDIMA*) algorithm was introduced and described. The algorithm works on a multimodal network with transit, walking, and vehicular network links. The TDIMA* algorithm finds paths for the three major modes (transit, walking, driving) and any feasible combination thereof (e.g., PNR). For improved realism, turn penalties, progressive transfer penalties, upper bounds on waiting, walking, and driving (where necessary) are incorporated. The algorithm is validated on the large-scale Chicago Regional network using real-world trips, and the results are compared with two online services (Google Directions API and RTA Trip Planner). The results provided by the TDIMA* for PNR trips are significantly better than the ones provided by either service.

As a next step, the algorithm will incorporate seating and standing capacity penalties, as well as monetary costs, such as tolls, parking fees, and transit fares. Although this algorithm can serve as a stand-alone routing tool, it is designed as an integral component of POLARIS, software for integrated activity-based modeling and traffic-transit simulation. This integration will allow for comprehensive modeling capabilities such as supply-side interactions between different modes, abandoning the use of skims for mode choice and activity scheduling, and en-route mode switching and rerouting.

Acknowledgments

Argonne National Laboratory's work was supported by the U.S. Department of Transportation, Federal Transit Administration under interagency agreement, through U.S. Department of Energy contract DE-AC02-06CH121357. The submitted manuscript has been created by UChicago Argonne, LLC, Operator of Argonne National Laboratory ("Argonne"). Argonne, a U.S. Department of Energy Office of Science laboratory, is operated under Contract No. DE-AC02-06CH11357. The U.S. Government retains for itself, and others acting on its behalf, a paid-up, nonexclusive, irrevocable worldwide license in said article to reproduce, prepare derivative works, distribute copies to the public, and perform publicly and display publicly, by or on behalf of the Government.

Author Contributions

The authors confirm contribution to the paper as follows: study conception and design: Ömer Verbas, Joshua Auld, Randy Weimer; data collection: Ömer Verbas, Hubert Ley, Shon Driscoll; analysis and interpretation of results: Ömer Verbas, Joshua Auld; draft manuscript preparation: Ömer Verbas. All authors reviewed the results and approved the final version of the manuscript.

References

1. Dijkstra, E. W. A Note on Two Problems in Connexion with Graphs. *Numerische Mathematik*, Vol. 1, No. 1, 1959, pp. 269–271.
2. Ford, L. R Jr. *Network Flow Theory*. DTIC Document, 1956.
3. Bellman, R. On a Routing Problem. *Quarterly of Applied Mathematics*, Vol. 16, No. 1, 1958, pp. 87–90.

4. Cooke, K. L., and E. Halsey. The Shortest Route through a Network with Time-Dependent Internodal Transit Times. *Journal of Mathematical Analysis and Applications*, Vol. 14, 1966, pp. 493–498.
5. Glover, F., D. Klingman, and N. Phillips. A New Polynomially Bounded Shortest Path Algorithm. *Operations Research*, Vol. 33, 1985, pp. 65–73.
6. Ziliaskopoulos, A. K., and H. S. Mahmassani. Time-Dependent, Shortest Path Algorithm for Real-Time Intelligent Vehicle Highway Systems Applications. *Transportation Research Record: Journal of the Transportation Research Board*, 1993. 1408: 94–100.
7. Ziliaskopoulos, A. K. *Optimum Path Algorithms on Multidimensional Networks: Analysis, Design, Implementation and Computational Experience*. University of Texas at Austin, Austin, TX, 1994.
8. Ziliaskopoulos, A. K., and H. S. Mahmassani. A Note on Least Time Path Computation Considering Delays and Prohibitions for Intersection Movements. *Transportation Research Part B: Methodological*, Vol. 30, 1996, pp. 359–367.
9. Ziliaskopoulos, A. K., and W. Wardell. An Intermodal Optimum Path Algorithm for Multimodal Networks with Dynamic Arc Travel Times and Switching Delays. *European Journal of Operational Research*, Vol. 125, 2000, pp. 486–502.
10. Abdelghany, K. F., and H. S. Mahmassani. Dynamic Trip Assignment-Simulation Model for Intermodal Transportation Networks. *Transportation Research Record: Journal of the Transportation Research Board*, 2001. 1771: 52–60.
11. Mahmassani, H. S., and K. Abdelghany. Dynasmart-IP: Dynamic Traffic Assignment Meso-Simulator for Intermodal Networks. *Advanced Modeling for Transit Operations and Service Planning*, Pergamon, Turkey, pp. 201–229.
12. Abdelghany, K. F., H. S. Mahmassani, and A. F. Abdelghany. A Modeling Framework for Bus Rapid Transit Operations Evaluation and Service Planning. *Transportation Planning and Technology*, Vol. 30, 2007, pp. 571–591.
13. Chriqui, C., and P. Robillard. Common Bus Lines. *Transportation Science*, Vol. 9, 1975, pp. 115–121.
14. Nguyen, S., and S. Pallottino. Equilibrium Traffic Assignment for Large Scale Transit Networks. *European Journal of Operational Research*, Vol. 37, 1988, pp. 176–186.
15. Nguyen, S., and S. Pallottino. Hyperpaths and Shortest Hyperpaths. In *Combinatorial Optimization* (Simeone, B., ed.), Springer Berlin Heidelberg, Germany, 1989, pp. 258–271.
16. Spiess, H., and M. Florian. Optimal Strategies: A New Assignment Model for Transit Networks. *Transportation Research Part B*, Vol. 23, 1989, pp. 83–102.
17. De Cea, J., and E. Fernández. Transit Assignment for Congested Public Transport Systems: An Equilibrium Model. *Transportation Science*, Vol. 27, 1993, pp. 133–147.
18. Cominetti, R., and J. Correa. Common-Lines and Passenger Assignment in Congested Transit Networks. *Transportation Science*, Vol. 35, 2001, pp. 250–267.
19. Schmöcker, J.-D., M. G. H. Bell, and F. Kurauchi. A Quasi-Dynamic Capacity Constrained Frequency-Based Transit Assignment Model. *Transportation Research Part B: Methodological*, Vol. 42, 2008, pp. 925–945.
20. Schmöcker, J.-D., A. Fonzone, H. Shimamoto, F. Kurauchi, and M. G. H. Bell. Frequency-Based Transit Assignment Considering Seat Capacities. *Transportation Research Part B: Methodological*, Vol. 45, 2011, pp. 392–408.
21. VerbasÖ. *Transit Network Assignment, Simulation and Frequency Setting: Integrated Approaches and Large Scale Application*. Northwestern University, Evanston, IL, 2014.
22. Verbas, İ. Ö., and H. S. Mahmassani. Finding Least Cost Hyperpaths in Multimodal Transit Networks. *Transportation Research Record: Journal of the Transportation Research Board*, 2015. 2497: 95–105.
23. Friedrich, M., I. Hofsaess, and S. Wekeck. Timetable-Based Transit Assignment Using Branch and Bound Techniques. *Transportation Research Record: Journal of the Transportation Research Board*, 2001. 1752: 100–107.
24. Noh, H. *Capacitated Schedule-Based Transit Assignment Using a Capacity Penalty Cost*. University of Arizona, Tuscon, AZ, 2013.
25. Hamdouch, Y., and S. Lawphongpanich. Schedule-Based Transit Assignment Model with Travel Strategies and Capacity Constraints. *Transportation Research Part B: Methodological*, Vol. 42, 2008, pp. 663–684.
26. Hamdouch, Y., and S. Lawphongpanich. Congestion Pricing for Schedule-Based Transit Networks. *Transportation Science*, Vol. 44, 2010, pp. 350–366.
27. Hamdouch, Y., H. W. Ho, A. Sumalee, and G. Wang. Schedule-Based Transit Assignment Model with Vehicle Capacity and Seat Availability. *Transportation Research Part B: Methodological*, Vol. 45, 2011, pp. 1805–1830.
28. Noh, H., M. Hickman, and A. Khani. Hyperpaths in Network Based on Transit Schedules. *Transportation Research Record: Journal of the Transportation Research Board*, 2012. 2284: 29–39.
29. Lu, C.-C., H. S. Mahmassani, and X. Zhou. A Bi-Criterion Dynamic User Equilibrium Traffic Assignment Model and Solution Algorithm for Evaluating Dynamic Road Pricing Strategies. *Transportation Research Part C: Emerging Technologies*, Vol. 16, 2008, pp. 371–389.
30. Lu, C.-C., and H. S. Mahmassani. Modeling User Responses to Pricing: Simultaneous Route and Departure Time Network Equilibrium with Heterogeneous Users. *Transportation Research Record: Journal of the Transportation Research Board*, 2008. 2085: 124–135.
31. Lu, C.-C., and H. S. Mahmassani. Dynamic Pricing with Heterogeneous Users: Gap-Driven Solution Approach for Bicriterion Dynamic User Equilibrium Problem. *Transportation Research Record: Journal of the Transportation Research Board*, 2009. 2090: 75–85.
32. Lu, C.-C., and H. S. Mahmassani. Modeling Heterogeneous Network User Route and Departure Time Responses to Dynamic Pricing. *Transportation Research Part C: Emerging Technologies*, Vol. 19, 2011, pp. 320–337.
33. Hart, P. E., N. J. Nilsson, and B. Raphael. A Formal Basis for the Heuristic Determination of Minimum Cost Paths. *IEEE Transactions on Systems Science and Cybernetics*, Vol. 4, No. 2, 1968, pp. 100–107.
34. Chabini, I., and S. Lan. Adaptations of the A* Algorithm for the Computation of Fastest Paths in Deterministic Discrete-Time Dynamic Networks. *IEEE Transactions on*

Intelligent Transportation Systems, Vol. 3, No. 1, 2002, pp. 60–74.
35. Zhao, L., T. Ohshima, and H. Nagamochi. A* Algorithm for the Time-Dependent Shortest Path Problem. *Proc., WAAC08: The 11th Japan-Korea Joint Workshop on Algorithms and Computation*, Fukuoka, Japan, 2008.
36. Bander, J. L., and C. C. White. A New Route Optimization Algorithm for Rapid Decision Support. *Proc. SAE/IEEE-VTS VNIS*, Detroit, MI, 1991, pp. 709–728.
37. Khani, A., M. Hickman, and H. Noh. Trip-Based Path Algorithms Using the Transit Network Hierarchy. *Networks and Spatial Economics*, Vol. 15, No. 3, 2015, pp. 635–653.
38. Google Developers. *General Transit Feed Specification (GTFS) Static Overview*, https://developers.google.com/transit/gtfs/. Accessed May 1, 2017.
39. Furth, P. G., and F. B. Day. Transit Routing and Scheduling Strategies for Heavy-Demand Corridors. *Transportation Research Record: Journal of the Transportation Research Board*, 1985. 1011: 23–26.
40. Furth, P. G. Alternating Deadheading in Bus Route Operations. *Transportation Science*, Vol. 19, 1985, pp. 13–28.
41. Furth, P. G. Zonal Route Design for Transit Corridors. *Transportation Science*, Vol. 20, 1986, pp. 1–12.
42. Auld, J., M. Hope, H. Ley, V. Sokolov, B. Xu, and K. Zhang. POLARIS: Agent-Based Modeling Framework Development and Implementation for Integrated Travel Demand and Network and Operations Simulations. *Transportation Research Part C: Emerging Technologies*, Vol. 64, 2016, pp. 101–116.
43. Google Developers. *Google Maps Directions API*, https://developers.google.com/maps/documentation/directions/. Accessed July 1, 2017.
44. Regional Transit Authority. *RTA Trip Planner*, http://www.rtachicago.org/plan-your-trip. Accessed July 1, 2017.

The Standing Committee on Traveler Behavior and Values (ADB10) peer-reviewed this paper (18-06765).

Article

TRR
JOURNAL OF THE TRANSPORTATION RESEARCH BOARD

Transportation Research Record
2018, Vol. 2672(47) 231–241

DOI: 10.1177/0361198118774750
journals.sagepub.com/home/trr
SAGE

Investigating Behavioral Differences in Heterogeneous Decision Rule Segments: An Empirical Analysis

Parthan Kunhikrishnan[1,2] and Karthik K. Srinivasan[1]

Abstract

Conventional and contemporary models of travel choice make the restrictive assumption of homogeneity in decision rules. Recent literature has shown empirical evidence for potential heterogeneity in decision rules with regard to utility maximization and regret minimization. Notwithstanding these advances in modeling decision rules, behavioral understanding in the differences in these alternative decision rule segments has not been sufficiently understood. Moreover, the factors which influence the choice of these decision rules have not received significant attention. This study proposes a framework which considers decision makers to have both utility maximizing and regret minimizing tendencies. The variation in these tendencies across decision makers renders the framework heterogeneous. A heterogeneous decision rule model is developed assuming the decision rule adopted to be a latent construct. The study characterizes the regret minimizing and utility maximizing segments based on average values of the segmental attributes. The empirical findings show evidence to confirm that utility maximizers tend to be predominantly captive to personal vehicle usage while regret minimizers might be non-captive to any particular mode. The nature and extent of influence of factors affecting the choice of decision rule is also examined.

Choice of travel mode is one of the most important decisions for urban commuters. When choosing their travel mode, most decision makers face a set of alternatives. Comparing these alternatives, the decision maker makes his or her choice of travel mode. This choice is made by considering a variety of factors including the attributes of the alternatives, trip characteristics, personal and household characteristics, etc. Every rational decision maker is assumed to adopt certain heuristics to process the information regarding these influential factors in order to arrive at his or her choice. This set of heuristics is called decision rule(s).

Decision rules can be broadly categorized as: fully compensatory, non-compensatory, and semi-compensatory in nature (*1*). A fully compensatory decision rule is one where the poor performance of an alternative on one attribute can be completely compensated by good performance on an equally important attribute (*2*). In a non-compensatory rule, poor performance on one attribute cannot be overcome by an improvement in another, regardless of the magnitude of improvement. In semi-compensatory decision rules, this type of compensation is partially possible.

The classic example of a fully compensatory decision rule is that of random utility maximization (RUM). Owing to the ease of its operationalization and its simplicity in representing choice behavior, the utility maximization framework has been in use for quite a long time. Especially with regard to travel-related decisions, utility maximization is the most popular decision rule assumption. However, the utility maximization framework fails to acknowledge the risk-averse nature of travel-related decision making. Recognition of this drawback has led analysts to focus on alternative decision rule frameworks. Random regret minimization (RRM) is one such decision rule, which has received greater attention in the recent past (*2–6*).

The regret theory is based on a decision maker's experiences with an alternative due to poor performance on an attribute. This regret, although it could be compensated to a certain extent by good performance of an

[1]Transportation Engineering Division, Department of Civil Engineering, Indian Institute of Technology, Madras, Chennai, India
[2]Department of Civil Engineering, B.M.S. College of Engineering, Bengaluru, Karnataka, India

Corresponding Author:
Address correspondence to Parthan Kunhikrishnan:
kparthan86@gmail.com

equally important attribute, cannot be nullified completely. This forces the individual to settle for a trade-off between these attributes in making the choice. Decision makers who adopt semi-compensatory decision rules might potentially regret choosing a particular alternative if they eventually come to know that the performance of another alternative was better on an attribute to which they are very sensitive. Therefore, a regret minimizing decision maker attempts to minimize the potential for regret in his or her chosen alternative.

The propensity to minimize potential regret is founded on the natural human tendency of risk aversion. Risk aversion is the reluctance of a person to accept an alternative with an unreliable performance on one or more attributes as compared with another alternative which is more reliable but perhaps has a slightly poorer performance on those attributes. Decision makers tend to become risk averse when faced with situations of high uncertainty (*3*). Regret minimizing frameworks thus, in some sense, account for the risk-averse nature of a decision maker (*3*, *7*). On the contrary, fully compensatory decision rules are conventionally operationalized in a manner that does not account for the risk-averse nature of decision makers. In fact, they present decision makers as risk neutral; for example, the linearly additive linear-in-parameters utility structure in a RUM framework does not account for the decision maker's risk aversion to certain factors. Further, decision makers may associate similar characteristics with respect to different alternatives (e.g., crowding in bus and train) making them correlated. However, a multinomial logit model in a RUM framework assumes independence among irrelevant alternatives (IIA). McFadden (1976) and Ben-Akiva (1977) independently developed a choice set generation model, called the "Dogit" model, which considers decision makers to be either captive to a particular alternative or completely free to choose from a set of alternatives (*8*, *9*). This relaxes the restriction of IIA. Some of the other models which relax the restriction of IIA include the generalized extreme value (GEV) family of models. The most popular among the GEV family is the nested logit model (*10*) which involves nesting of similar alternatives within a nest. The probability of an alternative within a nest is affected by alternatives in other nests and hence IIA does not hold well in these models.

Conventional models assume that all the decision makers adopt the same decision rule, which may not be true. For example, a risk-neutral decision maker is more likely to adopt utility maximization than a risk-averse decision maker. Similarly, a risk-averse decision maker might be more of a regret minimizer than a utility maximizer. The current study is motivated by this potential heterogeneity with regard to decision rules. It is important to relax the decision rule homogeneity assumption since it could constrain the sensitivities to various factors influencing choice. Such restrictive assumptions might lead to incorrect understanding of behaviors. Decision rule heterogeneity might help to capture differential sensitivity to these factors. Understanding of the factors which influence the adoption of these decision rules is also limited (*11*) and hence merits further investigation. Such an exercise might help analysts to identify segments which are potential regret minimizers or utility maximizers. In light of the above discussion, the following objectives have been identified for the study:

- To propose a framework which allows decision makers to be heterogeneous with respect to the decision rule adopted.
- Benchmark the proposed decision rule framework against homogeneous decision rule frameworks like RUM and RRM.
- Capture and analyze the differential sensitivities in the utility maximizing and regret minimizing tendencies of a decision maker.
- Identify the factors influencing the latent decision rule adopted by the decision maker.

The scope of the study is limited to choice travel mode for commuting to work based on household interview survey data collected in Chennai city. The data was found to be representative of the worker population in Chennai. This study contributes to research on mode choice by comparing alternate behavioral frameworks representing distinct decision rules adopted by individuals. It establishes that decision makers tend to adopt different decision rules. Further, it attempts to segment decision makers based on the decision rules adopted and to compare their behavior with regard to travel mode choice. In addition, the study also captures the factors influencing the latent decision rule adopted by decision makers.

Literature Review

Heterogeneity in choice behavior refers to differences in various segments in the population with respect to the elements of decision making: decision maker; alternatives; attributes related to the decision maker and alternatives; and decision rule. In travel behavior analysis, considerable research has been carried out in capturing heterogeneity with regard to: (a) consideration/availability of alternatives in the choice set (*12*, *13*); (b) perceptions of service levels offered by these alternatives and sensitivity to the various attributes of the alternatives which influence their choice (*14*, *15*); and (c) both systematic and unobserved components of attributes related to the decision maker (*16*). However, heterogeneity with regard to the decision rules adopted by decision makers, especially in the context of travel choice, is very sparsely investigated and hence merits investigation. The following discussion highlights the research contributions with regard to decision rules, especially utility maximization and regret minimization, as well as critically evaluating them.

One of the important properties of the utility maximization decision rule is its fully compensatory nature. It is founded on very strong theoretical backgrounds of consumer behavior and random utility theory. Its extensive usage in representing decision making concepts has made it very popular among behavioral analysts. However, one of its key limitations is its inability to represent the risk-averse behavior of decision makers. In this regard, one of the decision rules which has recently gained attention in the literature is regret minimization. It seems to be a promising alternative to utility maximization in terms of its logical intuitiveness, statistical testability, mathematical tractability, and implementability. In addition, regret minimization, through binary comparisons of alternatives, in a subtle sense attempts to capture the risk-averse nature of the decision makers.

Regret theory was originally proposed and independently developed by Loomes and Sugden (*7*) and Bell (*17*) as an alternative theory of rational choice under uncertainty. The theory is based on the fact that an individual has the ability to anticipate feelings of regret and of satisfaction and the model proposed takes these factors into consideration. Founded on regret theory, Chorus et al. (*3*) proposed the random regret minimization approach and emphasized the key differences between RUM and RRM. Chorus (*2*) also presented various theoretical properties of the RRM framework and compared them with the properties of RUM. Following this, a series of regret minimization models were developed to model various dimensions of travel, like mode choice, recreation site choice, revealed shopping choice, stated route choice, etc. (*3*, *6*, *18*, *19*). Some of these experiments also attempted to quantify and compare the willingness-to-pay measures in the RRM and RUM framework using empirical data (*18*, *19*). While some of the empirical results provide evidence for the superior performance of RRM models in relation to RUM models, there is also evidence to argue otherwise. Chorus et al. (*18*) proposed an attribute-dependent decision rule framework in which each influential attribute was categorized as regret based or utility based (*18*). Note that the hybrid framework still considers the decision rule to be homogeneous across decision makers.

Most of the literature on RRM attempts to understand the behavioral implications of theoretical properties of the RRM framework, explores the possibilities of application of RRM framework to various travel choice dimensions, or attempts to benchmark the empirical performance of RRM against the conventional RUM framework. However, very few attempts have been made to model heterogeneity in decision rules in the context of travel-related decisions, especially choice of travel mode. Srinivasan et al. (*11*) and Hess et al. (*20*) are the studies (to the best of the authors' knowledge) which have attempted to capture the potential heterogeneity in decision rules, namely RUM and RRM (or random disutility minimization [RDM]) frameworks. Although these studies establish the improvement in statistical consistency and predictive ability of the heterogeneous decision rule model, they fail to compare the behavioral differences in the two decision rule segments.

The factors influencing the choice of decision rule are not well understood and hence not much can be derived about them from the literature. While Hess et al. (*20*) used a constants-only binary logit model to explain the choice of decision rule, Srinivasan et al. (*11*) explain it using socio-economic factors. So, other than socio-economic characteristics, not much is known from previous studies in particular with regard to risk tolerance levels or expectations of service attributes which influence the choice of decision rule.

Methodology

Formulation and Estimation Details

Random Utility Maximization. The utility of alternative j as perceived by decision maker i is given by

$$U_j = V_j + \varepsilon_j, \; j = 1, 2, 3, \ldots K \tag{1}$$

if the systematic component is V_j and the unobserved component is ε_j. The decision maker chooses the alternative with the greatest utility. The error terms are assumed to be identically and independently Gumbel distributed with mean 0. The resulting probability expression is a multinomial logit model (MNL) as in Equation 2.

$$P_j^{util} = \frac{\exp(V_j)}{\sum_{k=1}^{K} \exp(V_k)}, \; j = 1, 2, \ldots K \tag{2}$$

Random Regret Minimization. The RRM has a non-linear regret expression with a binary comparison of attributes as in Equation 3. The error term in RRM is also assumed to be Gumbel distributed and hence the MNL probability expression is used. The deterministic component of the potential regret that an individual experiences in choosing alternative j with respect to an attribute m is given by

$$R_{jm} = \sum_{k=1}^{K} \ln\left[1 + \exp\left(\beta_{km}X_{km} - \beta_{jm}X_{jm}\right)\right] \quad (3)$$

and so the regret experienced on j with respect to all attributes is given by

$$R_j = \sum_{\substack{k=1 \\ k\neq j}}^{K} \sum_{m=1}^{M} \ln\left[1 + \exp\left(\beta_{km}X_{km} - \beta_{jm}X_{jm}\right)\right] \quad (4)$$

where:

R_j is the deterministic component of regret experienced on choosing jth alternative,

M is the set of attributes influencing the alternatives, and

K is the total number of alternatives.

Regret minimization is operationalized using a systematic "regret" component involving binary comparison of attributes. Hence, regret minimization involves performing binary operations on attributes. However, the subjective factors were measured on a Likert scale of 1 to 5. Since the objective function for regret minimization involves binary operations, ordinal values cannot be used in these models. Hence, these ordinal ratings need to be converted into a continuous measure on which binary operations are possible. This was done using ordered probit models. The model represents the ordinal indicator of subjective factor as a continuous latent measure making it consistent with a hybrid choice modeling approach. However, for simplicity of estimation and with the focus on the decision rule process, the ordinal factor was estimated independently, making it a sequential likelihood estimation process. These continuous latent measures of perception rating could be used in the regret minimization framework in Equation 4. In an ordered choice model,

$$y_j^* = \theta_j S_j + \omega_j, \quad (5)$$

the latent "preference" variable y_j^* is not observed. The observed component of y_j^* is S_j. The ordered probit model is based on normally distributed error term ω_j assumptions.

$$\begin{aligned} S_j &= 0 \text{ if } y_j^* \leq \phi_0, \\ &= 1 \text{ if } \phi_0 \leq y_j^* \leq \phi_1, \\ &= 2 \text{ if } \phi_1 \leq y_j^* \leq \phi_2, \\ &= 3 \text{ if } \phi_2 \leq y_j^* \leq \phi_3, \\ &\ldots \\ &= L \text{ if } y_j^* \geq \phi_{L-1} \end{aligned} \quad (6)$$

The probability that S_j takes a value l is given by

$$P(S_j = l) = P(y_j^* \text{ is in the } l^{th} \text{ range}) \quad (7)$$

The final probability structure of an RRM–MNL is as follows:

$$P_j^{reg} = \frac{\exp(-R_j)}{\sum_{k=1}^{K} \exp(-R_k)}, \; j = 1, 2, \ldots K \quad (8)$$

Heterogeneous Decision Rule (HDR). As already discussed, it is intuitive to consider that decision makers are heterogeneous with regard to the decision rules they adopt. The current study assumes decision makers choose between regret minimization and utility maximization. However, the decision rule adopted by any decision maker is unknown and hence considered to be latent in the proposed framework. The probability that an individual adopts one of these decision rules is assumed to have a binary logit model structure. Further, the conditional probability that the alternative j is chosen, given the decision maker follows a particular decision rule, is assumed to have an MNL structure. Consequently, the overall probability that an individual i chooses an alternative j is given by

$$P_j = p(util)\, P_j^{util} + p(reg)\, P_j^{reg} \quad (9)$$

where: $p(util)$ is the marginal probability that individual i is a utility maximizer,

P_j^{util} is the conditional probability that alternative j is chosen given that individual i is a utility maximizer,

$p(reg)$ is the marginal probability that individual i is a regret minimizer,

P_j^{reg} is the conditional probability that alternative j is chosen given that individual i is a regret minimizer.

Since the framework assumes the decision rule to be latent, decision makers are probabilistically assigned between utility maximization and regret minimization. These probabilities are $p(util)$ and $p(reg)$ respectively. This latent decision rule is dependent on a continuous random variable, rv_j, representing the propensity to adopt a decision rule. If the value of this random variable exceeds a threshold, the decision maker is assumed to adopt regret minimization, and utility maximization otherwise. A set of factors is assumed to influence the continuous propensity to choose the decision rule.

$$W_j^{util} = \sum_m \alpha_m Z_m \quad (10)$$

$$p(util) = \frac{\exp\left(W_j^{util}\right)}{1 + \exp\left(W_j^{util}\right)} \quad (11)$$

$$p(reg) = 1 - p(util) \quad (12)$$

Table 1. Descriptive Statistics of Sample Data

Average household income: Rs 15,527			
Gender: Male: 84% Female: 16%		Average age: 36.9	
Respondents with:			
Only two-wheeler	Only car	Both two-wheeler and car	No vehicle
61%	2%	15%	22%
Bicycle ownership: 56%		Driving knowledge: 78%	
Work duration (hours/work day)			
<= 8	8–10		>= 10
35%	43%		22%
Work days/week			
3	4	5	6
1%	1%	31%	67%
Respondents with:			
Flexible time schedule: 49%		Employed spouse: 27%	
Travel mode chosen at least twice in the last 3 months			
Two-wheeler	Car	Bus	Train
69%	21%	64%	51%
Autorickshaw	Shared autorickshaw	Company bus	
52%	8%	28%	
Travel mode share			
Two-wheeler 42.0%	Car 6.0%	Bus 20.5%	
Train 16.6%	Autorickshaw 2.2%	Shared autorickshaw 1.4%	
Company bus 5.6%	NMT 5.7%		

where:

W_j^{util} = Utility associated with the choice of RUM as the decision rule

Z_m = Set of individual and contextual factors which influence the choice of decision rule

α_m = Corresponding coefficients of Z_m

Maximum likelihood estimation technique is used to obtain the coefficients in the three models. The likelihood expression of all three models is as follows ($\delta_k = 1$, if alternative j is chosen and $= 0$, if not):

$$L(\beta) = \prod_{i}^{nobs} \prod_{k}^{K} (P_k^{\delta_k}) \tag{13}$$

Data Description

This section describes the empirical data on work travel mode choice obtained from a sample of 872 workers in Chennai city, India (*11*). Data was collected using face-to-face interviews conducted at randomly sampled households. Table 1 highlights the descriptive statistics of the sample data collected.

The data was found to be reasonably representative of the worker population in Chennai city. In this regard, average household size for the sample and the population were 4.37 and 4.51 respectively. The average age of a worker was 36.9 years for the sample and around 38 for the city. The average household income of the sample (Rs 15,527) was also consistent with the population (Rs 14,500) (*21*).

The alternative modes of travel to work include: motorized two-wheeler, private car, bus, train, intermediate public transport (IPT) (here: autorickshaw, shared autorickshaw, and company bus), and non-motorized transport (NMT) (walk and bicycle). The shares for the modes are shown in Table 1. Two-wheeler has by far the largest share, probably due to its maneuverability and flexibility in trip scheduling. The relatively smaller share of 6% in private car could be due to low car ownership rates and greater travel cost of this mode. A larger proportion of workers chose to travel by bus (20%) than by train (17%). Longer commuting distances and increased physical strain could explain the low share of NMT (5.7%).

Lack of door-to-door services and limited fleet size might have inhibited the usage of shared autorickshaw (1.4%). The choice to travel by company bus (5.6%) is affected by (lack of) service availability depending on the employer organization, and its operation in limited corridors. Usage of autorickshaw (2.2%) may be hindered by operators' lack of compliance with a regulatory fare structure.

Respondents were asked to rate subjective factors of the alternative travel modes—like comfort, safety, reliability, stress, cost, flexibility in departure time, and multiple destinations—on a Likert scale of 1 to 5, where 1 indicates poor and 5 indicates excellent performance. From these ratings, it was found that decision makers perceive the travel cost of personal vehicles and IPT to be high. These ratings also indicate that users give high priority to reliability and comfort in public transit. These ratings may also reflect the risk-averse behavior of decision makers.

The models which are developed account for unavailability of alternatives in the choice set. Certain heuristics were considered to define unavailability of alternatives in the choice set. The utilities of the following alternatives are set to negative infinity so that the probability of choosing the alternative is 0 if

- **Two-wheeler/car:** the household does not own a two-wheeler or car.
- **Bus/train:** distance from home to nearest bus stop or railway station is more than 2 km.
- **NMT:** distance of commute is more than 2.5 km.
- **Company bus/shared autorickshaw:** the person has not chosen it for the last 3 months at least twice.

Results and Discussion

Statistical Evidence for Decision Rule Heterogeneity

Three different models were developed: pure RUM, pure RRM, and HDR consisting of a utility maximizing segment and a regret minimizing segment. The models were developed using a data set of 872 valid observations. The pure RUM model had a likelihood of −870.89, the pure RRM model had a likelihood of −871.13, and the HDR model had a likelihood of −738.05. Since the two homogeneous decision rule models are special cases of HDR, chi-squared likelihood ratio test could be used to compare the goodness-of-fit measures of both pure RUM (21 coefficients) and pure RRM (21 coefficients) models separately with the HDR model (54 coefficients). The results of likelihood ratio tests show that both pure RUM (test statistic of 265.68 against a critical value of 47.40) and pure RRM (test statistic of 266.16 against a critical value of 47.40) perform worse than the HDR model. Hence the test, at 95% confidence level, confirms that the assumption of decision rule homogeneity needs to be relaxed.

Model Validation

The pure RUM, pure RRM, and HDR models are validated using a holdout data set of 30%. The model validation results in Table 2 show that the predicted and actual likelihood ratio index (ρ^2values) are reasonably close to each other (within 10% to 15%). This could be considered as a valid representation of heterogeneity in decision rules. In particular, the relative difference in likelihood ratio indices was the lowest in the case of HDR, which indicates its better predictive ability. The above results confirm that there could be segments in the population which are heterogeneous in the decision rules adopted.

Analyzing Behavioral Differences between Utility Maximizing and Regret Minimizing Segments

Table 2 presents the coefficients for the variables discussed above for all three model structures. The following inferences are made with regard to these models.

Level-of-Service Factors—Travel Time and Travel Cost. The two level-of-service factors that were included in the model specification are travel time and travel cost. The sensitivity to travel time in both the utility maximizing and regret minimizing segments in the HDR model was consistent with that in the pure RUM and pure RRM models. The sensitivity to travel cost in all the modes was found to be significant in the pure RUM model. This sensitivity was highest for travel cost of train, followed by bus, shared autorickshaw, autorickshaw, car, and two-wheeler. The same hierarchy was observed in the pure RRM model except that the decision makers were found to be insensitive to autorickshaw travel cost. On the contrary, the utility maximizing segment in HDR was found to be most sensitive to travel cost in shared autorickshaw, followed by train and bus. The segment was insensitive to travel cost in two-wheeler, car, and autorickshaw. However, the regret minimizing segment of HDR was most sensitive to travel cost in train, followed by bus, shared autorickshaw, car, and two-wheeler. The segment was insensitive to travel cost in autorickshaw. Hence, unlike the utility maximizing segment of HDR, the regret minimizing segment of HDR was consistent in its hierarchy of travel cost sensitivities with the pure RRM model. These findings with respect to travel cost sensitivities exhibit a clear hierarchy from public transport to personal vehicle. This ordering is consistent with increasing privacy, comfort levels, and accessibility provided by the alternative modes. The intermediate range of sensitivity to IPT travel cost is a reflection of (a) the role of IPT modes in urban transport in Chennai city and (b) the service levels IPT offers compared with personal and public transport modes. The high sensitivity to travel cost in train may be due to lower accessibility of trains, and longer waiting times—both while purchasing tickets for travel as well for the service arrival—and difficulty in reaching railway stations.

The HDR model is able to show the differences in travel cost sensitivities in the segments classified based on a decision rule; for example, the regret minimizing segment was sensitive to travel cost in two-wheeler and car while the utility maximizing segment was insensitive to it. However, both the segments were sensitive to travel cost in shared autorickshaw, bus, and train. This is contrary to the findings in the pure RUM and pure RRM model, where sensitivities to travel cost in all the modes were significant.

Subjective Factors. The subjective factors that were included in the model are continuous estimates of perceptional ratings for comfort, safety, reliability, and accessibility of different modes. In order to allow comparison across sensitivities to these different subjective factors, the continuous ratings were normalized by their

Table 2. Comparison of Alternative Decision Rule Models

Variable description	Pure RUM		Pure RRM		HDR-RUM		HDR-RRM	
Alternate specific constant	Coeff.	*t*-stat.	Coeff.	*t*-stat.	Coeff.	*t*-stat.	Coeff.	*t*-stat.
Car	1.24	2.10	0.51	2.07	6.34	0.79	0.75	2.78
Bus	−1.95	−0.71	−0.99	−1.47	10.85	2.48	5.19	0.34
Train	−1.30	−0.42	−0.17	−0.18	9.09	1.82	−1.88	−0.95
Autorickshaw	−3.69	−0.59	−1.87	−1.76	8.92	0.91	−6.32	−1.13
Shared autorickshaw	−3.14	−0.54	0.19	0.13	20.29	2.11	−5.17	−1.98
Company bus	−8.59	−1.76	−1.36	−1.35	9.52	1.37	−11.16	−2.18
NMT	5.75	3.19	4.36	2.33	11.41	3.00	7.21	2.06
Level of service factors								
Travel time (generic)	−0.02	−2.48	−0.01	−2.80	−0.03	−2.76	−0.01	−1.54
Travel cost (two-wheeler)	−0.18	−2.24	−0.06	−2.32	0.15	0.68	−0.14	−3.28
Travel cost (car)	−0.28	−2.96	−0.09	−3.72	−1.33	−0.83	−0.14	−4.01
Travel cost (bus)	−1.28	−5.96	−0.32	−6.04	−0.87	−2.92	−0.84	−1.43
Travel cost (train)	−2.07	−4.46	−0.47	−4.87	−1.02	−2.19	−1.71	−3.57
Travel cost (autorickshaw)	−0.32	−1.30	0.16	0.46	−0.01	−0.01	0.13	0.86
Travel cost (share autorickshaw)	−0.95	−1.83	−0.24	−2.60	−2.22	−1.69	−0.20	−1.67
Subjective factors								
Comfort of personal vehicle	0.58	4.39	0.25	3.38	0.86	3.27	0.52	3.75
Safety of bus (bus)	0.50	4.48	0.49	4.47	0.16	1.15	−0.14	0.00
Reliability of train or IPT	0.40	2.68	0.09	2.45	0.01	0.06	0.36	2.76
Accessibility of train (train)	0.78	2.40	0.24	1.86	1.07	2.63	−0.03	−0.17
Trip-related factors								
Distance to workplace (train)	0.08	3.65	0.02	3.42	0.09	3.06	0.00	−0.11
Distance to workplace (company bus)	0.21	8.23	0.09	3.81	0.20	6.30	−0.15	−0.07
Work-related travel (all modes except personal vehicle)	−0.66	−3.55	−0.19	−3.50	−0.33	−1.00	−0.19	−1.54
Binary logit decision rule model								
Constant					−1.01	−1.41		
Education level – school					−1.23	−1.69		
Young age (<40)					1.50	2.75		
Joint trip					−1.99	−2.09		
High congestion rating					−1.93	−3.61		
Road condition rating is bad					−1.49	−2.59		
Bus frequency					1.97	4.19		
Log-likelihood summary	Pure RUM		Pure RRM		Latent HDR model			
Log-likelihood (0)	−1528.84		−1528.84		−1528.84			
Log-likelihood (convergence)	−870.89		−871.13		−738.05			
Likelihood ratio index (ρ^2)	0.43		0.43		0.52			
No. of observations	872		872		872			
Model validation								
Hold-out validation data set (30%)	−290.89 (0.38)		−291.59 (0.38)		−230.55 (0.51)			
Calibration model coefficients applied on validation data set	−313.32 (0.33)		−315.01 (0.33)		−252.61 (0.46)			
Relative difference (%) in Likelihood ratio index (ρ^2) values	12.5%		13.1%		9.2%			

respective standard deviations. The subjective factors which were found to be significant in the models are: the effect of comfort in personal vehicles, safety in bus, reliability of train and IPT, and accessibility of train. All the coefficients were positive, indicating that increase in perceptive rating for these subjective factors leads to increase in utility and decrease in regret for the corresponding alternatives. A similar trend was also observed in the HDR model. The pure RUM model showed a higher sensitivity to accessibility in train, followed by equal sensitivity to all other factors. However, pure RRM showed a higher sensitivity for safety in bus followed by almost comparable sensitivity on all other factors. The utility maximizing segment in HDR was insensitive to reliability in train and safety in bus, however, the remaining sensitivities were consistent with those of a pure RUM model. Similarly, the regret minimizing segment were insensitive to bus safety and to reliability in train in their choices of bus and train respectively.

Trip-Related Factors. Performing work-related trips to places other than workplace and distance to the workplace from home were the trip-related factors that were significant in influencing the mode choice behavior of the decision makers. With regard to work-related trips, all the models showed a negative coefficient. This is indicative of the fact that the decision makers who perform work-related trips outside the workplace have limited travel alternatives other than personal vehicle. This effect was absent in the utility maximizing segment but evident in the regret minimizing segment of the HDR model.

The effect of distance to the workplace from home is intuitive in nature. The likelihood of choosing train and company bus increased in comparison to other modes as distance increased. This effect was observed in all the models. Further, the intensity of this effect was greater in the case of company bus than train. This could be because the establishments (like IT companies) which provide transport facilities for their employees are located on the outskirts of the city. In the case of train, the decision maker might choose this mode only if the distance to the workplace is beyond a minimum threshold distance and hence the usage of train might be associated with longer commuting distances. However, this effect of work distance was absent in the regret minimizing segment of the HDR model. This was also confirmed by the elasticity values in the respective segments with respect to work distance (not shown here) .

Factors Influencing the Choice of Decision Rule

The choice of the latent decision rule between utility maximization and regret minimization was modeled assuming the probability structure of a binary logit model. The following factors were found to be significant in the model. In this model, the positive coefficient indicates that, with an increase in the value of these factors, the probability that an individual considers utility maximization as a decision rule increases. Graduates and postgraduates are less likely to be regret minimizers than those commuters who are not graduates. Similarly, younger commuters tend to be utility maximizers more than regret minimizers. Decision makers who travel to work with other household members are less likely to adopt utility maximization as their decision rule. Commuting as a pair or as a group imposes more constraints on travel, like departure time, mode usage, etc., and hence this causes some impedances to the commuters in their travel choice. However, most commuters who travel with their family members tend to anticipate this impedance and try to minimize the regret associated with it. Decision makers who feel that congestion is high or road conditions are poor are more likely to be regret minimizers. Such ratings for these factors itself is reflective of higher expectations and comparison with higher levels of service which is characteristic of regret minimizing behavior. With the increase in the number of workers among members in the household, the probability of the decision maker being a utility maximizer increases. With a decrease in frequency of buses (number of services per hour), the decision maker tends to be a utility maximizer rather than a regret minimizer.

Segmental Analysis

In this section, the characteristics of these two segments (or classes) in the HDR model are compared. These segments are compared in terms of their individual/household characteristics and trip-related attributes. This comparison is done based on the values of the attributes computed based on the segmental analysis exercise proposed by Bhat (*22*). The average value of any given attribute in the segment D is computed using the following expression,

$$\overline{A_D} = \frac{\sum_{i}^{N} p_D^i \cdot A^i}{\sum_{i}^{N} p_D^i} \tag{14}$$

where:

A^i is the value of the socio-economic attribute for individual i,

p_D^i is the segment membership probability of individual I, and

$\overline{A_D}$ is the segmental average of an attribute.

The segment-wise average values of these attributes characterize these segments and could aid in differentiating the tendencies of these segments. The average of various attributes computed for both these segments is shown in Table 3. In addition to attributes, the mode shares in each of these segments are computed using the same expression as in Table 3. While differences were observed for some of the attributes in these segments, differences in certain other attributes were negligible. However, these attributes with differential in mean values could be used to understand and characterize these segments.

In the RUM segment, the share of two-wheeler was highest (55%), followed by car (12%) and train (9%). This was followed by NMT (8%), bus (7%), company bus (5%), autorickshaw (3%), and shared autorickshaw (1%). In the RRM segment, bus had the highest share (37%), followed by train (26%). This was followed by two-wheeler (19%), company bus (8%), NMT (4%), autorickshaw, and shared autorickshaw (3% and 2% each). The lowest share was for car (1%). The RUM segment had the highest share for personal vehicles (two-

Table 3. Average Values of Attributes in Each Segment

Variables	HDR – RUM	HDR – RRM
Low income	0.31	0.37
Medium income	0.49	0.49
High income	0.20	0.16
Male	0.79	0.81
Distance	10.15	13.3
Young	0.47	0.51
Middle age	0.43	0.36
Old age	0.10	0.13
Captive by vehicle ownership	0.15	0.31
Captive by driving knowledge	0.08	0.16
Semi-captive	0.22	0.26
Choice segment	0.55	0.27
Solo trip	0.85	0.97
Joint trip	0.15	0.03
Short distance (<= 7 km)	0.45	0.34
Medium distance (7–13 km)	0.20	0.18
Long distance (>= 13 km)	0.35	0.48
Bus frequency	1.64	12.12
No. of transfers	0.08	0.18
Employed spouse	0.24	0.20
Flexibility in time	0.46	0.43
Mode shares		
Two-wheeler	55%	19%
Car	12%	1%
Bus	7%	37%
Train	9%	26%
Autorickshaw	3%	3%
Shared autorickshaw	1%	2%
Company bus	5%	8%
NMT	8%	4%

wheeler and car) and the lowest share for IPT modes (autorickshaw, shared autorickshaw, and company bus). Public transport had relatively lower shares than personal vehicles. The RRM segment, on the other hand, had the highest shares for public transport. It also showed a considerable preference for two-wheeler but very little preference for car. The segment showed relatively greater shares for company bus (8%) but very low shares for autorickshaw, shared autorickshaw (2% each), and NMT (4%). The relatively very high shares for two-wheeler and car in the RUM segment suggest that the segment is mostly captive to the personal vehicle (67% shares of personal vehicle). On the contrary, the RRM segment has very high public transport shares (63%) as well as high two-wheeler shares (19%) and relatively greater company bus shares (8%). This means that the segment is not captive to any particular mode. Based on the above observations, it is considered intuitive to characterize the RUM segment as a group which is captive to personal vehicle usage and the RRM segment to be a non-captive group.

The proportion of commuters who have restricted or no access to personal vehicles (either due to lack of driving knowledge or vehicle ownership) is very high (23% in RUM segment versus 47% in RRM segment) in the non-captive group (RRM segment) when compared with the personal vehicle captive group (RUM segment). Further, the proportion of commuters who have full access to personal vehicles is also very high in the RUM segment when compared with the RRM segment (58% in RUM segment and 22% in RRM segment). This implies that the RUM segment is mostly captive to personal vehicle usage. The captivity of personal vehicle in the RUM segment could also be motivated by the greater proportion of joint trips in the segment (15% in segment1 versus 3% in segment 2). This is because, owing to its greater flexibility, the personal vehicle is very useful in making joint trips when compared with other modes. Commuters with flexible work time are more likely to be in the personal vehicle captive group than in the non-captive group. This could also have led to the higher personal vehicle shares in the personal vehicle captive group than in the non-captive group.

The proportion of long-distance trips in the non-captive group is significantly greater (48%) than in the personal vehicle captive group (35%). This could be the reason for higher train and company bus shares in the non-captive group (37% and 8%) than the personal vehicle captive group (9% and 4%). The percentage of short and medium distance trips is considerable in the non-captive group (34% short distance, 18% medium distance) but less than in the personal vehicle captive group (45% short distance, 20% medium distance). This might be the reason for greater share of two-wheelers when compared with the other modes in the non-captive group.

The relatively higher shares of bus in the non-captive group (37% versus 5% in the personal vehicle captive group) are possibly due to them recording very high bus frequency (12.12 services/hr against 1.64 service/hr).

Summary and Conclusions

This empirical study attempts to capture potential heterogeneity in decision rules adopted by decision makers with respect to random utility maximization and random regret minimization. The specific objectives of the study were to (a) propose a framework which relaxes the restriction of decision rule homogeneity; (b) compare its performance with RUM and RRM frameworks; (c) capture the behavioral differences between the utility maximizing segments and regret minimizing segments in the population; and (d) identify the factors influencing the choice of the latent decision rule. In order to achieve these objectives, empirical models were developed using the data collected from working commuters through household interview survey in Chennai city. MNL models were used to model the probabilities in pure RUM and pure RRM models while a latent class model was developed for representing decision rule heterogeneity. Model specification included variables related to level-of-service factors, subjective factors, and trip-related factors. The following findings from the study are noteworthy.

The proposed HDR model performed statistically better than the homogeneous decision rule models, the pure RUM and pure RRM models. This could be considered as primary evidence for decision rule heterogeneity in work travel mode choice. The effects observed in homogenous decision rule frameworks might be normalized effects and reflect the sensitivity in one of the decision rule sub-segments.

On performing the segmental analysis of the two segments, RUM and RRM, the following differences are observed. It was observed that the RUM segment displayed a stronger "inertia, captivity or loyalty" to the personal vehicle compared with the RRM segment. The personal vehicle shares were very high in the RUM segment. No particular mode type had an advantage over all other modes, although public transport modes had the highest shares for the RRM group. The RRM segment had a relatively larger proportion of long-distance trips while the RUM segment had more short and medium trips. The segmental analysis thus supports the model results regarding critical differences between the RUM and RRM segments.

In two-stage simultaneous models, modelers are often faced with the dilemma of specification of suitable variables at the appropriate stage, decision rule process, selection of the chosen mode or both, in the context of heterogeneous decision rule models. Existing models show evidence of heterogeneity due to socio-demographics at the mode level. Tthis may be misleading, however, as such models assume homogeneous decision rules, and thus heterogeneity at a decision rule level may be wrongly reflected as heterogeneity at the mode level.

Some variables were tested at both levels and found to be significant only at the latent decision rule level (e.g. age, education level, bus frequency, etc.). However, in other cases, some variables were included only at the decision rule level and not at the choice level to avoid collinearity, confounding, and error correlations across the two levels (e.g. distance to workplace, joint trip, reliability in train, safety in bus etc.). In such cases, inclusion of variables at both levels led to unstable or counter-intuitive estimates. Since very few studies have explored heterogeneity in decision rules, no explicit attempt has been made to segregate the effect of attributes at both stages. Hence, classification of these attributes and their specification requires more scientific analysis and further investigation.

To the authors' knowledge, data regarding decision rules adopted by users in travel surveys is very sparse, particularly with regard to revealed preference data, mainly because of focus on the alternative travel modes chosen rather than the decision process. However, researchers in marketing, psychometrics, and behavioral economics have attempted to study decision rules in other contexts using stated preference and conjoint experiments. Although a number of decision rules have been modeled in the context of travel behavior for different choice dimensions, almost all these studies have been based on the assumption of homogeneous decision rule.

It is a challenge to frame questions to study the decision process and attribute trade-offs with regard to revealed preference data as it involves providing a number of scenarios to elicit such preferences. There is a possibility of framing and selectivity bias with regard to such data, particularly as the choice process may be latent or partially formed or not well articulated by the decision maker. For these reasons, the reliability, robustness, and validity of such data and instruments to capture decision rules and choice processes directly need to be ensured.

In the absence of such data, and the fact that the choice process is latent, the latent and hybrid choice approaches offer a useful methodological construct to infer the decision making rules. However, the corroboration and validation of these insights with direct measurement remains an important and immediate next step for the generalization and practical use of the models.

Acknowledgments

The support rendered by the Centre of Excellence in Urban Transport at IIT Madras, funded by the Ministry of Urban Development, Government of India, is gratefully acknowledged.

Author Contributions

Both authors contributed to all aspects of the paper.

References

1. De Palma, A., M. Ben-Akiva, P. Moffatt, N. Picard, K. Train, P. Wakker, and J. Walker. Risk, Uncertainty and Discrete Choice Models. *Marketing Letters*, Vol. 19, No. 3–4, 2008, pp. 269–285.
2. Chorus, C. G., and G. C. de Jong. Modelling Experienced Accessibility for Utility Maximizers and Regret-Minimizers. *Journal of Transport Geography*, Vol. 19, 2011, pp. 1155–1162.
3. Chorus, C. G., T. Arentze, and H. J. P. Timmermans. A Comparison of Regret-Minimization and Utility-Maximization in the Context of Travel Mode-Choices. Presented at 87th Annual Meeting of the Transportation Research Board, Washington, D.C., 2008.

4. Chorus, C. A New Model of Random Regret Minimization. *Ejtir*, Vol. 10, No. 2, 2010, pp. 181–196.
5. Chorus, C. G. Logsums for Utility-Maximizers and Regret-Minimizers, and Their Relation with Desirability and Satisfaction. *Transportation Research Part A: Policy and Practice*, Vol. 46, No. 7, 2012, pp. 1003–1012.
6. Boeri, M., A. Longo, E. Doherty, and S. Hynes. Site Choices in Recreational Demand: A Matter of Utility Maximization or Regret Minimization. *Journal of Environmental Economics and Policy*, Vol. 1, No. 1, 2012, pp. 32–47.
7. Loomes, G., and R. Sugden. Regret Theory: An Alternative Theory of Rational Choice Under Uncertainty. *Economic Journal*, Vol. 92, No. 368, 1982, pp. 805–824.
8. Ben-Akiva. *Choice Models with Simple Choice Set Generating Processes*. Working paper, Massachusetts Institute of Technology, Cambridge, Mass., 1976.
9. McFadden, D. *The Theory and Practice or Disaggregate Demand Forecasting for Various Modes for Urban Transportation*. Urban Travel Demand Forecasting Project Working Paper No. 7623. Institute of Transport Studies, UC Berkley, Calif., 1976.
10. Train, K. E. *Discrete Choice Methods with Simulation*, 2nd ed. Cambridge University Press, 2004.
11. Srinivasan, K. K., G. M. Naidu, and T. Sutrala. Heterogeneous Decision Rule Model of Mode Choice Incorporating Utility Maximization and Disutility Minimization. *Transportation Research Record: Journal of the Transportation Research Board*, 2009. 2132: 59–68.
12. Cascetta, E., and A. Papola. Random Utility Models with Implicit Availability/Perception of Choice Alternatives for the Simulation of Travel Demand. *Transportation Research Part C: Emerging Technologies*, Vol. 9, No. 4, 2001, pp. 249–263.
13. Swait, J. Choice Set Generation within the Generalized Extreme Value Family of Discrete Choice Models. *Transportation Research Part B: Methodological*, Vol. 35, No. 7, 2001, pp. 643–666.
14. Yang, C. Modeling Multiple Sources of Heterogeneity in Mode Choice Model. *Journal of the Eastern Asia Society for Transportation Studies*, Vol. 6, 2005, pp. 2031–2045.
15. Dissanayake, D., and T. Morikawa. Investigating Household Vehicle Ownership, Mode Choice and Trip Sharing Decisions Using a Combined Revealed Preference/stated Preference Nested Logit Model: Case Study in Bangkok Metropolitan Region. *Journal of Transport Geography*, Vol. 18, No. 3, 2010, pp. 402–410.
16. Kunhikrishnan, P., and K. K. Srinivasan. Choice Set Variability and Contextual Heterogeneity in Work Trip Mode Choice in Chennai City. *Transportation Letters*, Vol. 7867, 2017, pp. 1–16.
17. Bell, D. E. Regret in Decision Making under Uncertainty. *Operations Research*, Vol. 30, No. 5, 1982, pp. 961–981.
18. Chorus, C. G., J. M. Rose, and D. A. Hensher. Regret Minimization or Utility Maximization: It Depends on the Attribute. *Environment and Planning B: Urban Analytics and City Science*, Vol. 40, No. 1, 2013, pp. 154–169.
19. Thiene, M., M. Boeri, and C. G. Chorus. Random Regret Minimization: Exploration of a New Choice Model for Environmental and Resource Economics. *Environmental and Resource Economics*, Vol. 51, No. 3, 2012, pp. 413–429.
20. Hess, S., A. Stathopoulos, and A. Daly. Allowing for Heterogeneous Decision Rules in Discrete Choice Models: An Approach and Four Case Studies. *Transportation*, Vol. 39, No. 3, 2012, pp. 565–591.
21. National Council of Applied Economic Research Report. 2006. http://www.ncaer.org/Lbrry.asp Date accessed: March 2017.
22. Bhat, C. R. An Endogenous Segmentation Mode Choice Model with an Application to Intercity Travel. *Transportation Science*, Vol. 31, No. 1, 1997, pp. 34–48.

The Travel Analysis Methods Section (ADB00) peer-reviewed this paper (18-06557).

Sage Publications USA
sagepub.com

ISBN 978-1-5108-8253-9

9 781510 882539

CURRAN ASSOCIATES INC.
proceedings.com